Solid–Liquid Separation Technologies

Solid–Liquid Separation Technologies
Applications for Produced Water

Edited by
Olayinka I. Ogunsola and Isaac K. Gamwo

CRC Press
Taylor & Francis Group
Boca Raton London New York

CRC Press is an imprint of the
Taylor & Francis Group, an **informa** business

First edition published 2022
by CRC Press
6000 Broken Sound Parkway NW, Suite 300, Boca Raton, FL 33487-2742

and by CRC Press
2 Park Square, Milton Park, Abingdon, Oxon, OX14 4RN

ISBN: 978-0-367-89328-6 (hbk)
ISBN: 978-0-367-54880-3 (pbk)
ISBN: 978-1-003-09101-1 (ebk)

DOI: 10.1201/9781003091011

Typeset in Times
by codeMantra

Contents

Preface

Water generally coexists with deposits of fossil energy resources (coal, oil, and gas) and is subsequently coproduced with these resources during production. The market value of produced water is very low compared to the energy resource it is coproduced with; in fact, produced water is typically considered a by-product or waste and is generally disposed. However, several methods are available for handling produced water, including reinjection for disposal and treatment for reuse.

The economic and environmental importance of produced water has recently gained increased attention. Injecting large volumes of produced water underground over a long period of time and within a conducive geological setting/environment has been found to cause induced seismicity. At the same time, produced water is also a potential source of value-added products, such as minerals and other elements the water dissolved during the thousands of years it lay buried in its unique geologic setting. The water/produced water also contains chemicals added during oil and gas exploration and production operations and native bacteria in the original geologic setting.

These constituents can be recovered and used for manufacturing value-added products, while the treated water can be beneficially used for other purposes. New advances in solid–liquid separation technologies offer promising means for cost-effective production of the various products while reducing the environmental impacts associated with produced water injection.

This book provides information on produced water and water related technologies, including an overview of characteristics, treatment, and beneficial reuse of produced water; some policy and regulatory considerations for produced water; and an overview of various solid–liquid separation technologies (such as membrane, filtration, crystallization, desalination, etc.).

This book also covers recent research advances in solid–liquid separation not related to produced water, as well as overview of the amenability of produced water to varying treatment technologies. The authors anticipate this book will be particularly useful for produced water/water treatment plant engineers, designers, and operators, as well as for researchers, professionals, technology developers, and policy/regulatory entities. The book is also a valuable resource for graduate and undergraduate courses in solid–liquid separation, process design engineering, and environmental engineering and related fields.

Editors

Olayinka I. Ogunsola, Ph.D.

Dr. Olayinka (Yinka) Ogunsola is a Senior Program Manager/Environmental Engineer in the Department of Energy (DOE) Office of Fossil Energy and Carbon Management, where he manages the Department's onshore oil and gas and produced water research and development programs. Prior to joining DOE in 2000, Dr. Ogunsola was a National Research Council Senior Research Associate at the U.S. Department of Energy, National Energy Technology Laboratory (NETL), Morgantown, West Virginia. While at NETL, Dr. Ogunsola planned and conducted research related to solid–fluid separation and transport reactor system in general.

Before working at NETL, he was a Research Associate Professor in the School of Mineral Engineering, University of Alaska Fairbanks. Dr. Ogunsola also worked at the then Western Research Laboratory, Canada Center for Energy & Mineral Technology (CANMET), Devon, Alberta, Canada, as NSERC Research Fellow. Prior to joining CANMET, he was a Senior Lecturer in the Department of Petroleum Engineering, University of Port Harcourt, Port Harcourt, Nigeria.

Dr. Ogunsola has published over 55 peer-reviewed scientific papers in reputable journals, symposia, and proceedings and has coedited two books in the areas of fuel and energy engineering. He has organized and chaired/co-chaired a number of symposia and technical sessions at technical conferences/meetings of American Chemical Society, National Organization for the Advancement of Black Chemists and Chemical Engineers, and American Institute of Chemical Engineers. He holds a Bachelor of Science (Honors) in Fuel and Energy Engineering from Leeds University, Leeds, England, and a Ph.D. from the Pennsylvania State University, University Park, Pennsylvania, USA.

Isaac K. Gamwo, Ph.D., P.E., AIChE Fellow

Dr. Isaac K. Gamwo is a Senior Research General Engineer at the United States Department of Energy's National Energy Technology Laboratory (Pittsburgh), where he leads a multi-institutional research group as both the Technical Coordinator and Principal Investigator to extend thermodynamic mineral scale models to high-temperature, high-pressure conditions. Dr. Gamwo has mentored several post-doctoral researchers, doctorate candidates, and summer interns.

Dr. Gamwo is serving a five-year term as a Director of the American Institute of Chemical Engineer's (AIChE) Separation division. He is also the current Technical Program Chair of the AIChE's Fluid-Particle Separations area. Dr. Gamwo has recently been elected as the 2024 AIChE's Separations Division Chair.

Dr. Gamwo earned his M.S. and Ph.D. in chemical engineering from the Illinois Institute of Technology (Chicago). He is a Licensed Professional Engineer, a Fellow of AIChE, and a Member of the National Organization of Black Chemists and Chemical Engineers (NOBCChE). Dr. Gamwo's work and expertise has garnered recognition through several awards including the 2020 American Institute of Chemical Engineers—MAC Eminent Chemical Engineer award, the 2017 NOBCChE Cannon award for excellence in chemical engineering, and two Gold Excellence in Government awards (2011 Gold Award for Outstanding Professional Employee; 2002 Gold Award for Rookie). Dr. Gamwo previously served as an Assistant Professor at the University of Akron (Ohio) and Tuskegee University (Alabama), and as an Affiliate Graduate Faculty Member at Virginia Commonwealth University (Richmond). Dr. Gamwo coauthored the book *Design and Understanding of Fluidized Bed Reactors* (Verlag 2009) and coedited the book *Ultraclean Transportation Fuels* (Oxford University Press, 2007). He is credited on over 150 articles and presentations.

Contributors

Hossain M. Azam
Department of Civil Engineering
University of the District of Columbia
Washington, D.C.

Hseen O. Baled
U.S. Department of Energy
Research and Innovation Center
National Energy Technology Laboratory
Pittsburgh, PA
and
Department of Chemical and Petroleum
    Engineering
University of Pittsburgh
Pittsburgh, PA

Meghan Brandi
U.S. Department of Energy
Research and Innovation Center
National Energy Technology Laboratory
Pittsburgh, PA

B.J. Carney
Northeast Natural Energy
Charleston, WV

Jeffrey J. Chalmers
Department of Chemical and
    Biomolecular Engineering
The Ohio State University
Columbus, OH

Xiaoyi Chen
Department of Chemical and Biological
    Engineering
University at Buffalo, The State
    University of New York
Buffalo, NY

Jia W. Chew
School of Chemical and Biomedical
    Engineering
Nanyang Technology University
Singapore, Singapore
and
Singapore Membrane Technology
    Center
Nanyang Environmental and Water
    Research Institute
Singapore, Singapore

Isaac K. Gamwo
U.S. Department of Energy
Research and Innovation Center
National Energy Technology Laboratory
Pittsburgh, PA

James Gardiner
U.S. Department of Energy
Research and Innovation Center
National Energy Technology Laboratory
Pittsburgh, PA

Jenifer Gómez-Pastora
Department of Chemical and
    Biomolecular Engineering
The Ohio State University
Columbus, OH

McMahan Gray
U.S. Department of Energy
Research and Innovation Center
National Energy Technology Laboratory
Pittsburgh, PA

J. Alexandra Hakala
U.S. Department of Energy
Research and Innovation Center
National Energy Technology Laboratory
Pittsburgh, PA

Derek M. Hall
U.S. Department of Energy
Research and Innovation Center
National Energy Technology Laboratory
Pittsburgh, PA
and
Department of Energy and
    Mineral Engineering
The EMS Energy Institute
The Pennsylvania State University
University Park, PA

Jacob Hutfles
Department of Mechanical Engineering
University of Colorado Boulder
Boulder, CO

Tuo Ji
National Energy Technology Laboratory
NETL Support Contractor
Pittsburgh, PA

Barbara Kutchko
U.S. Department of Energy
Research and Innovation Center
National Energy Technology Laboratory
Pittsburgh, PA

Alison E. Lewis
Crystallization and Precipitation Unit,
    Department of Chemical
University of Cape Town
Cape Town, South Africa

Haiqing Lin
Department of Chemical and Biological
    Engineering
University at Buffalo, The State
    University of New York
Buffalo, NY

Christina Lopano
U.S. Department of Energy
Research and Innovation Center
National Energy Technology Laboratory
Pittsburgh, PA

Serguei N. Lvov
U.S. Department of Energy
Research and Innovation Center
National Energy Technology Laboratory
Pittsburgh, PA
and
Department of Energy and
    Mineral Engineering
The EMS Energy Institute
The Pennsylvania State University
University Park, PA

Justin Mackey
U.S. Department of Energy
Research and Innovation Center
National Energy Technology Laboratory
Pittsburgh, PA

Elena Subia Melchert
U.S. Department of Energy
Washington, DC

John Pellegrino
Department of Mechanical Engineering
University of Colorado Boulder
Boulder, CO

Valentina Prigiobbe
Department of Civil, Environmental,
    and Ocean Engineering
Stevens Institute of Technology
Hoboken, NJ

**Sankaranarayanan Ayyakudi
Ravichandran**
Department of Mechanical Engineering
University of Colorado Boulder
Boulder, CO

Kevin Resnik
National Energy Technology Laboratory
NETL Support Contractor
Pittsburgh, PA

Fan Shi
National Energy Technology Laboratory
NETL Support Contractor
Pittsburgh, PA

Nicholas Siefert
U.S. Department of Energy
Research and Innovation Center
National Energy Technology Laboratory
Pittsburgh, PA

Yee Soong
U.S. Department of Energy
National Energy Technology Laboratory
Pittsburgh, PA

Torsten Stelzer
Department of Pharmaceutical Sciences
Crystallization Design Institute,
    Molecular Sciences Research Center
University of Puerto Rico
San Juan, PR

Henry J. Tanudjaja
School of Chemical and Biomedical
    Engineering
Nanyang Technology University
Singapore, Singapore

Thomas J. Tarka
U.S. Department of Energy
National Energy Technology Laboratory
Pittsburgh, PA

Madison Wenzlick
U.S. Department of Energy
National Energy Technology Laboratory
NETL Support Contractor
Albany, OR

Walter Christopher Wilfong
National Energy Technology Laboratory
NETL Support Contractor
Pittsburgh, PA

Xian Wu
Department of Chemical and
    Biomolecular Engineering
The Ohio State University
Columbus, OH

Wei Xiong
U.S. Department of Energy
Research and Innovation Center
National Energy Technology Laboratory
Pittsburgh, PA

Zi Ye
Department of Civil, Environmental,
    and Ocean Engineering
Stevens Institute of Technology
Hoboken, NJ

Shouliang Yi
National Energy Technology Laboratory
NETL Support Contractor
Pittsburgh, PA

1 Produced Water Treatment Technologies
An Overview

Isaac K. Gamwo
National Energy Technology Laboratory

Hossain M. Azam
University of the District of Columbia

Hseen O. Baled
National Energy Technology Laboratory
and University of Pittsburgh

CONTENTS

DOI: 10.1201/9781003091011-1

1

1.1 INTRODUCTION

Produced water (PW) is the wastewater separated from production fluid during oil and gas (O&G) production (Larson, 2018; WEF, 2018; Jiménez et al., 2018). PW is generated from both conventional and unconventional sources such as the coal bed methane, tight sands, and gas shale (Jiménez et al., 2018). PW includes formation/connate water, flowback water (injected water), and condensation water. Amount of PW generated during production of crude oil and natural gas can be as high as ten times the volume of hydrocarbon produced. PW volume can rise to as much as 98% of production fluids (e.g., at late stage of oil (gas) production), when production is no longer economical (Larson, 2018; Gray, 2020; Lusinier et al., 2019). Thus, the ratio of PW to oil varies from well to well, and over the life of the well. Typically, PW to oil volume ratio is over 3:1 and can be as high as over 20:1 (Larson, 2018; Jiménez et al., 2018). The global PW production is approximately 10.44 billion gallons/day (Jiménez et al., 2018), whereas the U.S. produced an estimated 890 billion gallons/year of PW in 2012 (GWPC, 2019).

   PW contains numerous chemicals, some of which are toxic organic and inorganic compounds (Jiménez et al., 2018). Physical and chemical properties of PW vary, depending on the geographic location of the field, the geological formation, the extraction method, and the type of hydrocarbon product being produced. Furthermore, PW may include chemical additives, which are dosed in during drilling to treat or prevent operational problems and to enhance subsequent oil/water separation (Jiménez et al., 2018). Thus, both the flow rate and PW composition change over time, leading to varying PW management strategies (WEF, 2017). Multiple separation steps are typically required to separate oil and water from PW (WEF, 2017). Most regulatory policies and technical requirements focus on treatment of O&G content; salt content is also critical in onshore operations (Jiménez et al., 2018). The major PW constituents of concern may be categorized in the following groups: salts, expressed as salinity, total dissolved solids (TDS), or electrical conductivity; oil and grease; BTEX (benzene, toluene, ethylbenzene, and xylenes); PAHs (polyaromatic hydrocarbons); organic acids; phenol; natural inorganic and organic compounds, e.g., chemicals that cause hardness and scaling such as calcium, magnesium, barium, carbonate, and sulfates; and chemical additives used in drilling, fracturing, and operating the well (e.g., biocides and corrosion inhibitors) (Arthur et al., 2011).

   The degree of PW management depends on the site's treatment requirements and typically includes deep well injection/disposal, reinjection, evaporation ponds, surface water discharge, treatment, and reuse (WEF, 2017; Dores et al., 2012). Local water scarcity, legislation, risk of formation plugging, high costs associated with PW disposal, quality of water used in enhanced oil recovery (EOR), and increasing demand for water in production operations are some of the drivers for appropriate PW management techniques. Due to scarcity of water resulting from climate change-induced drought, regulations have become more stringent, disposal method costs have increased, and beneficial reuse is becoming a more viable option (Larson, 2018; WEF, 2018). PW disposal includes deep well injection and discharge into surface water, which requires treatment to remove dispersed and dissolved oil, solids, and toxic compounds. In offshore operations, the common practice is to discharge treated PW to the sea. Hence the main treatment objective is to reduce oil and grease to levels required to meet discharge regulations and environmental standards (Dores et al., 2012).

Reinjection into petroleum formations for hydraulic fracturing, waterflooding to maintain the pressure in the reservoir and displace the petroleum fluids, and EOR are the most widely used PW management strategies practiced in the industry. Reinjection of PW is generally considered the most environmentally friendly option because it substantially reduces the freshwater or seawater consumption (Lusinier et al., 2019). Reinjection of PW requires removal of suspended solids (SS) to avoid formation plugging. In addition, scale forming constituents such as barium (Ba) and calcium (Ca) must also be removed to minimize scaling.

Water injection is usually utilized as a secondary oil recovery technique in oil fields when reservoirs deplete. By contrast, water is not typically injected in gas reservoirs; hence, PW from gas fields is mostly formation water and condensed water. PW from gas reservoirs is generally much less in volume than that produced from oil fields (Ahmadun et al., 2009). However, due to the higher concentrations of volatile hydrocarbons, PW discharged from gas fields is much more toxic than the PW from oil wells (Duraisamy et al., 2013; Jiménez et al., 2018).

Currently, the majority of PW generated at onshore O&G facilities is reinjected underground either for disposal or for EOR processes. Thus, the major focus of onshore facilities is the types of treatment technologies mainly designed for dispersed O&G and SS to avoid plugging and pumps damage (WEF, 2017, 2018). The common practice for offshore operations is to discharge the treated PW to the sea, leading to the main treatment objective of reducing O&G to acceptable levels and mitigating toxicity impacts on aquatic fauna and flora. Moreover, the requirement for fracturing fluid has changed over the years, leading to different treatment requirements (WEF, 2017). Depending on the location of the onshore O&G facilities, different types of treatment technologies are available, including primary (e.g., hydrocyclone, corrugated plate separator, American Petroleum Institute (API) separator, or similar) and secondary (e.g., flotation units, such as induced gas flotation [IGF], dissolved gas flotation [DGF], dissolved air flotation [DAF], dissolved nitrogen flotation [DNF], and compact flotation unit [CFU]), to support the goal of reducing O&G concentrations in treated PW to 30 or 40 mg L^{-1} (Dores et al., 2012; Veil et al., 2004). Nonetheless the combination of these primary and secondary treatment technologies is unable to produce an effluent that meets the quality standard for beneficial reuse in irrigation or industrial processes (Dores et al., 2012).

There is an increasing push for PW recycling for irrigation, livestock watering, aquifer storage, and municipal and other industrial uses due to climate-induced water scarcity (Al-Ghouti et al., 2019). In addition, highly treated PW may be used for other beneficial uses such as irrigation and industrial processes.

There is need for tertiary treatment of PW or a polishing treatment for the reduction of O&G content, TDS, and other concerning substances depending on the end use. Apart from the O&G and Total suspended solids (TSS) concentrations, those tertiary/polishing treatment technologies focus on treatment of micro- and nanoscale particles, salinity (9% or greater), volatile compounds, extractable organics (acidic, basic, and neutral), ammonia, and hydrogen sulfide. API has assessed several proven tertiary or polishing treatment technologies to reduce the pollutants in PW to desirable effluent quality or almost undetectable levels. These technologies include carbon adsorption (modular granular-activated carbon systems), air stripping (packed tower with air bubbling through the PW stream), membrane filtration (nanofiltration and reverse osmosis polymeric membranes), ultraviolet light (irradiation by UV lamps), chemical oxidation (ozone and/or hydrogen peroxide oxidation), and biological treatment (aerobic system with fixed-film bio-tower or suspended growth) (Igwe et al., 2013). The types of primary, secondary, and tertiary treatment applicable for PW treatment are shown in Figure 1.1. Overall, the specific treatment process or train depends on the characteristics of PW and desired end use of the treated PW. Typical onshore and offshore treatment trains, focused on O&G and TSS removal, are shown in Figure 1.2.

1.2 CHARACTERISTICS OF PRODUCED WATER

PW is a very complex mixture of water and several thousand other constituents similar to those found in crude oil. The physical and chemical properties of PW are variable (Al-Ghouti et al., 2019; Jiménez et al., 2018), and the complex composition, concentrations, and toxicity of PW are influenced by geographic location of the field, composition of the fracking fluid, the geological formation, extraction method, the lifetime of the reservoir, reservoir conditions (e.g., pressure and temperature), and the chemical characteristics of the hydrocarbon being produced. In the O&G industry, the O&G content is generally regulated along with salt contents, total suspended solids (TSS), and other constituents (Jiménez et al., 2018). The toxicity of PW discharged from gas platforms is many times higher than the toxicity of discharge from oil wells, but the volumes of PW are less than those from oil production. These constituents can be (1) organic compounds including oil and grease, (2) suspended solids (SSs), (3) dissolved solids/salts, (4) heavy metals, (5) radioactive materials, (6) bacteria, (7) dissolved gases, etc.

Typical concentrations of constituents found in PW are shown in Table 1.1. Dissolved and dispersed oil compounds are composed of hydrocarbons such as BTEX, naphthalene, phenanthrene, dibenzothiophene (NPD), polyaromatic hydrocarbons (PAHs), phenols, and organic acids (Al-Ghouti et al., 2019; Jiménez et al., 2018). Most of the hydrocarbons do not dissolve in water and mainly disperse as an emulsion or clearly separate into two phases. Therefore, O&G in PW can be in the form of free, dispersed, and emulsified oil. Suspended solids (SSs)/insoluble produced solids include sand, clays, slit, proppants, carbonate and sulfate scales, corrosion products, etc. Some other inorganic crystalline substances such as SiO_2, Fe_2O_3, and Fe_3O_4 can also be found in PW. Large amounts of SSs could lead to serious problems such as clogging flow lines and plugging the well bore downhole, thereby reducing production. The concentration of TSS ranges from a few milligrams per liter up to ~ 5,000 mg L^{-1} (Al-Ghouti et al., 2019). PW may also contain deposited high-molecular-weight components as solid precipitates, such as paraffin waxes and asphaltenes.

Dissolved natural salts and minerals are present in PW as cations and anions such as Na^+, K^+, Ca^{2+}, Mg^{2+}, Ba^{2+}, Cl^-, SO_4^{2-}, and CO_3^{2-}. Sodium and chloride are the

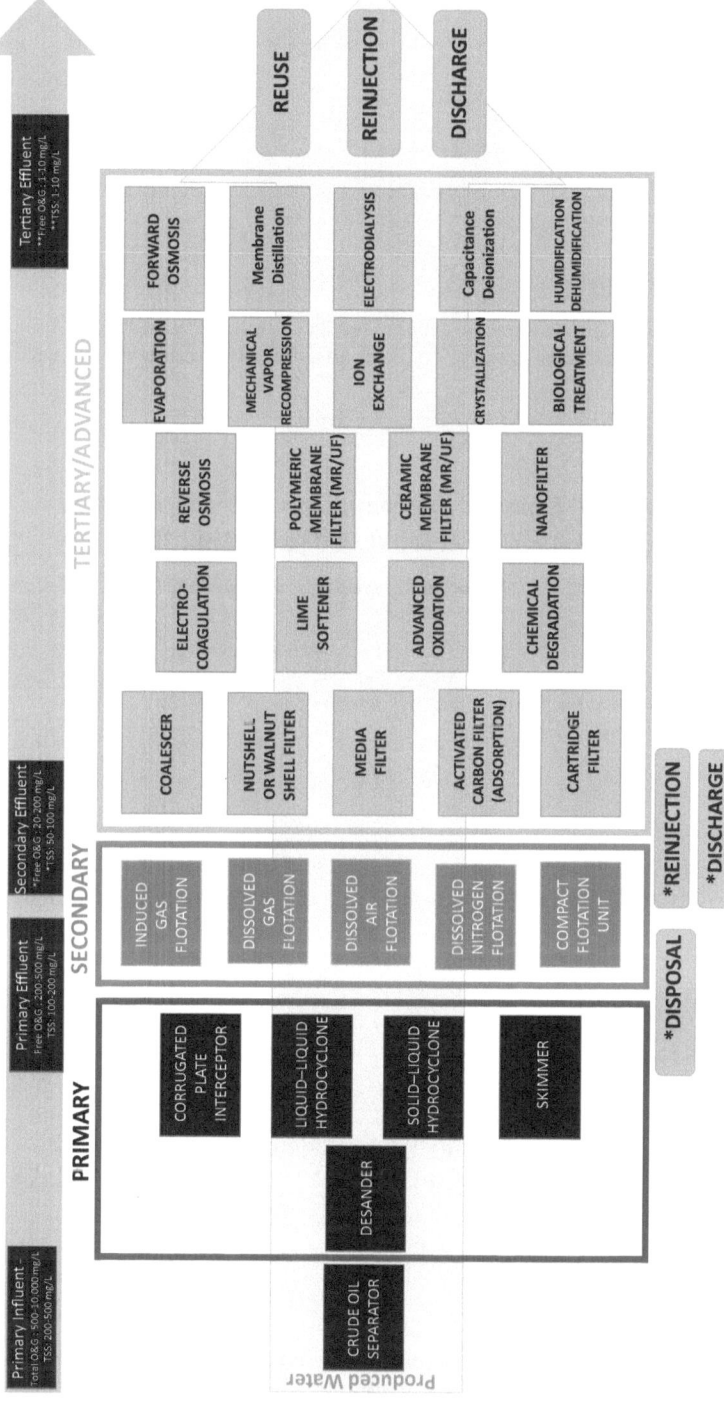

FIGURE 1.1 Primary, secondary, and tertiary treatment technologies applicable for PW treatment (Larson, 2018; WEF, 2018; Jimenez, 2018).

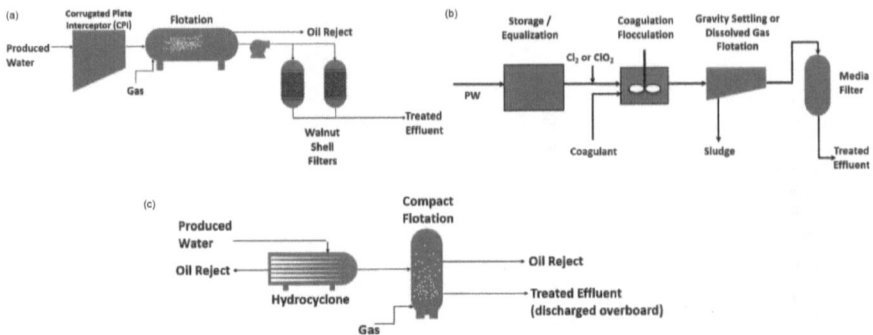

FIGURE 1.2 (a) Typical onshore conventional produced water (PW) treatment, (b) typical unconventional PW treatment, and (c) typical offshore PW treatment. (Adopted from Larson, 2018.)

TABLE 1.1

Main Components and Reported Concentrations in Produced Water (Al-Ghouti et al., 2019; Jiménez et al., 2018; Nasiri et al., 2017)

Parameter/Heavy Metals (mg L^{-1}-Otherwise Shown)	Reported Ranges of Values	Heavy Metals (mg L^{-1})	Reported Ranges of Values
Density (kg m^{-3})	1,014–1,140	Calcium (Ca)	13–25,800
Conductivity (µS cm^{-1})	4,200–58,600	Sodium (Na)	132–97,000
Surface tension (dyn cm^{-1})	43–78	Potassium (K)	24–4,300
pH (unitless)	4.3–10	Magnesium (Mg)	8–6,000
COD	1,220–2,600	Iron (Fe)	<0.01–100
TOC	0–1,500	Aluminum (Al)	310–410
TSS	1.2–1,000	Boron (B)	5–95
TDS	100–400,000	Barium (Ba)	1.3–650
Total oil (IR) (O&G)	2–565	Cadmium (Cd)	<0.005–0.2
Benzene	0.032–778.51	Copper (Cu)	<0.02–1.5
Ethylbenzene	0.026–399.84	Chromium (Cr)	0.02–1.1
Toluene	0.058–5.86	Lithium (Li)	3–50
Xylene	0.01–1.29	Manganese (Mn)	<0.004–175
Volatile compounds (BTEX)	0.39–35	Lead (Pb)	0.002–8.8
Chloride	800–200,000	Strontium (Sr)	0.02–1,000
Bicarbonate	77–3,990	Titanium (Ti)	<0.01–0.7
Sulfate	<2–1,650	Zinc (Zn)	0.01–35
Ammonium (as N)	10–300	Arsenic (As)	<0.005–0.3
Sulphite	10	Mercury (Hg)	<0.005–0.3
Phenol	0.009–23	Silver (Ag)	<0.001–0.15
Total organic acids	0.001–10,000	Beryllium (Be)	<0.001–0.004
Volatile fatty acids (VFA)	0.009–4,900	Palladium (Pd)	0.008–0.88

COD, chemical oxygen demand; TSS, total suspended solids; TDS, total dissolved solids; O&G, oil and gas; BTEX, benzene, toluene, ethylbenzene, and xylenes.

main ions responsible for the salinity of PW. The TDS varies considerably and is usually higher than seawater, ranging from a few parts per million (ppm) to approximately 400,000 ppm (Ahmadun et al., 2009; Jiménez et al., 2018). The high salinity (i.e., high TDS) of PW makes it unsuitable for reuse and generally requires an expensive and energy-intensive treatment to reduce the TDS to acceptable levels where the PW can be reused. In addition, concentrated brine often results in the formation of scales such as calcite ($CaCO_3$) and barite ($BaSO_4$) upon temperature and pressure changes, causing serious problems such as plugging of reservoir rock pores, production losses, and equipment damage.

PW may contain trace quantities of heavy metals, such as iron, nickel, copper, zinc, arsenic, cadmium, mercury, and lead (Ahmadun et al., 2009), which are classified as dissolved inorganic compounds. Naturally occurring radioactive materials/radionuclides such as ^{226}Ra and ^{228}Ra may also be present in oilfield PW (Ahmadun et al., 2009). Like heavy metals, naturally occurring radioactive materials are also classified as dissolved inorganic compounds. There can be bacteria/viruses present that require treatment, such as sulfur oxidizing anaerobic bacteria, which can cause corrosion and scaling and thereby clog the pipelines and formation pores. Large quantities of dissolved gases are contained in oilfield brines, mostly volatile hydrocarbons, but also CO_2, O_2, and H_2S are commonly found in PW.

In addition to its natural components, PW may include chemical additives dosed in drilling to treat or prevent operational problems and enhance oil/water separation. Such additives include gas hydrate inhibitors, corrosion inhibitors, oxygen scavengers, scale inhibitors, biocides to mitigate bacterial fouling, asphaltene dispersants, paraffin inhibitors, defoamers, emulsion breakers, clarifiers, coagulants, flocculants, etc. (Daigle et al., 2012). Some of these chemicals are highly toxic even at low concentrations (Table 1.2).

TABLE 1.2

Typical Concentrations of Produced Water Chemical Additives (Ahmadun et al., 2009)

Chemical Name	Concentrations in Oil Field		Concentrations in Gas Field	
	Typical (mg L⁻¹)	Range (mg L⁻¹)	Typical (mg L⁻¹)	Range (mg L⁻¹)
Corrosion inhibitor (contains amide/imidazoline compounds)	4	0.3–10	4	0.3–10
Scale inhibitor (contains phosphate ester/ phosphate compounds)	10	0.2–30	...	...
Demulsifier (contains oxylated resins/ polyglycol ester/alkyl aryl sulphonates	1	1–2	...	...
Polyelectrolyte (e.g., polyamine compounds)	2	0–10	...	...
Methanol	...	...	2,000	1,000–15,000
Glycol (DEG)	...	...	1,000	7.7–2,000

1.3 TREATMENT METHODS FOR PRODUCED WATER

Since PW contains several different contaminants with varying concentrations, numerous treatment technologies with a series of individual unit processes are required to remove contaminants that might not be removed through a single process. Prior to disposal or any form of PW reuse, proper contaminant removal treatment is required to comply with environmental regulations and to meet the requirements and standards for reuse applications. The treatment required depends on the PW composition and how the PW is disposed or reused. Onshore PW is usually discharged into deep disposal wells, and only dispersed hydrocarbons and SS are removed to prevent formation plugging (Hussain et al., 2014). On the other hand, PW in offshore operations is often discharged to sea, and only hydrocarbons are treated to acceptable concentrations to meet the environmental regulations and standards. Reuse in oilfield operations, such as in waterflooding, drilling, and hydraulic fracturing, may require only limited PW treatment to meet the needs for these operations. However, reuse in beneficial applications such as in agriculture irrigation and industrial processes might require more extensive treatment to comply with more restrictive limitations and meet the quality required (Gray, 2020).

A typical PW treatment process has three main stages: (1) primary treatment, (2) secondary treatment, and (3) tertiary/advanced treatment steps with a pretreatment step (Figure 1.1). The pretreatment step is done to remove large oil droplets, coarse particles, and gas bubbles with the goal of reducing the amount of dispersed contaminants that would otherwise pass through the crude oil separator. The primary treatment step involves removing small oil droplets and particles using desanders, skim tanks, plate pack interceptors, API separators, and/or liquid–liquid or solid–liquid hydrocyclones. The secondary treatment involves removal of much smaller oil droplets and particles using gas flotation (e.g., IGF and DGF.) and, sometimes, hydrocyclones and centrifuges. The tertiary/advanced treatment step is usually employed to remove ultrasmall droplets and particles and dispersed hydrocarbons ($<10\,\mathrm{mg\,L^{-1}}$) using techniques such as dual media filters, cartridge filters, and membranes (WEF, 2017, 2018; Al-Ghouti et al., 2019).

Different physical, chemical, and biological processes are employed at different stages (e.g., primary, secondary, and tertiary/advanced step) of PW treatment. A well-designed combination (hybrid method) of two or more treatment technologies is commonly used to achieve a high degree of treatment and to reduce energy consumption. In general, a viable treatment method should have low operating costs and high efficiency. Additionally, in offshore uses, the technology should also be compact to accommodate space and weight limitations (Nonato et al., 2018). Typical onshore and offshore O&G and TSS treatment trains are shown in Figure 1.2.

The treatment methods can be broadly classified into basic separation methods designed to remove suspended solids and dispersed oil and grease, and more advanced techniques tailored for the removal of dissolved solids and hydrocarbons to achieve a higher degree of treatment (Lin et al., 2020). Basic separation methods include gravity separation, media filtration, flotation, coagulation-flocculation, and cyclonic/centrifuge separation. Commonly used advanced treatment methods include membrane filtration, adsorption, distillation, ion exchange, advanced oxidation processes, etc.

The detailed description of treatment methods, their advantages, and drawbacks can be found in several recent excellent reviews on the treatment of PW (Al-Ghouti et al., 2019; Jiménez et al., 2018; Nasiri et al., 2017; Nonato et al., 2018; Wei et al., 2020).

In general, the treatment technologies are selected and recommended based on the following factors: (1) source of PW: onshore and offshore, (2) PW composition and concentration of pollutants, (3) regulations and environmental standards associated with discharge and reuse, (4) space requirements, and (5) cost of treatment. An overview of the separation technologies in use or with potential to treat PW is presented in Table 1.3 and Figure 1.1.

This chapter briefly describes some treatment techniques, including physical, chemical, or biological processes, for separating different types of contaminants from PW. Biological methods such as the activated sludge process, aerated filtration, and membrane bioreactors are not extensively utilized in PW treatment, but interest is increasing due to recycling and beneficial reuse of PW. Biological treatment processes are generally mostly used in refineries, petrochemical, and other downstream facilities to remove dissolved organic compounds by biodegradation, in which aerobic or

TABLE 1.3

Summary of Existing and Emerging Technologies for Produced Water

Treatment Technology	Dispersed Oil & Grease	Dissolved Hydrocarbons	Suspended Solids	Dissolved Solids
Physical Methods				
Gravity separator	X		X	
Hydrocyclones	X		X	
Microfiltration			X	
Ultrafiltration	X	X	X	
Nanofiltration	X	X	X	X
Reverse osmosis	X	X	X	X
Membrane distillation	X	X	X	X
Thermal separators				X
Flotation	X		X	
Activated carbon adsorption	X	X	X	X
Chemical Methods				
Chemical precipitation	X		X	X
Ion exchange				X
Advanced oxidation processes		X		
Electrodialysis				X
Electrochemical processes		X		
Biological Methods				
Aerated filtration	X	X		
Activated sludge		X		
Membrane bioreactors	X	X	X	X

anaerobic microorganisms decompose the dissolved hydrocarbons into smaller molecules that can then be converted into water, CO_2, and biomass through biological oxidation. In general, when compared to physical and chemical treatments, biological treatments have higher removal efficiencies for dissolved hydrocarbons and are relatively less expensive. However, they suffer from serious challenges such as large footprints, which make biological treatments unsuitable for offshore applications. Other major challenges are the toxicity of some dissolved compounds, such as BTEX, and the high salinity of PW, which may strongly limit biological activity. Interested reader is referred to the comprehensive recent reviews on biological treatments of PW (Lusinier et al., 2019; Camarillo and Stringfellow, 2018; Wei et al., 2003).

As discussed earlier, PW treatment processes focus on the removal of oil and grease and other contaminants. PW treatment equipment (e.g., API gravity separator, corrugated plate separator, and IGF) have different capacities for particles size removal. Table 1.4 shows the list of different de-oiling technologies with respect to their particle size.

Treatment technologies such as corrugated plate separator, centrifuge, hydrocyclone, and gas flotation can be used effectively to recover oil from emulsions and/or water with high oil content prior to discharge. PW from water-drive reservoirs and water flood production are the most likely feedstocks, containing oil and grease in excess of 1,000 mg L^{-1} (Arthur, 2005). Treatment processes such as extraction, ozone/hydrogen peroxide, oxygen, and adsorption can remove oil from water with low oil and grease content (<1,000 mg L^{-1}) or remove trace quantities of oil and grease prior to membrane processing. Oil reservoirs and thermogenic natural gas reservoirs usually contain trace amounts of liquid hydrocarbons. Biogenic natural gas, such as coal-based natural gas (CBNG), may contain no liquids in the reservoir but when pumped to the surface, the water takes up lubricating fluids from the pumps. The basic description of de-oiling technologies, their respective advantages and disadvantages, together with their types of waste stream are described in Table 1.5.

TABLE 1.4

Summary of Different Oil Removal Technologies for Produced Water (Arthur et al., 2005; WEF, 2019)

Oil Removal Technology	Minimum Size of Particles Removed (µm)
American Petroleum Industry (API) gravity separator	150
Corrugated plate separator	40
Induced gas flotation (no flocculants)	25
Induced gas flotation (with flocculants)	3–5
Hydrocyclone	10–15
Mesh coalescer	5
Media filter	5
Centrifuge	2
Membrane filter	0.01

TABLE 1.5

Description, Advantages, Disadvantages, and Waste Streams of Different De-oiling Technologies

Treatment	Description	Advantages	Disadvantages	Waste Stream
Corrugated plate separator	Separation of free oil from water under gravity effects enhanced by flocculation on the surface of corrugated plates	a. No energy required b. Cheaper and effective for bulk oil removal c. No moving parts d. Robust and resistant to breakdowns	a. Inefficient for fine oil particles b. Requirement of high retention time c. Maintenance	Suspended particles slurry at the bottom of the separator
Centrifuge	Separation of free oil from water under centrifugal force generated by spinning the centrifuge cylinder	a. Efficient removal of smaller oil particles and suspended solids b. Lesser retention time c. High throughput	a. Energy requirement for spinning b. High maintenance cost	Suspended particles slurry as pretreatment waste
Hydrocyclone	Free oil separation under centrifugal force generated by pressurized tangential input of influent stream	a. Compact modules b. Higher efficiency and throughput for smaller oil particles	a. Energy requirement to pressurize inlet b. No solid separation c. Fouling d. Higher maintenance cost	
Gas Flotation	Oil particles attached to induced gas bubbles and float to the surface	a. No moving parts b. Higher efficiency due to coalescence c. Easy operation d. Robust and durable	a. Generation of large amount of air b. Retention time required for separation c. Skim volume	Skim oW volume, lumps of oil
Extraction	Removal of free or dissolved oil soluble in lighter hydrocarbon solvent	a. No energy required b. Easy operation c. Removes dissolved oil	a. Use of solvent b. Extract handling c. Regeneration of solvent	Solvent regeneration waste
Ozone/hydrogen peroxide/oxygen	Strong oxidizers oxidize soluble contaminant and remove them as precipitate	a. Easy operation b. Efficient for primary treatment of soluble constituents	a. On-site supply of oxidizer b. Separation of precipitate c. Byproduct C 2	Solids precipitated in slurry form
Adsorption	Porous media adsorbs contaminants from the influent stream	a. Compact packed bed modules b. Cheaper and efficient	a. High retention time b. Less efficient at higher feed concentration	Used adsorbent media, regeneration waste

Adapted from Arthur et al. (2005).

Removal of bacteria, viruses, microorganisms, algae, etc., from PW is necessary to prevent scaling, water contamination (to protect potability), or fouling of the reservoir, tubulars, and surface equipment. Microorganisms can occur naturally in PW or may be added during de-oiling treatments. Advanced filtration techniques are one effective technology used to remove microorganisms. UV light treatment, chorine or iodine reaction, ozone treatment, and pH reduction are other treatments available to disinfect PW (Arthur et al., 2005). The basic description, advantages, disadvantages, and waste streams of major disinfection techniques are shown in Table 1.6.

Removal of dissolved solid, salts, or impurities is one of the key functions of the water treatment systems. TDS in PW ranges from <2,000 to >150,000 ppm. The choice of desalination method depends on TDS content and the treatment system's compatibility to function in the presence of extra contaminants in the PW. O&G operators have attempted evaporation, distillation, membrane filtration, electric separation, and chemical treatments to remove TDS from PW. Microfiltration (MF), ultrafiltration (UF), nanofiltration (NF), and reverse osmosis (RO) utilize high pressure across the membranes to accomplish filtration of contaminants from PW. Cations such as Na^+, K^+, Ca^{2+}, Mg^{2+}, Ba^{2+}, Sr^{2+}, and Fe^{2+} and anions such as Cl^-, SO_4^{2-}, CO_3^{2-}, and HCO_3^- affect PW chemistry in terms of buffering capacity, salinity, and scale potential as well as subsequent removal efficiency of the treatment technologies. PW also contains trace quantities of various heavy metals such as cadmium, chromium, copper, lead, mercury, nickel, silver, and zinc, mostly from natural origins, that affect relevant treatment technologies. Tables 1.7 and 1.8 provide descriptions, advantages, disadvantages, and waste streams of different desalination technologies and membrane processes. Technologies shown in Table 1.7 typically require less power and less pretreatment than membrane technologies. Suitable PW feed will have TDS value between 1,000 and 10,000 mg L^{-1}. Some of these treatment processes remove oil and grease contaminants while others require oil and grease contaminants to be reduced

---

TABLE 1.6

Description, Advantages, Disadvantages, and Waste Streams of Different Disinfection Technologies

Treatment	Description	Advantages	Disadvantages	Waste Stream
UV light/ ozone	Passing UV light or ozone to produce hydroxyl ions that kills microbes	a. Simple and clean operation b. Highly efficient disinfection	a. On-site supply of ozone b. Other contaminants reduce efficiency	Small volumes of suspended particles at the end of the treatment
Chlorination	Chlorine reacts with water to produce hypochlorous acid which kills microbes	a. Cheaper and simplest method	a. Does not remove all types of microbes	

Adapted from Arthur et al. (2005).

TABLE 1.7

Description, Advantages, Disadvantages, and Waste Stream of Different Desalination Technologies

Treatment Methods/ Technology	Description	Advantages	Disadvantages	Waste Stream
Lime softening	Addition of lime to remove carbonate, bicarbonate, and hardness	Cheaper, accessible, can be modified	Chemical addition, posttreatment necessary	Used chemical and precipitated waste
Ion Exchange	Dissolved salts or minerals are ionized and removed by exchanging ions with ion exchangers	Low energy required, possible continuous regeneration of resin, efficient, mobile treatment possible	Pre- and posttreatment required for high efficiency, produce effluent concentrate	Regeneration chemicals
Electrodialysis	Ionized salts attract and approach to oppositely charged electrodes passing through ion exchange membranes	Clean technology, no chemical addition, mobile treatment possible, less pretreatment	Less efficient with high concentration influent, require membrane regeneration	Regeneration waste
Electro- deionization	Enhanced electrodialysis due to presence of ion exchange resins between ion exchange membranes	Removes weakly ionized species, high removal rate, mobile treatment possible	Regeneration of ion exchange resins, pre/ posttreatment necessary	Regeneration waste, filtrate waste from posttreatment stage
Capacitive deionization	Ionized salts are adsorbed by the oppositely charged electrodes	Low energy required, higher throughput	Expensive electrodes, fouling	Regeneration waste
Electrochemical Activation	Ionized water reacts with ionized chloride ion to produce chlorite that kills microbes	Simultaneous salt and microbial removal, reduce fouling	Expensive electrodes	Regeneration waste
Rapid spray evaporation	Injecting water at high velocity in heated air evaporates the water which can be condensed to obtained treated water	High-quality treated water, higher conversion efficiency	High energy required for heating air, required handling of solids	Waste in sludge form at the end of evaporation

(Continued)

TABLE 1.7 (*Continued*)
Description, Advantages, Disadvantages, and Waste Stream of Different Desalination Technologies

Treatment Methods/ Technology	Description	Advantages	Disadvantages	Waste Stream
Freeze-thaw evaporation	Utilize natural temperature cycles to freeze water into crystals from contaminated water and thaw crystals to produce pure water	No energy required, natural process, cheaper	Lower conversion efficiency, long operation cycle	N/A

Adapted from Arthur et al. (2005).

TABLE 1.8
Description, Advantages, Disadvantages, and Waste Stream of Membrane Technologies

Treatment	Description	Advantages	Disadvantages	Waste Stream
Microfiltration	Membrane removes micro particles from the water under the applied pressure	Higher recovery of fresh water, compact modules	High energy required, less efficiency for divalent, monovalent salts, viruses, etc.	Concentrated waste from membrane backwash during membrane cleaning, concentrate stream from the filtration operation
Ultrafiltration	Membrane removes ultraparticles from the water under the applied pressure	Higher recovery of fresh water, compact modules, viruses, and organics removal	High energy, membrane fouling, low MW organics, salts, etc.	
Nanofiltration	Membrane separation technology removes species ranging between ultrafiltration and reverse osmosis (RO)	Low MW organics removal, hardness removal, divalent salts removal, compact module	High energy required, less efficient for monovalent salts and lower MW organics, membrane fouling	
Reverse Osmosis	Pure water is squeezed from contaminated water under pressure differential	Removes monovalent salts, dissolved contaminants, and compact modules	High pressure requirements, even trace amounts of oil and grease can cause membrane fouling	

Adapted from Arthur et al. (2005).

before their operation. Removal of trace oil and grease, microbial, soluble organics, divalent salts, acids, and trace solids are possible via membrane-based technologies. Contaminants can be targeted by the selection of the membrane. Removal of sodium chloride, other monovalent salts, and other organics can be achieved via a RO membrane, although some organic species may require pretreatment. While energy costs increase with higher TDS, RO can efficiently remove salts in excess of 10,000 mg L^{-1}.

PW softening, sodium adsorption ratio (SAR) adjustments, and removal of trace contaminants, pollutants, naturally occurring radioactive materials (NORM), etc., are part of PW treatment in some regions, depending on the PW composition. Different biological treatment technologies (e.g., fixed-film treatment, membrane bioreactors, wetlands and ponds, activated sludge treatment, anaerobic treatment, and bio-electrochemical treatment) are also emerging, though not used widely yet. The desire to recycle and reuse PW has led to increased interest in its biological treatment. Technical details and their relevance to PW treatment are described below for various widely used physical and chemical PW treatment processes. Some physical treatment processes included are (1) hydrocyclones, (2) API separator and corrugated plate separator/interceptor, (3) media filtration (e.g., nutshell filter), (4) gas flotation (5) membrane filtration, (6) membrane distillation, (7) thermal separators, and (8) activated carbon adsorption. Some chemical treatment processes included are (1) chemical precipitation, (2) ion exchange, (3) advanced oxidation, (4) electrodialysis, and (5) electrochemical processes.

1.3.1 HYDROCYCLONES

In the petroleum industry, cyclones are often used for desanding, for instance at the wellhead, to protect the downstream equipment. Hydrocyclones are also widely used to treat PW. A cyclone uses centrifugal acceleration to mechanically reduce or increase, depending on the process objectives, the concentration of a dispersed phase (aggregates, particles, droplets, etc.) within a dispersant media (Jiménez et al., 2019). Hydrocyclones can be classified as liquid–liquid, liquid–solid, or gas–liquid separation types (Liu et al., 2015). Hydrocyclones are mainly used to remove suspended solid particles and dispersed oil droplets based on the density difference and centrifugal force. As shown in Figure 1.3, a hydrocyclone has two sections: a cylindrical section, where the feed stream enters under pressure tangentially at the top, and a conical section. While the heavier phase is forced toward the wall of the hydrocyclone and discharged at the bottom (underflow), the lighter phase flows toward the center and leaves at the top (overflow). Three-phase cyclonic separators have also been designed to remove solids and oil from PW (Ahmadun et al., 2009). Hydrocyclones do not require any chemicals or pretreatment; however, hydrocyclones cannot remove dissolved components. A typical cyclone removal efficiency for dispersed oil is approximately 50%–70% (Ahmadun et al., 2009).

1.3.2 API SEPARATOR AND CORRUGATED PLATE SEPARATOR/INTERCEPTOR

The API separator (Figure 1.4) is a gravity-based device designed using Stokes law. Most SS will settle to the bottom of the separator as a sediment layer, the oil will rise to top of the separator, and the wastewater will compose the middle layer. Any settled

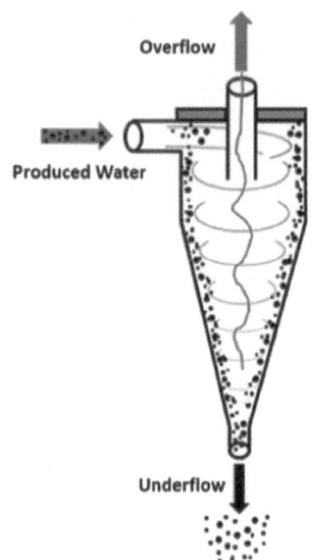

FIGURE 1.3   A general scheme of a hydrocyclone separator.

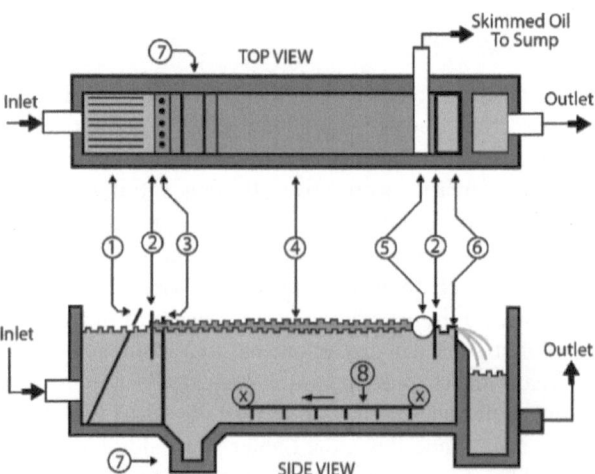

FIGURE 1.4   A general scheme of an API separator (Robinson, 2013): (1) trash trap (inclined rods), (2) oil retention baffles, (3) flow distributors (vertical rods), (4) oil layer, (5) slotted pipe skimmer, (6) adjustable overflow weir, (7) sludge sump, and (8) Chain and flight scraper.

gross solids and trash must be periodically removed from the trash screen in the inlet chamber (Duraisamy et al., 2013; Judd et al., 2014; Han et al., 2017). Whereas conventional oil–water separators can only remove free oil, API separators are designed to remove oil droplets with diameters as small as 0.015 cm (150 μm). Under most operating conditions, the API separator will remove both free oil and SS down to a concentration between 50 and 200 mg L⁻¹ (WEF, 2017, 2018). Chemical oxygen demand (COD) removals in the range of 16%–45% and TSS removals in the range of 33%–68% have been reported. Removing the bulk of free oils, greases, and SS from the wastewater reduces overloading and other problems in downstream treatment processes (Duraisamy et al., 2013; Judd et al., 2014; Han et al., 2017). Plate separators, or coalescing plate separators (CPI), are similar to API separators and are also based on Stokes law principles but include inclined plate assemblies (parallel packs). The underside of each parallel plate provides more surface for suspended oil droplets to coalesce into larger globules (Boraey, 2018; Ahmadun et al., 2009). Separation of free oil from water under gravity is enhanced by flocculation on the surface of corrugated plates. CPI is widely used for oil recovery from emulsions or water with high oil content prior to discharge. Water may contain oil and grease in excess of 1,000 mg L^{-1} (Ahmadun et al., 2009).

1.3.3 MEDIA FILTRATION

A relatively simple technique used in O&G treatment process, filtration, is based on the use of porous filter media to allow only water and not the impurities (e.g., oil and grease) to pass through it. Filtration technology is extensively used to remove oil and grease and total organic carbon (TOC) from PW (more than 90% efficiency). Various porous materials can be used as filter media, such as sand, gravel, anthracite, and walnut shell. However, sand is the most widely used material due to its availability, low cost, and efficiency. Walnut shell filters are commonly used for PW treatment because they are not affected by water salinity and might be applicable to any type of PW. Filter efficiency can be further enhanced if coagulants are added to the feed water prior to filtration. Media regeneration and solid waste disposal are setbacks to this process (Igunnu and Chen, 2014; Ahmadun et al., 2009).

1.3.4 ACTIVATED CARBON ADSORPTION

Adsorption is considered one of the best techniques used in a tertiary/advanced step to achieve high water quality with nearly undetectable levels of pollutants. Activated carbon is particularly effective in removing contaminants, thanks to its unique characteristics, including high surface reactivity, high adsorption ability, large surface area, and microporous structure (Al-Ghouti et al., 2019). In addition to suspended particles and insoluble free hydrocarbons, activated carbon can also be used to remove dissolved organic compounds, heavy metals, and radioactive materials. Installation and maintenance costs are the major disadvantages of activated carbon adsorption. As in other adsorption processes, the activated carbon must be regenerated after a few runs. Various chemicals such as acids and organic solvents can be used to regenerate the activated carbon, which results in liquid waste disposal and an increase in treatment costs (Al-Ghouti et al., 2019).

1.3.5 GAS FLOTATION

This widely used treatment process for oilfield PW can be used to remove volatile organics, oil, and grease from PW (Igunnu and Chen, 2014). A gas such as nitrogen or air is injected into the PW to remove suspended particles and dispersed oil droplets.

DGF and IGF are two subdivisions of the gas flotation technology based on the method used to generate gas bubbles and the resultant bubble size (Al-Ghouti et al., 2019). The process efficiency mainly depends on the contaminants to be removed, liquids density differences, temperature, and oil droplet size. Particles of 25 μm can be removed by dissolved air flotation, and 3–5 μm particles can be removed when coagulation is used as pretreatment step (Al-Ghouti et al., 2019). Fine solid particles and small oil droplets attach to the micro gas bubbles and rise together to the surface due to an increase in buoyancy or a diminished aggregate density. As a result, foam forms at the water surface, which can then be removed by skimming, and the clarified water is collected at the bottom of the flotation zone. This process is simple, robust, and requires no moving parts. The disadvantages include a large amount of gas and a large skim volume (Al-Ghouti et al., 2019).

1.3.6 MEMBRANE FILTRATION

Membrane systems can compete with more complex treatment technologies for treating water with high oil content, low mean particle size, and flowrates greater than $150\,m^3h^{-1}$ and is, consequently, suitable for medium and large offshore platforms (Ahmadun et al., 2009). A membrane is a thin semi-permeable layer of organic (e.g., polymeric membranes) or inorganic (e.g., ceramic membranes) material that separates a pollutant from PW when an external pressure is applied across the membrane. As shown in Figure 1.5, pressure-driven membrane separation technologies are classified according to pore size (i.e., MF, UF, NF, and RO). Whereas MF and UF membranes primarily remove bacteria, viruses, proteins, colloidal particles, and SS particles, NF membranes and RO can reject molecules and ions. This is because water flows through the pores of MF and UF membranes, whereas in NF and RO membranes water moves through the molecular structure (Thomas, 2019). In RO membranes, an external hydraulic pressure suppresses the osmotic pressure and forces the permeate to diffuse through the membrane. While NF membranes can remove multivalent ions such as calcium, magnesium, and sulfate, RO can retain monovalent ions, such as sodium and chloride, in addition to multivalent ions (Dores et al., 2012; Thomas, 2019). RO osmosis membranes can achieve 99% salt rejection (Ahmad et al., 2020) and 99.9% organic rejection (Ahmad et al., 2020).

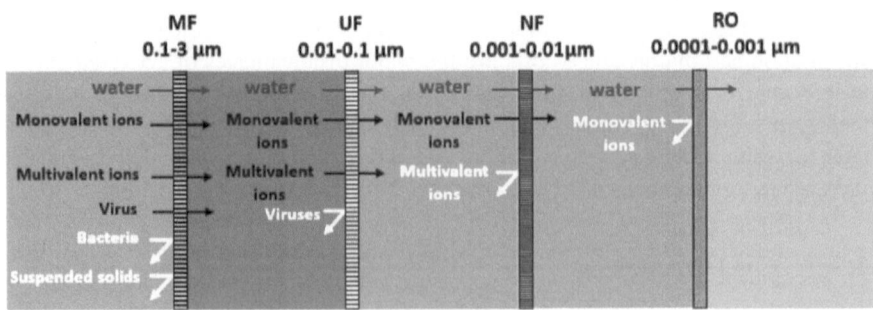

FIGURE 1.5 Membrane filtration technologies (MF, microfiltration; UF, ultrafiltration; NF, nanofiltration; RO, reverse osmosis.)

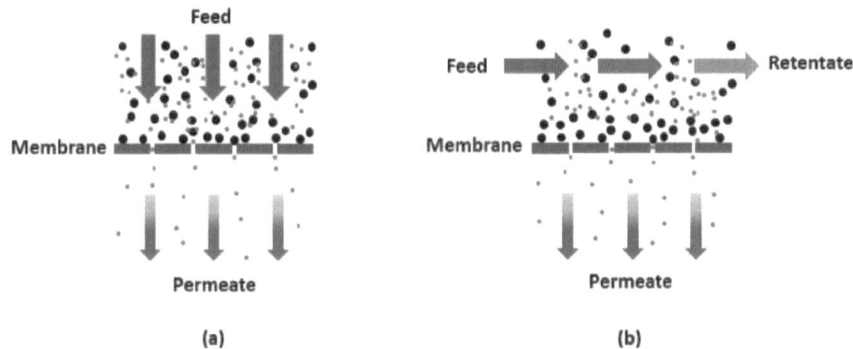

FIGURE 1.6 Operation modes for membranes: (a) dead-end filtration and (b) crossflow filtration.

Based on their material type, membranes can also be classified into polymeric, inorganic, and composite. Polymeric membranes are prepared from materials like polytetrafluoroethylene (PTFE), polyacrylonitrile (PAN), polysulfone (PS), and polyvinylidenedifluoride (PVDF). These membranes are highly efficient for removing dispersed oil and SS particles. Inorganic membranes include ceramic membranes, metallic membranes, glass membranes, and zeolitic membranes. They have better chemical and thermal stability than polymeric membranes, but they are generally more expensive (Duraisamy et al., 2013; Dickhout et al., 2017). Membranes can be operated in two modes, dead-end filtration and crossflow filtration, as shown in Figure 1.6.

Compared to traditional separation methods, UF is one of the most effective methods for oily wastewater treatment, especially for PW, because of its high oil removal efficiency. UF requires no chemical additives, incurs low energy costs, and has small space requirements (Ahmadun et al., 2009). The main advantages of membrane filtration technologies for O&G treatment include their (1) small footprint, which makes membrane filtration suitable for both onshore and offshore operations; (2) modularity, easy to upgrade capacity; (3) consistent and high-quality permeate; (4) ease of operation, fully automated; (5) little or no chemical requirements; (6) small sludge quantities; and (7) continuous processing. The major issue with membrane filtration technologies is membrane fouling caused by the complex contaminants in PW. In membrane fouling, a layer of solids, oil, and other PW constituents form on the membrane surface resulting in decreased permeate flux, selectivity, and membrane lifetime. Furthermore, fouled membranes require higher pressure during operation. Fouling in membranes can be either reversible or irreversible. Reversible fouling is due to deposited particles or dissolved components on the membrane and can be reversed by backwashing with pure water. Irreversible fouling is a result of strong sorption on the membrane surface and in the membrane pores (Duraisamy et al., 2013).

1.3.7 MEMBRANE DISTILLATION

Unlike pressure-driven membrane filtration processes, membrane distillation (MD) is a thermally driven process based on the temperature difference (or vapor pressure difference) between the hot brine and cold distillate streams. Hydrophobic membranes are used in MD to allow only water vapor to pass through. MD has four major configurations, including direct contact MD (DCMD), air gap MD (AGMD), vacuum MD (VMD), and sweeping gas MD (SGMD) with 100% (theoretical) solute rejection capacity (Wang and Chung, 2015; Nasiri et al., 2017). These configurations differ in how the driving force (the vapor pressure gradient) is applied. Among them, DCMD, which utilizes a hydrophobic microporous membrane, is the simplest to operate. Study results from Al-Salmi et al. show that DCMD has great potential for treatment of PW (Al-Salmi et al., 2020). MD exhibits several advantages, such as high selectivity, high salt rejection efficiency, no external pressure, and fewer fouling issues (Ahmad et al., 2020). The main drawbacks include high energy consumption, long-time operation instability, and membrane wetting (Ahmad et al., 2020).

1.3.8   Thermal Separators

Thermal separation processes are widely used for water desalting, particularly in regions where energy sources are readily available and are mainly used for large desalting plants, which include PW treatment processes (Nasiri et al., 2017). Thermal separation can be used for desalting water with high TDS, up to 40,000 mg L^{-1}. Some chemicals, such as EDTA and acids, are used in conjunction with thermal separation to prevent scaling (Nasiri and Jafari, 2017). Major thermal desalination techniques include multistage flash (MSF), multi-effect distillation (MED), and vapor compression distillation (VCD). In MSF distillation, water evaporation occurs by reducing the pressure of the feed stream instead of heating. MED generally uses steam to evaporate water in a series of evaporators. In VCD, compression of vapor provides the required heat. A combination of these thermal processes such as a hybrid MED–VCD can also be used to treat PW (Igunnu and Chen, 2014). This hybrid treatment method has some advantages over the other conventional thermal technologies such as reduced overall costs and less fouling (Jiménez et al., 2018). Various evaporator designs such as horizontal tube, vertical tube rising film, and vertical tube falling film are used to improve heat transfer rates. These evaporators offer several advantages; they are simple and require minimal pretreatment and substantially fewer chemicals. A major drawback is that evaporators increase the concentration of solids, which results in crystal precipitation and scaling (Dores et al., 2012; Nasiri et al., 2017). Another thermal technique is freeze-thaw evaporation (FTE®). This mature technology was developed in 1992 by Energy & Environmental Research Centre (EERC) and B.C. Technologies Ltd. (BCT). FTE is based on "freezing point depression," a phenomenon in which salts and other dissolved constituents in PW decrease the freezing point of the solution to a temperature below the freezing point of pure water. When PW is cooled below 32°F but above its freezing point, pure water crystallizes; the ice crystals are then collected and melted to obtain cleaner water. The concentrated solution remains unfrozen (Igunnu and Chen, 2014). This technology is easy to operate and robust, but it requires large ponds and only works in cold seasons with subfreezing temperatures.

1.3.9   Chemical Precipitation

Precipitation is considered a conventional chemical treatment processes of PW (Al-Ghouti et al., 2019). Chemical precipitation is used to remove suspended solids, dispersed oil droplets, and colloidal particles from PW using flocculation and coagulation chemicals. The basic idea is to increase the size of the solid particles so they can precipitate. In coagulation, the electrostatic repulsion between the particles is reduced by chemicals called coagulants, such as aluminum sulfate, ferric chloride, and lime. These coagulants react with the suspended particles to form precipitants. In flocculation, the particles are brought together by water-soluble polymeric agents. The addition of coagulation chemicals can remove almost 97% of SS and oil from PW (Al-Ghouti et al., 2019). Chemical precipitation is a simple technology for removing suspended particles, but it is ineffective in removing dissolved components. Another concern is the increased concentration of some toxic metals in the sludge that forms due to the use of chemicals (Duraisamy et al., 2013; Jiménez et al., 2018).

1.3.10 Ion Exchange

Another chemical technology widely used in industrial applications for PW treatment is ion exchange technology. This technology can remove various PW constituents such as dissolved heavy metals, arsenic, salts, radium, and uranium (Arthur et al., 2005). The method utilizes resins, in which cations or anions in the resin exchange similarly charged ions in the PW (Jiménez et al., 2018). Since the resin favors divalent ions (Ca, Mg, etc.) over monovalent (Na) ions for replacement, secondary treatment for SAR (sodicity) is required (Arthur et al., 2005). Ion exchange has been applied in many industrial operations including for the treatment of coal bed methane (CBM) PW (Igunnu and Chen, 2014). Resins can be especially suitable for eliminating monovalent and divalent ions and metals present in PW, with capacity to remove boron from RO permeate of PW (Jiménez et al., 2018). Ion exchange technology has a lifetime of approximately 8 years and requires pretreatment for solids removal, as well as the use of chemicals for resin regeneration and disinfection (Jiménez et al., 2018).

1.3.11 Advanced Oxidation Processes

In oxidation processes, oxidants, such as ozone (O_3), hydrogen peroxide (H_2O_2), chlorine, and ultraviolet (UV), or mixtures of these oxidants, are used to crack down dissolved organic contaminants into simple, less toxic molecules. Advanced oxidation processes (AOPs) have been extensively studied and are considered mature technologies. AOPs have received increasing interest for the treatment of PW in industrial-scale applications due to numerous advantages such as their capability to achieve complete mineralization of organic components and the minimal time (i.e., minutes) required for oxidation. Chemical oxidation (e.g., AOPs) is a well-known and consistent technology for the removal of color, odor, COD, biochemical oxygen demand (BOD), organics, and some inorganic compounds from PW (Jiménez et al., 2018). As recommended for wastewaters with COD below 5 g L^{-1}, the treatment of PW with a high organic load requires pretreatment operations, such as dilution, coagulation, and flocculation, as well as optimization of reagents and energy consumption and minimization of reaction time (Jiménez et al., 2018).

1.3.12 ELECTRODIALYSIS

In electrodialysis, an electrochemical charge drives the separation process, which is used to treat PW, particularly for the removal of dissolved salts. In this process, a stack of alternating anion and cation selective membranes separated by spacer sheets is used to remove salts from PW with low TDS concentrations. When an electrical current is applied to the cell, only anions (e.g., Cl⁻) can pass through the positively charged membrane (anode), and similarly, only cations (e.g., Na⁺) can migrate to the negatively charged membrane (cathode), thereby producing alternating cells of diluted and concentrated solutions between the selective membranes (Al-Ghouti et al., 2019). Like any other process with integrated membranes, fouling is a major limitation of electrodialysis technology. Electrodialysis was successfully applied to PW from a conventional well that contained H_2S, oil, organic acids, etc. (Jiménez et al., 2018).

1.3.13 OTHER ELECTROCHEMICAL PROCESSES

Other electrochemical technologies, including water electrolysis, electrodeposition, fuel cells, and photo-electrochemistry, can be used to treat PW through the use or generation of electricity. However, many of these treatment technologies are either rarely employed for PW or are mainly designed to treat dissolved organic compounds. Although, most of these processes have not yet been commercially applied to treat PW, results from several studies indicate that these relatively green and low-cost technologies have a great potential for produced water treatment (Dores et al., 2012; Hussain et al., 2014; Lin et al., 2020).

1.4 CONCLUSIONS

Current and emerging PW treatment technologies were briefly reviewed. These technologies enable the reinjection, safe disposal, and reuse of the enormous amount of PW generated by the O&G industry. PW is a complex mixture of water and many other constituents including suspended and dissolved materials. While the primary and the secondary treatment technologies may suffice for reinjection and offshore disposal, the tertiary or polishing technologies are critically essential for beneficial reuse of PW. Current research efforts in developing biological, electrochemical, and other emerging PW treatment technologies will enhance reuse and material recovery from produced water.

REFERENCES

Ahmad, N. A.; Goh, P. S.; Yogarathinam, L. T.; Zulhairun, A. K.; Ismail, A. F. Current advances in membrane technologies for produced water desalination. *Desalination* **2020**, 493, 114643.

Ahmadun, F. R.; Pendashteh, A.; Abdullah, L.C.; Biak, D. R. A.; Madaeni, S. S.; Abidin, Z. Z. Review of technologies for oil and gas produced water treatment. *Journal of Hazardous Materials* **2009**, 170, 530–551.

Al-Ghouti, M. A.; Al-Kaabi, M. A.; Ashfaq, M. Y.; Da'na, D. A. Produced water characteristics, treatment and reuse: A review. *Journal of Water Process Engineering* **2019**, 28, 222–239.

Al-Salmi, M.; Laqbaqbi, M.; Al-Obaidani, S.; Al-Maamari, R. S.; Khayet, M.; Al-Abri, M. Application of membrane distillation for the treatment of oil field produced water. *Desalination* **2020**, 494, 114678.

Arthur, J. D.; Langhus, B. G.; Patel, C. Technical summary of oil and gas produced water treatment technologies, **2005**. Technical Report, ALL Consulting, LLC. https://hero.epa.gov/hero/index.cfm/reference/details/reference_id/2224750.

Arthur, J.D.; Dillon, L.W.; Drazan, D.J. Management of produced water from oil and gas wells; *Working Document of the NPC North American Resource Development Study*, **2011**, 32.

Boraey, M. A Hydro-Kinematic approach for the design of compact corrugated plate interceptors for the de-oiling of produced water. *Chemical Engineering and Processing - Process Intensification* **2018**, 130, 127–133. https://www.sciencedirect.com/science/article/pii/S0255270118302794?casa_token=JXlb7hV5KeEAAAAA:tsTYz0OpqUCF1rywt6tVUunIzizq23qCiffQ2f9l43mWyxV1B1IiocrU3KLYIneg5AhWeFk916U

Camarillo, M. K.; Stringfellow, W. T. Biological treatment of oil and gas produced water: A review and meta-analysis. *Clean Technologies and Environmental Policy* **2018**, 20(6), 1127–1146. http://dx.doi.org/10.1007/s10098-018-1564-9. https://escholarship.org/uc/item/286323rr

Daigle, T.; Hantz, S.N.; Janjua, R.S.N. *Treating and Releasing Produced Water at the Ultra Deepwater Seabed*, doi:10.2118/165574-PA; Corpus ID: 109655910; Published 30 April 2012, Engineering; Oil and gas facilities. https://www.semanticscholar.org/paper/Treating-and-Releasing-Produced-Water-at-the-Ultra-Daigle-Hantz/ea517bd3304944ab3b716103af3be8529ccf2e45

Dickhout, J. M.; Moreno, J.; Biesheuvel, P. M.; Boels, L.; Lammertink, R. G. H.; de Vos, W. M. Produced water treatment by membranes: A review from a colloidal perspective. *Journal of Colloid and Interface Science* **2017**, 487, 523–534.

Dores, R.; Hussain, A.; Katebah, M.; Adham, S. Using advanced water treatment technologies to treat produced water from the petroleum industry. *SPE International Production and Operations Conference and Exhibition*, Doha, Qatar, 14–16 May **2012**, SPE 157108.

Duraisamy, R. T.; Beni, A. H.; Henni, A. State of the art treatment of produced water. In Elshorbagy, W., Chowdhury, R.K., (eds.). *Water Treatment*. InTech: Rijeka, Croatia, **2013**.

Gray, M. Reuse of produced water in the oil and gas industry. *SPE International Conference and Exhibition on Health, Safety, Environment, and Sustainability, Virtual Event Due to COVID-19*, 27–31 July **2020**, SPE-199498-MS.

GWPC. *Produced Water Report. Regulations, Current Practices, and Research Needs.* Groundwater Protection Council. 2019.

Han, L.; Tan, Y. Z.; Netke, T.; Fane, A. G.; Chew, J. W. Understanding oily wastewater treatment via membrane distillation. *Journal of Membrane Science* **2017**, 539, 284–294.

Hussain, A.; Minier-Matar, J.; Janson, A.; Gharfeh, S.; Adham, S. Advanced technologies for produced water treatment and reuse. *International Petroleum Technology Conference*, Doha, Qatar, 20–22 January, **2014**, IPTC 17394.

Igunnu, E. T.; Chen, G. Z. Produced water treatment technologies. *International Journal of Low-Carbon Technologies* **2014**, 9, 157–177.

Igwe, C. O.; Saadi, A. A.; Ngene, S. E. Optimal options for treatment of produced water in offshore petroleum platforms. *Journal of Pollution Effects & Control* **2013**, 1(2). https://pdfs.semanticscholar.org/37ef/5baa4a84780b7bc86c10e6989743fe93bb98.pdf

Jiménez, S.; Micó, M. M.; Arnaldos, M.; Medina, F.; Contreras, S. State of the art of produced water treatment. *Chemosphere* **2018**, 192, 186–208.

Jiménez, S.; Andreozzi, M.; Micó, M. M.; Álvarez, M. G.; Contreras, S. Produced water treatment by advanced oxidation processes. *Science of the Total Environment* **2019**, 666, 12–21. https://www.sciencedirect.com/science/article/pii/S0048969719306163

Judd, S.; Qiblawey, H.; Al-Marri, M.; Clarkin, C.; Watson, S.; Ahmed, A.; Bach S. The size and performance of offshore produced water oil-removal technologies for reinjection. *Separation and Purification Technology* **2014**, 134, 241–246.

Larson, A. Produced Water: Oil and Gas Terminology Glossary. *Water Environment Federation* **2018**, Factsheet. WSEC-2017-FS-013 IWWC.

Lin, L.; Jiang, W.; Chen, L.; Xu, P.; Wang, H. Treatment of produced water with photocatalysis: Recent advances, affecting factors and future research prospects. *Catalysis* **2020**, 10, 924.

Lusinier, N.; Seyssiecq, I.; Sambusiti, C.; Jacob, M.; Lesage, N. Biological treatments of oilfield produced water: A comprehensive review. *SPE Journal* **2019**, 24, 2135–2147.

Nasiri, M.; Jafari, I.; Parniankhoy, B. Oil and gas produced water management: A review of treatment technologies, challenges, and opportunities. *Chemical Engineering Communications* **2017**, 204, 990–1005.

Nonato, T. C. M.; De A Alves, A. A.; Sens, M. L.; Dalsasso, R. L. Produced water from oil - A review of the main treatment technologies. *Journal of Environmental Chemistry and Toxicology* **2018**, 2, 23–27.

Robinson, D. *Oil and Gas: Treatment and Discharge of Produced Waters Onshore (Part 1)*, 2013, https://www.filtsep.com/content/other/oil-and-gas-treatment-and-discharge-of-produced-waters-onshore-part-1

Thomas, A. *Produced Water treatment. Essentials of Polymer Flooding Technique*, First Edition, John Wiley & Sons Ltd, Hoboken, NJ, **2019**.

Veil, J.; Puder, M.G.; Elcock, D.; Redweik, R.J.J. *A White Paper Describing Produced Water From Production of Crude Oil, Natural Gas and Coal Bed Methane*, 2004, http://www.netl.doe.gov/publications/oilpubs/prodwaterpaper.pdf.

Wang, P.; Chung, T. S. Recent advances in membrane distillation processes: Membrane development, configuration design and application exploring. *Journal of Membrane Science* **2015**, 474, 39–56. https://www.sciencedirect.com/science/article/pii/S0376738814007091?casa_token=8uwUTIg3ltwAAAAA:LyyjNHuM2be63wy12dh3sLDtKBbxPqTY7o-5CUvRY3cpoKEiX5bWE8yZOJsSFrHAY7sVa_kt_Co

Water Environment Federation (WEF) Fact Sheet, *Produced Water: Oil and Gas Terminology Glossary*, 2017.

Water Environment Federation (WEF) Webinar, *Fundamentals of Produced Water Treatment in the Oil and Gas Industry*, 2019.

Water Environment Federation (WEF) Workshop. Fundamentals of Produced Water Treatment in the Oil & Gas Industry. Workshop 14. Water Environment Federation's Technical Exhibition and Conference, **2018**.

Wei, N.; Wang, X.H.; Li, F.K.; Zhang, Y.J.; Guo, Y. Treatment of high-salt oil field produced water by composite microbial culture. *Urban Environment & Urban Ecology* **2003,** 16, 10–12.

Wei, X.; Kazemi, M.; Zhang, S.; Wolfe, F. A. Petrochemical wastewater and produced water: Treatment technology and resource recovery. *Water Environment Research* **2020**, 92, 1695–1700.

2 Produced Water Overview

Characteristics, Treatment, and Beneficial Uses

Elena Subia Melchert
U.S. Department of Energy

CONTENTS

2.1 INTRODUCTION

For those unfamiliar with geologic formations, it should be understood that oil and gas deposits are generally colocated with water. The source of this water is the ancient seas and oceans that once covered the earth (Water Environment Federation 2019). Having been in contact with the hydrocarbon-bearing formation for centuries, the water may bear some of the chemical constituents of both rock and oil (Colorado School of Mines/Advanced Water Technology Center undated).

The processes of oil and gas exploration and production involve both the demand for and the supply of water. Fresh groundwater is generally part of the process to drill oil and gas wells and is used to cool the drill bit as it cuts into rock, and to carry to the surface the drill cuttings. After the well is drilled, constructed, completed, and finally put on production, the fluids that are produced comprise oil, natural gas, and water, hence the term "produced water". These three fluids together are constrained in the pore space of the rock in the geologic formation so when one is produced all

DOI: 10.1201/9781003091011-2

three are produced. At the surface, this multiphase fluid stream is separated into these three components and each is dealt with separately. In general, separation begins with the density differences between these fluids: the natural gas at the top is taken by pipeline to a natural gas processing plant and then to market, and the oil taken from the middle is taken by pipeline directly to market. The water taken from the bottom of the separation vessel, however, does not have an established market and so must be dealt with as a cost of doing business.

When produced water is brought to the surface it must be managed in a manner that is consistent with the business model of the oil and gas producer. Therefore, in some cases, it must be handled only as a waste that has a disposal cost. However, an alternate opportunity for revenue can be realized if a value proposition could be developed for the reuse of treated produced water. If the most profitable option is to dispose of this "waste" in a saltwater disposal (SWD) well that has been drilled into a suitable geologic formation deep in the earth, then it should be done. Alternatively, depending on quality, the water may be made available for reuse in oilfield operations such as drilling or for hydraulic fracturing, and it could be possible that on a limited basis the produced water may be treated for some other possibly beneficial use such as manufacturing or even nonedible crops.

In general, treating produced water can transform it from a waste to a valuable resource. The cost of treatment coupled with the cost of transporting treated or untreated produced water can be a challenge for developing the value proposition. However, as the demand for freshwater increases and the cost of treatment and transport decreases using advanced technologies, the value proposition may improve.

This discussion will focus on produced water and its role in oil and gas exploration and production, its general characteristics, and an overview of the types of treatment technologies. It will generally cover policy and regulatory issues related to treated and untreated produced water including beneficial use of treated produced water, and finally, lists a few stakeholders who have interest and standing in the reuse of treated produced water.

2.2 CHEMICAL, PHYSICAL, AND BIOLOGICAL CHARACTERISTICS AND PROPERTIES OF PRODUCED WATER

The suitability of produced water for reuse is dependent on its chemical characteristics and properties and the value proposition of changing these characteristics and properties to make treated produced water suitable for reuse. If the value proposition is weak, oil and gas producers will not consider treatment and the final disposition of produced water will be injection into SWDs. As a physical commodity involving pipelines, storage tanks, and other facilities of finite capacity, if produced water is not properly managed as asset or waste, then it creates a "bottleneck" and oil and gas production must stop until the produced water is properly handled.

The characteristics and properties of produced water vary significantly within and across oil and gas plays, and even within the same well over time. For example, the Southwest USA is known for drought while the Northeast is known for more abundant water resources. Treatment technologies can be used to change these characteristics and properties but at a cost that is compared to the cost of disposal.

Nonetheless, depending on the cost, the volume, and the availability of freshwater, a case may be made for the use of treated produced water. Therefore, it is important to consider those characteristics and properties of produced water that make it suitable or unsuitable for treatment for possible beneficial reuse.

Treatment technologies must address the changing nature of the chemical, physical, and/or biological profile of the produced water. It is for this reason that the reader is cautioned to have not only a basic understanding of produced water dynamics but also to update their understandings to include awareness of advancing trends in produced water treatment. Produced water is responsive not only to its nascent geologic environment but also to ongoing trends used in oil and gas production operations that may change the chemical, physical, or biological profile of the produced water.

As illustrated in Table 2.1, these ancient waters are in contact with the rocks that comprise the geologic formation and so can take on some of the characteristics of the rock. Geologic formations also include the shells from former sea life, plant debris, animal bones, algae, and bacteria that add to this "produced water soup".

The major components of produced water—besides the water—are oil, grease, and salt. Produced water may also contain other organic and inorganic compounds that vary geographically, geologically, and temporally even within the same well. The physical properties of produced water include solutions and suspensions, emulsions, and adsorbed particles and particulates (Veil et al. 2004). Veil et al. also noted that

TABLE 2.1

Oil, Gas, and Water Reside Together in Geologic Formations Comprised Primarily of Sedimentary Rocks

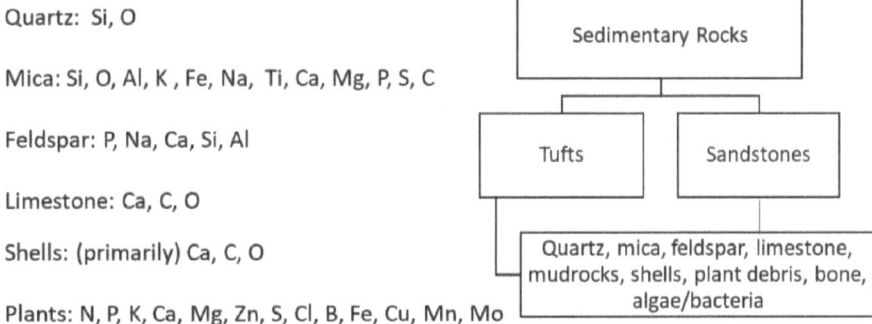

Quartz: Si, O

Mica: Si, O, Al, K , Fe, Na, Ti, Ca, Mg, P, S, C

Feldspar: P, Na, Ca, Si, Al

Limestone: Ca, C, O

Shells: (primarily) Ca, C, O

Plants: N, P, K, Ca, Mg, Zn, S, Cl, B, Fe, Cu, Mn, Mo

Bones: Ca, P, O, H

Algae/Bacteria: C, N, P

These rocks are made of different minerals that are made of elements on the periodic table. The oil comes from originally organic materials such as plants, bone, algae, and bacteria, and other debris each with its own chemical composition that also come in contact with the water.

Adapted from American Association of Petroleum Geologists (undated) and Veil et al. (2004).

produced water can also contain what is often called "source water", which is ground-water or seawater that has been injected into the oil and gas reservoir for pressure maintenance. Produced water may also contain bacteria and miscellaneous solids.

Also, to be found in produced water as noted by Veil et al. (2004) are the chemicals that were added to aid in the drilling, completions, stimulation, and production processes. These can include corrosion and scale inhibitors, emulsion breakers, coagulants, flocculants, clarifiers, solvents that can affect human toxicity and that can affect oil–water separation and microbiological populations.

Constituents in produced water from natural gas wells can differ slightly in that they also include condensed water. That is, water molecules that are in the vapor formed downhole will condense at the surface when subjected to ambient temperature and pressure. Veil et al. (2004) also note that water produced from natural gas wells may have benzene, toluene, and other low molecular-weight aromatic hydrocarbons and thus can be relatively more toxic than water produced from oil wells. Further, chemicals used for oilwell management such as those that inhibit the formation of hydrates will also be contained in the produced water from natural gas wells (Veil et al. 2004).

Veil et al. (2004) also note that there may be differences between water produced onshore versus offshore as related to pH and chlorides. All this variety in produce waters adds to the challenge of developing cost-effective produced water treatment trains needed to make the treated produced water suitable for reuse.

As illustrated in Table 2.2, there are constituents in produced water that similarly affect potential treatment options and as such can be grouped. For example, organic

---

TABLE 2.2

All the Constituents in Produced Water Reflect the Geologic Setting in Which They Reside and These Constituents Each Influence the Ability to Be Separated from H$_2$O at the Surface

Potential Produced Water Constituents

Dispersed oil	Small droplets suspended in the aqueous phase
Organics	Organic acids, polycyclic aromatic hydrocarbons, phenols, volatiles
Production well treatment chemicals	Biocides, reverse emulsion breakers, and corrosion inhibitors
Solids	Precipitated solids, sand and silt, carbonates, clays, proppant, corrosion products, and other suspended solids derived from the producing formation and from well bore operations
Precipitates	Calcium carbonate, calcium sulfate, barium sulfate, strontium sulfate, and iron sulfate
Metals	Zinc, lead, manganese, iron, and barium
Sulfates	Controls the solubility of several other elements in solution, particularly barium and calcium
NORM	Naturally occurring radioactive material

Adapted from Veil et al. (2004).

TABLE 2.3

This Table, Developed as an Illustration Only and Is *Not* Actual Data from a Specific Reservoir, Shows that Produced Water Can Differ Greatly Within the Same Geologic Basin

	Ion Concentration (mg L^{-1})	
Constituent	Water 1	Water 2
Chloride	110,000	20,000
Bromide	1,100	0
Iodide	50	0
Sulfate	1,350	1,600
Total Alkalinity	1,500	1,100
Sodium	60,000	13,500

constituents that are dispersed or dissolved include oil, grease, and some dissolved compounds (Veil et al. 2004).

Examples of produced water constituent profiles reveal the range of concentration for various ions (Table 2.3). These two cases, developed for illustration purposes only, show the potential extreme range of concentration for some constituents in produced water. While these two examples are not actual data from a specific reservoir, they illustrate the extremes that can occur within the same geologic basin. When designing a treatment train to examine a potential value proposition, there is no substitute for testing the actual produced water of interest to determine the actual constituent concentration profile.

Oil becomes dispersed in the produced water and affects how it should be pretreated. It must be removed from the produced water stream if it is to be reused in likely any application. Removal of oil is challenging because it can be emulsified in the water and is affected by the oil density and the efficiency of the physical separation equipment (Ali et al. 1999 as cited in Veil et al. 2004).

Organic components can also dissolve or become soluble in produced water. These include organic acids, polycyclic aromatic hydrocarbons (PAHs), phenols, and volatiles (Utvik 2003 as cited in Veil et al. 2004).

The chemicals used to treat the production wells can also add to the challenge of treating the produced water. These can include biocides, reverse emulsion breakers, and corrosion inhibitors (Glickman 1998 as cited in Veil et al. 2004). Corrosion inhibitors, essential for long-term wellbore integrity, can reduce the efficiency of the oil–water separation due to the formation of more stable emulsions (Veil et al 2004).

Solid particles can also become entrained in the produced water. This can include everything that exists naturally in the hydrocarbon reservoir such as precipitates, sand, silt, and clay plus any injected items such as proppant used in hydraulic fracturing and other wellbore operations (Cline 1998 as cited in Veil et al. 2004).

Metals including heavy metals can be present in low or high concentrations in produced water, and this depends entirely on the character of the geologic formation

TABLE 2.4

Oil and Natural Gas Exploration and Production Activities May Include the Use of Recycled Treated Produced Water

The taxonomic classification of bacteria in produced water can include:

Gamma-proteobacteria	Alpha-proteobacteria	Clostridia	Bacteroidetes
Synergistetes	Thermotogae	Spirochetes	

Prior to treatment, the produced water includes biological populations that must be addressed. The bacterial communities listed above were found during impoundment at the surface prior to reuse and/or disposal. Mohan et al. (2013).

(Utvik 2003 as cited in Veil et al. 2004). Metals typically found in produced waters include zinc, lead, manganese, iron, and barium.

Sulfates also affect treatment options and costs. This is because sulfate concentration can control the solubility of other elements such as barium and calcium (Utvik 2003 as cited in Veil et al. 2004).

Radioactive materials occur naturally in rock, soils, and produced water and are not just due to human activity. This is another feature unique to the hydrocarbon reservoir. The most common naturally occurring radioactive materials (NORM) compounds are radium 226 and radium 228 (Utvik 2003 as cited in Veil et al. 2004).

Salt content in produced water can range from nearly that of freshwater to salt levels much high than that of seawater—even up to ten times higher. Salinity reflects the total dissolved salts in the produced water.

The acidity of produced water is an important property because it affects the chemistry of the produced water and the treatment strategy needed to transform the produced water into a form suitable for reuse. The resulting treatment strategy can be affected by efforts to stabilize the pH which may increase the cost of the treatment.

Another important property of produced water is the microbial population. Geological formations are not sterile environments; therefore, the fluids that reside therein are teeming with microbiological life albeit anaerobic. As illustrated in Table 2.4, the microbiological community is diverse and can evolve as oil and gas production operations change the reservoir environment. These microbes are produced with the fluid and affect the quality of the produced water—as do the chemicals injected into the well to control the microbial population—including the ease of produced water treatment at the surface (Veil et al. 2004).

2.3   TYPES OF TREATMENT TECHNOLOGIES

Type of treatments used to remove constituents from produced water can be grouped into three key areas: clarifying, softening, and desalination. These then can further be defined to the more specific processes for removal of solids, oil and grease, soluble organics, divalent cations, and salts. Table 2.5 illustrates the various subprocesses that are used to achieve the desired removal of constituents from the produced water. Again, the choice of treatment and the design of the treatment train is a function

TABLE 2.5
Produced Water Treatment Train Compares Relative Cost of Treatment Depending on Number and Type of Constituents and the Potential Use of Treated Produced Water

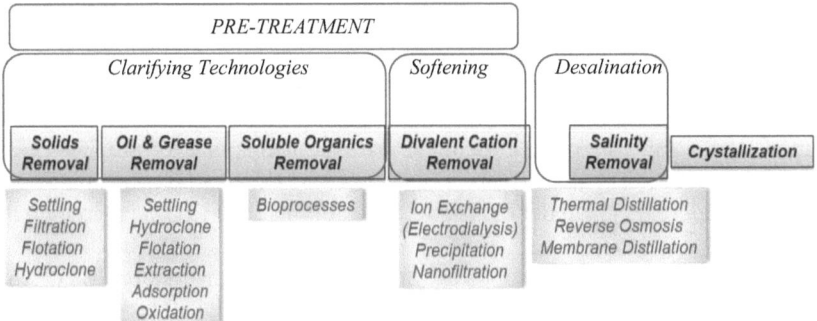

In general, cost of treatment moves from least cost on the left to most costly treatment on the right. Therefore, the quality of treated water needed for end use must be considered before treatment.

of the initial constituent profile of the produced water. This must be coupled with the target constituent concentration profile for the treated produced water which is dictated by the potential beneficial use and the value proposition for reuse of treated produced water.

The overall cost to treat produced water is a function of the degree of treatment needed to take the produced water from its initial state to a new state for a particular reuse. The more treatment needed to transform the produced water, the greater the sunk costs of the treated produced water, thus applying pressure to the value of the beneficial reuse. Pretreatment costs are generally less that those associated with desalination and crystallization. Obviously, the ideal value proposition from the perspective of the oil and gas producer would be one in which the costs of treatment are lower; from the perspective of the user the value proposition of the treated produced water would be one in which the cost of the treated produced water is less than the cost of freshwater. Further, as stated previously, any beneficial reuse of treated produced water must also compete with the low cost of disposal of untreated produced water.

2.3.1 PRETREATMENT TECHNOLOGIES

Clarifying technologies involve the removal of suspended solids including separators, settling, filtration, flotation, and hydrocyclone processes. The removal of oil and grease includes settling, hydrocyclone, flotation, extraction, adsorption, and oxidation. Removal of organics involves bioprocesses. Softening involves the removal of primarily Ca and Mg ions. These processes include ion exchange, precipitation, and nanofiltration.

2.3.2 Desalination Technologies

Desalination of produced water is an important part of the treatment train. Although there are some produced waters that may not need more than pretreatment, desalination represents an added cost of treatment. Removing salts is key to the reuse of most produced waters and can involve membrane processes, thermal process, and full system processes depending on the value proposition for the supply of treated produced water and the desired end use. Common processes used in desalination include thermal distillation, reverse osmosis, and other membrane distillation.

Membrane processes include reverse osmosis (RO), nanofiltration (NF), and ultrafiltration (UF) membranes when modified for use in treatment of produced water such as the application of an antifouling surface coating such as polydopamine which can increase salt rejection under high salinity conditions as in RO and increased flux in UF (Folio et al. 2018).

Thermal processes include multistage flash distillation: distillation, vapor compression, thermal compression, and mechanical compression. Other processes include solar distillation, humidification-dehumidification, membrane distillation, and freezing (Shatilla 2020).

Again, treating produced water can transform it from a waste to a valuable resource. The cost of treatment coupled with the cost of transporting treated or untreated produced water can be a challenge for developing the value proposition. However, as the demand for freshwater increases and the cost of treatment and transport decreases through the use of advanced technologies, the value proposition may improve.

2.4 BENEFICIAL USE OF TREATED PRODUCED WATER

Treated produced water may, in specific cases, substitute for freshwater especially as demand for freshwater increases or drought conditions restrict the use of freshwater for certain activities. Beneficial reuse of produced water includes oilfield reuse and non-oilfield use (Table 2.6).

The benefit of using treated produced water is associated with the reduced demand for freshwater in specific situations of scarcity of freshwater, such as drought conditions, or where the untreated produced water is of such quality that it requires little treatment and is, therefore, a cost-effective alternative to freshwater.

TABLE 2.6

As the Demand for Freshwater Increases or the Supply Decreases Due to Drought Conditions, Treated Produced Water May Offset the Use of Freshwater in Specific Applications

Potential non-oilfield uses of treated produced water:

Irrigation of nonfood crops or biomass for energy production or golf courses or parks

Suppression of fire or dust

Manufacturing processes (high water demand)

Ice rinks

In general, the first consideration of beneficial use of treated produced water is for use within the oilfield. Freshwater is used in the development of drilling sites and the roads leading to the drill site. It is used in the drilling process itself in fluids that cool the drill bit, or in the creation of a temporary barrier that holds up the sides of the wellbore and hold back the entry of water and other formation fluids that exist in the hole at depths near the surface, and to move the drill cuttings to the surface. Freshwater is also used to complete and stimulate the well such as in the process of hydraulic fracturing.

Historically, the oilfield reuse of produced water had not been considered as beneficial or even valuable. However, as a greater appreciation for the value of produced water has developed, an increasing number of oil and gas producers have developed and even accelerated their oilfield use of produced water. This has been supported in part by the development of the "midstream water" sector that is comparable in some respects to the oil and gas midstream sector that moves oil and gas from where it is produced to where it is sold. This midstream water sector has been an integral part of the transition of produced water from being viewed as just a waste needing disposal to a possible alternative to the use of freshwater in oilfield uses—even non-oilfield use.

The degree (and expense) of treatment necessary when considering reuse of produced water depends on the quality of the water needed for the intended use. Treatment is typically designed to be "fit for purpose". This "purpose" can include many applications depending on the quality of the produced water, the capability and cost to treat this produced water, and the final value of the treated produced water. This is in competition with the value of alternative options such as low-cost freshwater, the quality of the water needed for the beneficial use, or even the low cost of disposal of nontreated produced water.

Once the potential beneficial use of the treated produced water is determined, then the value proposition can be developed by examining the cost to treat the particular quality of this produced water, the cost to transport this treated produced water (which may be different from the cost to transport freshwater) versus the cost of freshwater, and the cost of untreated produced water disposal. However, the value proposition is not dependent on market value alone but also includes the benefit of use and the cost of regulatory compliance.

In 2019, the Groundwater Protection Council (GWPC) reported that almost half of the produced water in the USA is reused in the USA for conventional oil and gas operations including enhanced oil recovery and waterflooding and steam flooding. The remaining produced water is typically disposed of in saltwater disposal wells (Groundwater Protection Council 2019).

Further, the GWPC states that many factors including local use, initial produced water quality, volume of available produced water, long-term reliability of produced water volumes as an alternative to freshwater, logistics, infrastructure, and other considerations affect the decision to reuse produced water for oilfield and potential non-oilfield use. Reuse of treated produced water depends on its quality and suitability for reuse (Groundwater Protection Council 2019). Further, management and cost of produced water treatment waste as concentrated brine must also be considered (Groundwater Protection Council 2019).

2.5   POLICY AND REGULATION FRAMEWORK FOR BENEFICIAL USE OF TREATED PRODUCED WATER

The National Pollutant Discharge Elimination System (NPDES) governs the use of produced water, such as discharge directly to surface water, and water sent to a municipal wastewater treatment facility. The Underground Injection Control (UIC) program governs the produced water injection wells used for enhanced recovery of oil and gas.

This means that almost all activities related to produced water such as management practices, construction standards, and operational requirements are regulated. This includes federal, state, and local regulations, each with particular jurisdiction. Federal jurisdiction includes disposal, and State and local jurisdictions include various often numerous permits, licenses, and certificates (Groundwater Protection Council 2019).

Further, because the geologic boundaries of the oil- and gas-producing zones are not bound by surface geographic boundaries set by a single state or county or river basin more than one oil and gas operator may have to negotiate their respective responsibilities for managing the produced water and its ultimate disposition.

For example, determinations surrounding the source of the produced water and its ownership are subject to state water rights laws or regulations. Individual states may require water withdrawal permits from surface or subsurface sources. Water purchased from municipal sources will require contracts or other legal documents. The hauling of water by trucks may require permits and licenses. If pipelines are used, the purchase or lease of multiple easements from multiple landowners may be required. Further, additional permits may be needed if the pipelines cross streams, roads, rail, or other infrastructure—again state and local requirements may not align, which means additional legal procedures and costs will be incurred.

When it comes to storing the water, additional permits may be required to construct and manage storage pits—including the case of stormwater management. The use of water storage tanks may be covered by other previously obtained permits but in some cases may not be covered. It should be recognized that spilled water having the potential for environmental harm must be prevented, controlled, and/or mitigated, which may require additional planning and possibly approval.

2.6   PRODUCED WATER STAKEHOLDERS

With over 30 oil- and gas-producing states in the USA, the stakeholders interested in produced water are many and varied. They range from government at all levels including nongovernment organizations, to industry, to academia, and to members of the public (Table 2.7).

Industry stakeholders include oil and gas producers; service companies including those that gather, treat, and move produced water; innovators interested in developing new technologies for treating produced water; and possibly those non-oilfield industries that have a high demand for produced water that may be interested in using treated produced water. Those that provide logistics such as pipelines and trucks, those that provide treatment technologies, those that receive and use treated produced

TABLE 2.7

The Interests, Needs, and Preferences of Different Stakeholders Add a Level of Complexity to the Future Role of Produced Water in Society

Key Stakeholders for Produced Water

INDUSTRY—Water Supply: oil and gas producers, service companies, water management (including water treatment service and water transport service).

INDUSTRY—Water Disposal: wastewater transport, wastewater disposal, saltwater disposal wells

INDUSTRY—(Potential) Treated Water Demand: manufacturing, power plants, fire and dust suppression, agriculture, and biofuels

GOVERNMENT: federal, state, local, tribal

ACADEMIA: water profile, water option analyses, new treatment technology concepts, new treatment technologies

OTHER RESEARCHERS: private laboratories, oil and gas service company laboratories, National Labs

PUBLIC—Treated Water Demand: replacement for freshwater in nonhuman use

NONGOVERNMENT ORGANIZATIONS: approximately 1.5 million NGOs are operating in the USA[1] and some are interested in water

U.S. Department of State, Bureau of Democracy, Human Rights, and Labor (2021).

Disposal of produced water into saltwater disposal wells removes this resource from the water cycle which can be an important consideration for drought-prone regions. Risks—real or perceived—related to using treated produced water in oilfield operations and non-oilfield applications, and potential costs and benefits underpin the decisions related to the final disposition of produced water.

water in oilfields, and those that may be interested in using treated produced water for non-oilfield use are also key stakeholders. As the question of using treated produced water for non-oilfield uses is examined by government stakeholders and the public, industry must focus on the value proposition of doing so. Without such value, government may find that while it may be open to permitting the non-oilfield use of produced water, there may be no business case for doing so.

Government stakeholders include federal, state, local, and tribal. Government as a provider of basic services must ensure the health and safety of those who use these services. The use of treated produced water presents an opportunity to offset the use of freshwater; however, the strengths and weaknesses of that opportunity must be fully understood before the government can permit its use. Drought-prone states may need to turn to alternative sources of water to meet the needs of its citizens. Protecting the public is at the core of good government; hence, the risks of moving produced water out of the water cycle as is done when disposed of as waste must be carefully examined. Likewise, using treated produced water must be based on science so that no harm can be brought to humans if treated produced water is used outside the oilfield.

With respect to states, there are 32 oil- and gas-producing states in the USA (Statista 2020). The perspective of state agencies is different from that of industry with respect to the prudent spending of tax dollars, providing a high standard of public services and looking at the overall quality of life for the citizens of and visitors to that state. These are some of the questions that state agencies ponder, further some state agencies have overlapping responsibilities and authorities that must be coordinated because water affects so many aspects of life. Additionally, with respect to produced water, some states have produced water discharge permitting authorities while others may not. Reconciling this between states that share a single oil and gas reservoir can be a challenge especially on the question of who owns the water. This becomes a further challenge when a reservoir may cross three or four state lines and the negotiated position of two of the states does not easily reconcile with the laws and regulations of a third or fourth state.

Congress has jurisdiction over multiple water-related topics including water resources development, management, use, protection, water rights, and research. Federal agencies include those responsible for protection of human health which would include the Environmental Protection Agency, Department of Health and Human Services, Department of the Interior/Bureau of Reclamation, and Department of Agriculture to name a few (Congressional Research Service 2017).

Tribal nations largely share the same challenges as the states but as sovereign nations may have additional economic or financial goals and limitations that could come into play in the case of new financial opportunities or environmental vulnerabilities.

Academia and other research organizations provide needed analyses of produced water constituents that may affect the final disposition of produced water; development of commercial technologies and methodologies for the removal of non-water constituents; and invention of new technologies and methodologies for treating produced water. Academia may support innovators as they attempt to move new research from demonstration to commercialization. Academia may also support the government in its efforts to properly permit the final disposition of treated and untreated produced water. In general, this function is supported by either government or industry not academia. Therefore, a key role of academia is to respond to the needs of industry and government in addressing key questions of ultimate disposition of produced water.

Finally, the public depends on government for services including water supply and expects that service to be of highest quality at least cost. Public sentiment may or may not support the surface transport and use or the subsurface disposal of treated or untreated produced water. As taxpayers, the public is entitled to the highest level of service, and they hold the government accountable for the decisions made using taxpayer dollars and for the outcomes of those decisions. Therefore, some expect government to regulate and permit all aspects of the public services provided. This has led, in some cases, to important partnerships between government, industries, and academia to address the needs of the public and their quality of life especially for such fundamental needs as safe drinking water.

2.7 SUMMARY

Produced water represents a potential source of water that under specific conditions may substitute for the use of fresh groundwater in specific cases. These specific reuse opportunities relate to the value proposition associated with the quality and quantity of the produced water and the cost of any treatment that would be needed for that reuse opportunity consistent with regulatory considerations, if any.

Water is the "universal solvent" and, as stated earlier, it may take on the chemistry of the geologic formation such as dissolved mineral salts and organic compounds or it may be mixed with inorganic metals with traces of heavy metals and naturally occurring radioactive materials. These constituents and others would make produced water costly to treat for reuse (DOE 2020).

The cost and availability of treatment options are key barriers to the potential reuse of produced water, sometimes leaving disposal in SWDs as the only cost-effective option (Folio et al. 2018). This practice of disposal removes produced water from the water cycle and is under scrutiny by some stakeholders as drought states consider alternate sources of water supply, including the use of treated produced water.

REFERENCES

American Association of Petroleum Geologists, undated. Type of Reservoir Rocks. https://wiki.aapg.org/Reservoir#Type_of_Reservoir_Rocks (accessed April 3, 2021).

Congressional Research Service. 2017. Selected Federal Water Activities: Agencies, Authorities, and Congressional Committees. https://fas.org/sgp/crs/misc/R42653.pdf (accessed May 3, 2021).

Colorado School of Mines/Advanced Water Technology Center. Undated. About Produced Water: Produced Water 101. http://aqwatec.mines.edu/produced_water/intro/pw/ (accessed March 24, 2021).

DOE. 2020. Office of Fossil Energy. Produced Water: From a Waste to a Resource. https://www.energy.gov/fe/articles/produced-water-waste-resource (accessed March 24, 2021).

U.S. Department of State, Bureau of Democracy, Human Rights, and Labor. 2021. FACT SHEET, Non-Government Organizations (NGOs) in the United States. https://www.state.gov/non-governmental-organizations-ngos-in-the-united-states/#:~:text=Approximately%20 1.5%20million%20NGOs%20operate,development%2C%20and%20many%20other%20issues. (accessed May 13, 2021).

Folio, E., Ogunsola, O., Melchert, E., Frye, E. 2018. Produce Water Treatment R&D: Developing Advanced, Cost-Effective Treatment Technologies. URTEC 2886718. Presented at Unconventional Resources Technology Conference. https://library.seg.org/doi/10.15530/urtec-2018-2886718 (accessed March 24, 2021).

Groundwater Protection Council, 2019. Produced Water Report: Regulations, Current Practices, and Research Needs. https://www.gwpc.org/sites/gwpc/uploads/documents/Research/Produced_Water_Full_Report___Digital_Use.pdf (accessed March 24, 2021).

Mohan, A.M., Hartsock, A., Hammack, R., Vidic, R. 2013. Microbial communities in flowback water impoundments from hydraulic fracturing for recovery of shale gas. https://pubmed.ncbi.nlm.nih.gov/23875618/ (accessed December 18, 2021).

Shatilla, Y., 2020. Nuclear Reactor Technology Development and Utilization. https://www.sciencedirect.com/book/9780128184837/nuclear-reactor-technology-development-and-utilization (accessed December 18, 2021)

Statista. 2020. Crude oil production in the United States in 2020, by state. https://www.statista.com/statistics/714376/crude-oil-production-by-us-state/#:~:text=A%20total%20of%20 32%20of,for%20Defense%20Districts%20(PADD) (accessed May 13, 2021).

Veil, J.A., Puder, M.G., Elcock, D., Redweik, R., 2004. A White Paper Describing Produced Water From Production of Crude Oil, Natural Gas, and Coal Bed Methane. Prepared for the U.S. Department of Energy.

Water Environment Federation. 2019. https://www.wef.org/globalassets/assets-wef/3---resources/online-education/webcasts/presentation-handouts/presentation-handouts-25apr19.pdf (accessed June 26, 2021).

World Health Organization. 2003. Sodium in Drinking -Water. Background document for development of WHO Guideless for Drinking-water Quality. WHO/SDE/WSH/ 03.04/15. Originally published in guidelines for drinking water quality, 2nd.ed. Vol. Health criteria and other supporting information. World Health Organization, Geneva, 1996. Microsoft Word - GDWQ.2ndEdit.Sodium.doc (who.int) (accessed May 11, 2021).

3 Standard Water Treatment Techniques and Their Applicability to Oil and Gas Produced Brines of Varied Compositions

Nicholas Siefert
National Energy Technology Laboratory

Madison Wenzlick
National Energy Technology Laboratory,
NETL Support Contractor

CONTENTS

DOI: 10.1201/9781003091011-3

3.1   INTRODUCTION: MANAGING PRODUCED BRINE IN THE UNITED STATES

High-salinity brine is a by-product from most oil and gas wells across the globe, and the ratio of water to hydrocarbon tends to increase over time for any given well [1–3]. This literature review focuses on produced water generated in the United States and commercially available options for treatment and reuse. With a few exceptions in the Western United States, this produced water/brine is too saline to discharge into local water ways or to be used for agricultural purposes without significant treatment. Some studies on beneficial reuse with minimal treatment can be found in the following references [4–7]. Management of produced water is currently accomplished through a combination of evaporation in surface ponds, reinjection underground, and beneficial reuse. Reinjected brine is sometimes reinjected by itself and sometimes mixed with other fresh water or waste streams before reinjection, making the required treatment a function of both the produced brine and the mixed water streams. The management of produced water is subject to compliance with local, state, and federal regulations, which means that handling produced water will be unique to each state/region/basin.

The two main commercially practiced methods for the disposal of produced brine are (1) reuse for water flooding and enhanced oil recovery (EOR) and (2) deep-well injection into saltwater disposal (SWD) wells. In Table 3.1, we present the available data for produced water management in the United States from the year 2017, compiled by Veil [1]. In Table 3.2, we present the available data from 2012, also compiled by Veil [2]. As seen in Table 3.1, of the 24,483 MMbbl[1] produced water generated during oil and gas operations in 2017, 44% was managed by injection for enhanced oil recovery (EOR), 38% was injected for disposal, and 10% was injected at offsite commercial facilities for disposal [1,2]. Only 643 MMbbl[1], or 3%, went for beneficial reuse in 2017, although this is an increase from only 126 MMbbl[1] in 2012 (0.6% of produced water generated). For conventional reservoirs, reinjection for EOR can be a prudent choice because injection wells can be placed near the production well. Reinjection for EOR using produced water helps to maintain formation pressure, which tends to drop as hydrocarbon resources are extracted and to increase the hydrocarbon output of nearby production wells. For tight, unconventional shale reservoirs, this EOR operation can be difficult to achieve because of the difficulty of hydraulically connecting nearby injection wells to the production wells. For tight, unconventional shale reservoirs, produced water is typically only reinjected during fracturing operations, whereas for conventional hydrocarbon wells, EOR injection

---

[1] Note: 1 MMbbl = 1.59·10⁸ L, and bbl = barrel of oil, even for produced water. bbl is <u>not</u> U.S. liquid barrel. This allows for easy calculation of water to oil ratios.

TABLE 3.1

Top Ten States by Produced Water Volume in 2017: Management Methods and Volumes of Oil, Gas and Coalbed Methane Produced Water, 1 MMbbl = 1.59·10⁸L

State	Injection for EOR		Injection for Disposal		Surface Discharge		Evaporation		Offsite Commercial Disposal		Beneficial Reuse		Total Produced Water Managed	
	MMbbl/year	Fraction	MMbbl/year	Fraction	MMbbl/year	Fraction	MMbbl/year	Fraction	MMbbl/year	Fraction	MMbbl/year	Fraction	MMbbl/year	% of generated
U.S. Total	10,683	44%	9,303	38%	1,338	5%	102	0%	2,413	10%	643	3%	24,483	100%
Texas	4,558	46%	3,587	36%	34	0%	-	-	1,716	17%	-	-	9,895	100%
California	1,842	59%	694	22%	14	0%	29	1%	56	2%	471*	15%	3,105	99%
Oklahoma	1,277	45%	1,186	42%	-	-	-	-	382	13%	-	-	2,844	100%
Wyoming	802	46%	243	14%	648	37%	40	2%	2	0%	-	-	1,736	102%
Kansas	299	25%	906	75%	-	-	-	-	-	-	-	-	1,205	100%
Louisiana	71	7%	877	88%	-	-	-	-	50	5%	-	-	999	100%
New Mexico	351	40%	444	51%	-	-	-	-	-	-	79	9%	874	99%
Alaska	772	90%	85	10%	-	-	-	-	-	-	-	-	857	104%
North Dakota	41	8%	266	53%	-	-	-	-	199	39%	-	-	506	100%
Colorado	108	32%	157	47%	18	5%	20	6%	-	-	30	9%	333	107%

Source: Veil [1].

* Beneficial reuse in California includes 160 MMbbl/year for beneficial reuse in oilfields and 311 MMbbl/year for beneficial reuse outside oilfields.

Note: Total produced water managed may be more than 100% due to reported total volume of water used in EOR.

TABLE 3.2

Management Methods and Volumes of Oil, Gas, and Coalbed Methane Produced Water, U.S. Total and Top Ten States by Volume in 2012, 1 MMbbl = 1.59·10⁸L

	Injection for EOR		Injection for Disposal		Surface Discharge		Evaporation		Offsite Commercial Disposal		Beneficial Reuse		Total Produced Water Managed	
	MMbbl/year	Fraction	MMbbl/year	Fraction	MMbbl/year	Fraction	MMbbl/year	Fraction	MMbbl/year	Fraction	MMbbl/year	Fraction	MMbbl/year	% Total Generated*
U.S. Total	9,288	45%	8,010	39%	1,121	5%	691	3%	1,373	7%	126	0.6%	20,609	97%
Texas	3,718	48%	2,923	37%	371	5%	-	-	795	10%	NR	NR	7,807	105%
California	1,412	46%	623	20%	60	2%	649	21%	284	9%	46	2%	3,075	100%
Oklahoma	1,098	47%	1,087	47%	-	-	-	-	140	6%	-	-	2,325	100%
Wyoming	856	73%	313	27%	NR	NR	NR	NR	NR	NR	NR	NR	1,169	54%
Kansas	276	26%	785	74%	-	-	-	-	NR	NR	NR	NR	1,061	100%
Louisiana	31	3%	857	92%	-	-	-	-	39	4%	-	-	928	100%
Alaska	652	85%	85	11%	33	4%	-	-	-	-	-	-	769	100%
New Mexico	381	50%	381	50%	-	-	-	-	-	-	-	-	762	98%
Colorado	124	32%	124	32%	40	10%	35	9%	22	6%	48	12%	393	110%
North Dakota	52	18%	162	56%	-	-	-	-	77	26%	-	-	291	100%

Source: Veil [2].

* Note: Total produced water managed may be more than 100% due to reported total volume of water used in EOR, which may include fresh water as well as recycled produced water.

NR = not reported

Some of these values are estimated. Specifically, Veil estimated a 50%/50% split of water managed by EOR and disposal injection for Texas and New Mexico. For Texas, some of the 50% was assigned to offsite commercial disposal. For California, all water not managed by injection, discharge, evaporation, or commercial disposal was combined into an "Other" category; this was assumed to be mostly beneficial reuse and is shown as such [2].

TABLE 3.3

Produced Water Volumes from Selected U.S. States and U.S. Total; Total U.S. Oil Production, Gas Barrel of Oil Equivalent (BOE) Production; and the Ratio of Produced Water Volume to the Total Oil/Gas BOE Volume, 1 MMbbl = 1.59·10^8 L

	1985	1995	2002	2012	2017
State	MMbbl/ year	MMbbl/ year	MMbbl/ year	MMbbl/ year	MMbbl/ year
Texas	7,839	7,630	5,032	7,436	9,895
California	2,846	1,684	1,290	3,075	3,135
Oklahoma	3,103	1,643	1,253	2,325	2,844
Wyoming	785	1,401	2,119	2,178	1,705
Kansas	999	684	1175	1,061	1,205
Louisiana	1,347	1,346	1,080	928	999
New Mexico	445	706	113	769	880
Alaska	98	1,090	813	625	828
North Dakota	60	80	78	291	506
U.S. total produced water	20,609	17,922	14,160	21,181	24,392
U.S. total oil	3,862	3,037	2,783	3,258	4,794
U.S. total gas BOE[a]	2,636	2,962	3,009	3,810	4,392
U.S. total oil and gas BOE[a]	6,498	6,000	5,792	7,068	9,186
Ratio of U.S. produced water to total U.S. oil and gas equivalent	3.2	3.0	2.4	· 3.0	2.7

Sources: Produced water volumes from Veil et al. [8]; Veil [2]; Veil [1]. Oil and gas data from Looney [9].
[a] Total gas BOE calculated using a conversion factor of 6,000 cf of natural gas per barrel of oil equivalent.

into nearby wells to maintain formation pressure can be a continuous operation. In Table 3.3, we present historical trends for the historical volumes of total produced water generated in the top 10 producing states between 1985 and 2017 as well as the ratio of the U.S. total produced water volume to hydrocarbon volume, which has historically fluctuated around an average value of three.

There has been a growing demand for brine reinjection in the United States. For example, in 2019, there were approximately 180,000 Class II injection wells permitted by the Environmental Protection Agency (EPA) in the United States [10]. A Class II permit is required by the EPA for wells involved with the injection of fluids associated with oil and gas. In 2014, there were approximately 144,000 Class II injection wells in the United States. Historically, approximately 80% are used for EOR, and approximately 20% are designated as SWD wells [11]. In many regions of the United States, including Texas, North Dakota, and New Mexico, deep injection into SWD wells is currently the primary management practice for produced water from unconventional wells [12]. However, oil and gas well operators are experiencing increasingly strict governmental regulations on disposal limitations in some parts of the country.

Even when produced water is sent to a reinjection well or to a SWD well, a number of pretreatment steps are required to increase hydrocarbon recovery and/or to prevent mineral scaling and corrosion on the steel walls of above-ground pipes and underground piping for reinjection [13–16] or in the reservoir [17]. A produced water pretreatment system might be required to include all or several of the following steps: (1) large suspended solid removal (flocculation/settling) (2) oil/water separation (flotation), (3) fine suspended solid removal (filtration), (4) disinfection, (5) removal of dissolved organic and chemical oxygen demand (adsorption), and (6) divalent scale-forming ion removal (ion exchange/nanofiltration/chemical precipitation). Additionally, in some cases, removal of naturally occurring radioactive material (NORM) may be necessary to prevent release to the environment in cases of beneficial reuse [18,19], and this is discussed in conjunction with other treatment methods later in this chapter.

Due to the cost of the steps listed above as well as the costs associated with additional transportation and reinjection, the management of produced water from a well can compose a significant portion of the overall cost. For example, the capital cost of managing flowback water can range from 5% to 19% of a well's capital cost and managing produced water over time ranges from 7% to 52% of a well's lease operating expenses (LOE), depending on location [20]. For a number of environmental and economic reasons, the management of produced brine presents challenges for industry, government, and the public [21]. In addition, the use of SWDs has also been linked in some cases to increased seismic activity and possible contamination of drinking water aquifers, which are significant environmental risks [22–24]. Therefore, in recent years regulatory agencies have investigated alternatives to disposal by injection [25,26]. Some alternatives currently in use include beneficial reuse for agriculture, livestock or wildlife propagation ([4–7,27]), and deicing and/or dust suppressing on roads [19,28]; other potential uses considered include fire protection and use in power plant cooling towers [29].

Matching industries that need brine with those companies that produce brine could be a crucial way to reduce the amount of water sent to SWDs. Specifically, the capacity and capabilities of local water treatment infrastructure could be increased so that specific types of brine or salt required by local industries can be generated from local-produced brines. Another way of reducing dependence on SWDs is to dewater and concentrate brines to reduce the volume of brine headed for disposal by injection. In both scenarios, management by dewatering can concentrate brine that would be injected for disposal into a valuable stream for reuse or sale and create pure water (Figure 3.1). For example, dewatering can generate a commodity called ten-pound brine (10-lb. brine), which is used within the oil and gas industry and by the chemical industry to generate goods purchased globally [30]. Below is a list of potential products from produced brines if the streams are treated to recover resources [31,32].

1. Ten-pound brine: saturated NaCl brine (with limited scaling ions), a feed-stock for the chloralkali industry, and a high-density fluid used in the oil and gas industry
2. Road salt: NaCl, $MgCl_2$, and/or $CaCl_2$ for road salt and deicing applications
3. Lithium and/or rare earth elements

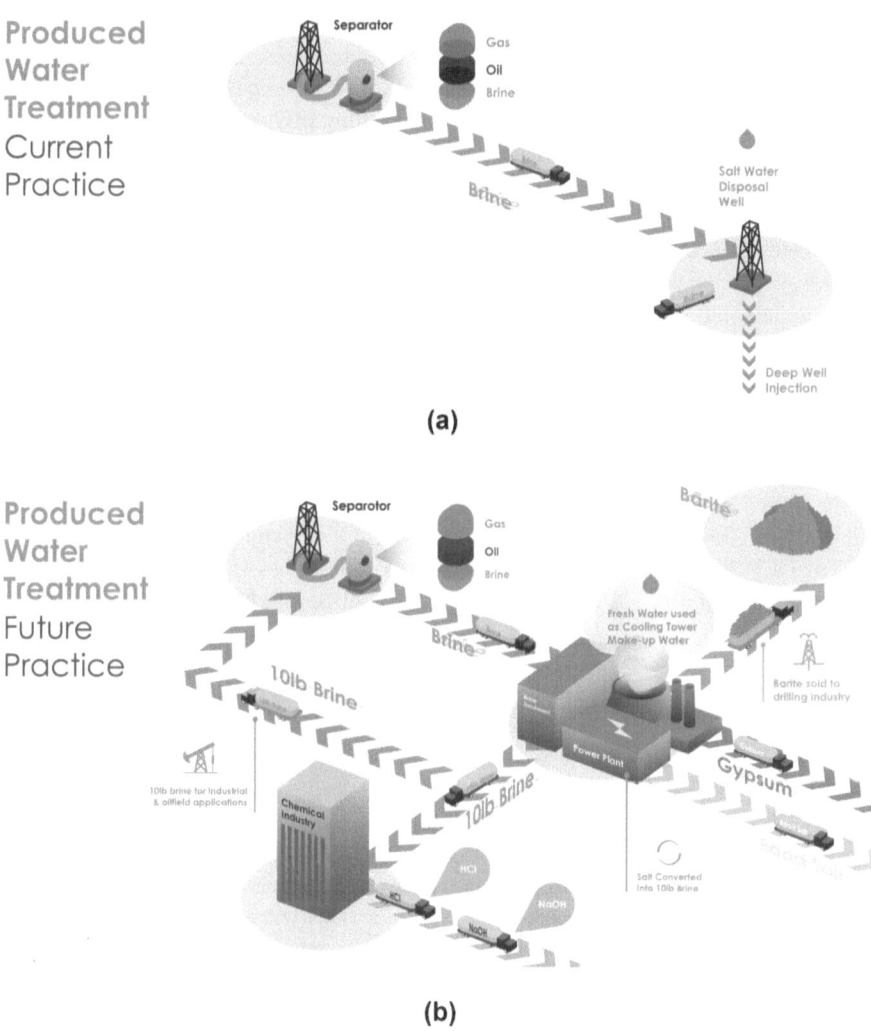

FIGURE 3.1 (a) Current practice of reinjection of produced brines in saltwater disposal wells. (b) Potential future application where produced brines are treated at power plants and converted into valuable resources.

4. Barite: solid barium sulfate with density $>4.2\,g\,cm^{-3}$, a weighting agent and drilling additive with low corrosion
5. Gypsum: calcium sulfate, used as wallboard and fertilizer
6. Low-salinity water for agriculture or power plant cooling tower make-up water

Note that (4) and (5) would imply mixing of the produced water with a sulfate-rich stream, such as acid mine drainage, coal power plant effluent, or sulfate-rich river water [18,33,34].

There have been several informative literature reviews on the topic of produced water management techniques and treatment technologies. Some of the most relevant literature reviews for the current discussion are listed below in Table 3.4. Also included in Table 3.4 are literature reviews on treating high-salinity brines extracted during CO_2 management, which have similar compositions as oil- and gas-produced brines, though often with lower organic and hydrocarbon concentrations.

One main difficulty to overcome regarding resource recovery from produced waters is to develop a cost-effective system to handle the broad spectrum of the time-varying suspended and dissolved materials in the produced brine. These may include heavy metals, natural hydrocarbons, water hardness, biological contaminants, and NORM. Systems integration and optimization for the region-specific waste streams from different basins and plays are needed to determine the best setup for a region. Additionally, the requirements for end use in terms of concentration constraints will inform the parameters of the chosen treatment and management process. The requirements for reuse scenarios may include environmental limits for discharge set by the EPA or industry-specific limits. It is in modeling these specific processes using advanced software, such as OLI Flowsheet (OLI Systems Inc., Parsippany, NJ [35]) or OLI chemistry in Aspen Plus [36], that efficiencies in operation and maximum reduction of costs can be compared with baseline processes.

TABLE 3.4
Relevant Literature Reviews on Produced Water Treatment

Author	Year	Title	Ref
Fakhru'l-Razi et al.	2009	Review of technologies for oil and gas produced water treatment	[37]
Igunnu and Chen	2012	Produced water treatment technologies	[38]
Shaffer et al.	2013	Desalination and reuse of high-salinity shale gas produced water: drivers, technologies, and future directions	[26]
Duraisamy et al.	2013	State of the art treatment of produced water	[39]
Munirasu et al.	2016	Use of membrane technology for oil field- and refinery-produced water treatment—a review	[40]
Giwa et al.	2017	Brine management methods: Recent innovations and current status	[41]
Pramanik et al.	2017	A review of the management and treatment of brine solutions	[42]
Arena et al.	2017	Management and dewatering of brines extracted from geologic carbon storage sites[a]	[43]
Kaplan et al.	2017	Assessment of desalination technologies for treatment of a highly saline brine from a potential CO_2 storage site[a]	[44]
Semblante et al.	2018	Brine pre-treatment technologies for zero liquid discharge systems	[45]
Jiménez et al.	2018	State of the art of produced water treatment	[46]
Chang et al.	2019	Potential and implemented membrane-based technologies for the treatment and reuse of flowback and produced water from shale gas and oil plays: A review	[47]
Al-Ghouti et al.	2019	Produced water characteristics, treatment and reuse: A review	[48]
Conrad et al.	2020	Fit-for-purpose treatment goals for produced water in shale oil and gas fields	[49]

[a] Literature focused on extracted brine treatment.

3.2 BRINE COMPOSITIONS

Depending on the basin, oil- and gas-produced brines vary in composition and in total dissolved solids (TDS). The values for TDS can range between 1 and 400 g L⁻¹ [26], although they generally range between 10 and 300 g L⁻¹. The typical salinity is significantly higher than the specification for drinking water and agricultural irrigation (0.5–2 g L⁻¹ TDS) [48]. As such, each produced brine will have its own set of treatment challenges and/or reuse scenarios. To provide some case studies for the wide variety of brine compositions and to provide context for the required treatment steps, we present in this section the characteristics of brines from several regions and formations across the United States. Brine composition data were collected from literature as well as national laboratory testing projects. The data were analyzed using OLI Stream Analyzer [50], which enables charge balance as well as the calculation of several brine attributes including osmotic pressure, density, viscosity, and ionic species in aqueous and solid forms. When pH and/or alkalinity values were measured, the pH and alkalinity were reconciled using OLI Stream Analyzer. The resulting brine attributes are shown in Table 3.5. Note that if a solid mineral forms, we present the concentration in mg of solid per L of total volume. For those mineral species that are undersaturated, we present the Q/K value, which is the ratio of the reaction activity quotient, Q, to the solubility product constant, K. The locations of the formations included in Table 3.5 are shown in Figure 3.2. This subsample of formations was chosen to demonstrate the geographical and compositional variation in produced brines across the United States. For more produced water compositional data, see the USGS Produced Waters Geochemical Database [51].

TABLE 3.5
Example Brine Compositions from Across the U.S. Using OLI Stream Analyzer V10.0

Formation:		Eagle Ford	Eagle Ford	Frio	Frio	Frio	Wolfcamp
Source ID:		Engle [58]	Slutz [59]	Morton [60]	Kharaka [61]	Knauss [62]	Stuckman Permian [63]
pH		6.6	6.7	7.6	6.6	6	–
Osmotic pressure[a]	atm	49	29	52	27	96	96
Activity of water	[-]	0.9653	0.9782	0.9618	0.9805	0.9313	0.9310
Ionic strength (x-based)	mol mol⁻¹	0.019	0.013	0.023	0.011	0.035	0.035
Specific elect. cond.	μmho cm⁻¹	83,978	59,811	92,936	53,979	142,572	139,829
Viscosity, abs	cP	0.990	0.956	1.015	0.946	1.100	1.094
Density (liquid only)	mg L⁻¹	1,038,390	1,025,740	1,045,752	1,022,080	1,074,288	1,073,190

(Continued)

TABLE 3.5 (*Continued*)
Example Brine Compositions from Across the U.S. Using OLI Stream Analyzer V10.0

Formation:		Eagle Ford	Eagle Ford	Frio	Frio	Frio	Wolfcamp
Source ID:		Engle [58]	Slutz [59]	Morton [60]	Kharaka [61]	Knauss [62]	Stuckman Permian [63]
Concentration of H_2O	mg L^{-1}	979,440	985,716	977,348	987,660	961,268	962,249
Total dissolved solids	mg L^{-1}	58,935	40,022	68,393	35,600	110,158	110,878
SiO_2(aq)	mg L^{-1}	91	–	74	61	25	–
CO_2(aq)	mg L^{-1}	6	46	<1	72	18	–
$B(OH)_3$(aq)	mg L^{-1}	520	–	53	36	–	–
Sodium (Na$^+$)	mg L^{-1}	*18,532*	*13,658*	21,824	12,683	39,825	*38,287*
Potassium (K$^+$)	mg L^{-1}	232	192	265	108	403	440
Lithium (Li$^+$)	mg L^{-1}	24	–	–	4	–	23
Ammonium $\left(NH_4^+\right)$	mg L^{-1}	–	–	32	12	–	–
Calcium (Ca^{2+})	mg L^{-1}	3,140	1,270	3,961	681	2,160	3,361
Iron (Fe^{2+})	mg L^{-1}	–	4	8	6	36	–
Manganese (Mn^{2+})	mg L^{-1}	–	–	4	1	–	–
Magnesium (Mg^{2+})	mg L^{-1}	195	111	151	88	450	593
Barium (Ba^{2+})	mg L^{-1}	3	–	–	8	48	–
Strontium (Sr^{2+})	mg L^{-1}	520	412	289	78	107	176
Chloride (Cl$^-$)	mg L^{-1}	35,192	23,797	*41,545*	*20,625*	*67,047*	66,320
Fluoride (F$^-$)	mg L^{-1}	–	–	1	1	–	1
Bromide (Br$^-$)	mg L^{-1}	245	–	127	55	–	497
Iodine (I$^-$)	mg L^{-1}	–	–	–	17	–	90
Bicarbonate $\left(HCO_3^-\right)$	mg L^{-1}	34	343	15	430	36	–
Acetate $\left(C_2H_3O_2^-\right)$	mg L^{-1}	138	–	–	649	–	–
Sulfate $\left(SO_4^{2-}\right)$	mg L^{-1}	16	254	49	8	3	1,084
Total suspended solids	mg L^{-1}	17	250	1,151	76	17	102
Barium sulfate (barite)	mg L^{-1}	17	25		57	17	–
Strontium sulfate (celestite)	mg L^{-1}	(Q/ K=0.19)	86	(Q/ K=0.07)	(Q/ K=0.01)	(Q/ K<0.01)	–
Calcium carbonate (calcite)	mg L^{-1}	(Q/ K=0.44)	2	1,132	(Q/ K=0.51)	(Q/ K=0.08)	–
Strontium carbonate (strontianite)	mg L^{-1}	(Q/ K=0.22)	88	(Q/ K=0.21)	(Q/ K=0.19)	(Q/ K=0.01)	13

(Continued)

TABLE 3.5 (*Continued*)
Example Brine Compositions from Across the U.S. Using OLI Stream Analyzer V10.0

Formation:		Eagle Ford	Eagle Ford	Frio	Frio	Frio	Wolfcamp
Source ID:		Engle [58]	Slutz [59]	Morton [60]	Kharaka [61]	Knauss [62]	Stuckman Permian [63]
Iron carbonate (siderite)	mg L^{-1}	–	49	19	19	($Q/$ $K=0.49$)	–
Silicon dioxide (lechatelierite)	mg L^{-1}	($Q/$ $K=0.91$)	–	($Q/$ $K=0.73$)	($Q/$ $K=0.56$)	($Q/$ $K=0.30$)	–
Calcium fluoride (fluorite)	mg L^{-1}	–	–	($Q/$ $K=0.03$)	($Q/$ $K=0.01$)	–	86
Sodium chloride (halite)		($Q/$ $K=0.01$)	($Q/$ $K=0.005$)	($Q/$ $K=0.01$)	($Q/$ $K=0.004$)	($Q/$ $K=0.05$)	($Q/$ $K=0.04$)

Formation:		Williston	Bakken	Utica	Marcellus	Marcellus
Source ID:		UND EERE [64]	Stuckman [63]	Spencer [65]	Phan#1 [66]	Phan#2 [66]
pH (median)		4.5	5.1	5.6	5.4	6.6
Osmotic pressure	atm	178	368	275	61	65
Activity of water	[-]	0.8775	0.7672	0.8192	0.9561	0.9533
Ionic strength (x-based)	mol mol^{-1}	0.060	0.097	0.088	0.0263	0.0276
Specific elect. cond.	µmho cm^{-1}	192,447	252,085	206,971	100,362	105,080
Viscosity, abs	cP	1.307	1.810	1.694	1.034	1.043
Density (liquid)	mg L^{-1}	1,121,450	1,196,150	1,166,280	1,051,930	1,055,020
Concentration of H$_2$O	mg L^{-1}	939,803	895,390	927,280	976,252	974,762
Total dissolved solids	mg L^{-1}	182,155	300,674	238,010	75,676	80,158
SiO$_2$(aq)	mg L^{-1}	<20	–	50	28	52
CO$_2$(aq)	mg L^{-1}	–	–	7	–	–
B(OH)$_3$(aq)	mg L^{-1}	1,850	–	8	90	90
Sodium (Na$^+$)	mg L^{-1}	49,900	81,332	49,680	18,995	20,498
Potassium (K$^+$)	mg L^{-1}	5,010	7,885	744	155	215
Lithium (Li$^+$)	mg L^{-1}	46	48	40	51	51
Rubidium (Rb$^+$)	mg L^{-1}	–	–	–	0.5	0.7
Ammonium (NH$_4^+$)	mg L^{-1}	–	2,948	–	–	–

(*Continued*)

TABLE 3.5 (*Continued*)
Example Brine Compositions from Across the U.S. Using OLI Stream Analyzer V10.0

Formation:		Williston	Bakken	Utica	Marcellus	Marcellus
Source ID:		UND EERE [64]	Stuckman [63]	Spencer [65]	Phan#1 [66]	Phan#2 [66]
Calcium (Ca^{2+})	mg L^{-1}	13,800	21,459	32,807	6,339	6,389
Iron (Fe^{2+})	mg L^{-1}	46	–	<1	133	132
Zinc (Zn^{2+})	mg L^{-1}	–	–	–	0.6	0.7
Nickel (Ni^{2+})	mg L^{-1}	–	–	–	0.1	0.1
Manganese (Mn^{2+})	mg L^{-1}	12	–	1	4.5	2
Magnesium (Mg^{2+})	mg L^{-1}	680	1,265	3888	655	688
Barium (Ba^{2+})	mg L^{-1}	2	6	2	2,340	2,400
Strontium (Sr^{2+})	mg L^{-1}	1,140	2,688	1997	1,317	1,399
Chloride (Cl^{-})	mg L^{-1}	*109,012*	*181,963*	*148,405*	*45,056*	*47,461*
Bromide (Br^{-})	mg L^{-1}	–	935	–	476	504
Fluoride (F^{-})	mg L^{-1}	<7	–	1	1	1
Iodide (I^{-})	mg L^{-1}	–	81	–	–	–
Bicarbonate (HCO$_3^{-}$)	mg L^{-1}	–	–	70	–	–
Acetate (C$_2$H$_3$O$_2^{-}$)	mg L^{-1}	–	–	–	–	–
Sulfate (SO$_4^{2-}$)	mg L^{-1}	152	63	125	<1	<1
Total suspended solids	mg L^{-1}	37	119	1467	36	106
Barium sulfate (barite)	mg L^{-1}	37	80	*(Q/K=0.62)*	<2	<2
Strontium sulfate (celestite)	mg L^{-1}	*(Q/K=0.75)*	39	949	<2	<2
Calcium sulfate (gypsum)		*(Q/K=0.16)*	*(Q/K=0.14)*	*(Q/K=0.27)*	–	–
Calcium carbonate (calcite)		–	–	*(Q/K=0.30)*	–	–
Iron disulfide (pyrite)	mg L^{-1}	–	–	277	19	–
Iron sulfide (pyrrhotite)	mg L^{-1}	–	–	<<<1	–	86
Silicon dioxide (lechatelierite)	mg L^{-1}	–	–	232	*(Q/K=0.66)*	*(Q/K=0.58)*
Calcium fluoride (fluorite)	mg L^{-1}	<14	–	–	17	8
Sodium chloride (halite)		*(Q/K=0.13)*	*(Q/K=0.64)*	*(Q/K=0.27)*	*(Q/K=0.013)*	*(Q/K=0.015)*

TABLE 3.5 (*Continued*)

Example Brine Compositions from Across the U.S. Using OLI Stream Analyzer V10.0

Formation:		Marcellus	Marcellus	Marcellus
Source ID:		Scenario 2 April 24[b]	Scenario 2 June 4[b]	Scenario 1[b]
pH		6	6	5.5
Osmotic pressure[a]	Atm	150 (145)	117 (113)	215 (206)
Activity of water	[-]	0.899054	0.920062	0.85939
Ionic strength (x-based)	mol mol^{-1}	0.0555	0.0464	0.0690
Specific elect. cond.	μmho cm^{-1}	167,295	147,794	197,262
Viscosity, abs	cP	1.271	1.185	1.425
Density (liquid)	mg L^{-1}	1,109,080	1,090,920	1,138,240
Concentration of H_2O	mg L^{-1}	953,107	961,584	936,291
Total dissolved solids	mg L^{-1}	155,958	129,336	201,924
SiO_2(aq)	mg L^{-1}	–	–	15
CO_2(aq)	mg L^{-1}	22	27	54
Sodium (Na^+)	mg L^{-1}	35,537	28,304	52,298
Potassium (K^+)	mg L^{-1}	273	261	116
Ammonium $(NH_4{}^+)$	mg L^{-1}	NA	NA	266
Calcium (Ca^{2+})	mg L^{-1}	16,900	14,500	18,400
Iron (Fe^{2+})	mg L^{-1}	376	101	66
Manganese ($Mn^{2+)}$	mg L^{-1}	20.6	–	8.1
Magnesium (Mg^{2+})	mg L^{-1}	1,680	1,480	1,550
Barium (Ba^{2+})	mg L^{-1}	3,120	3,230	2,919
Strontium (Sr^{2+})	mg L^{-1}	2,980	2,750	3,150
Chloride (Cl^-)	mg L^{-1}	93,800	77,200	122,000
Bromide (Br^-)	mg L^{-1}	1,100	1,340	919
Bicarbonate $(HCO_3{}^-)$	mg L^{-1}	154	147	184
[c]Citrate $(C_6H_5O_7{}^-)$	mg L^{-1}	4.7	4.7	4.7
Sulfate $(SO_4{}^{2-})$	mg L^{-1}	–	–	<0.1
Total suspended solids	mg L^{-1}	26	0	35
Barium sulfate (barite)	mg L^{-1}	–	–	35
Iron carbonate (siderite)	mg L^{-1}	26	–	–

Formation:		Wattenberg	Reef Ridge	Vedder	Vedder
Source ID:		Li Thesis (avg) [67]	(San Joaquin Basin) IDUSGS#3805 [51]	(San Joaquin Basin) IDUSGS#3813 [51]	(San Joaquin Basin) IDUSGS#46781 [51]
pH (median)		5.91	Measured 7.7	7.1	7.2
Osmotic pressure	Atm	14	9	31	15

(Continued)

TABLE 3.5 (Continued)
Example Brine Compositions from Across the U.S. Using OLI Stream Analyzer V10.0

Formation:		Wattenberg	Reef Ridge	Vedder	Vedder
			(San Joaquin Basin)	(San Joaquin Basin)	(San Joaquin Basin)
Source ID:		Li Thesis (avg) [67]	IDUSGS#3805 [51]	IDUSGS#3813 [51]	IDUSGS#46781 [51]
Activity of water	[-]	0.9896	0.993607	0.977533	0.988682
Ionic strength (x-based)	mol/ mol	0.0058	0.0032	0.0128	0.0060
Specific elect. cond.	μmho cm^{-1}	30,090	15,113	58,003	31,347
Viscosity, abs	cP	0.918	0.924	0.960	0.922
Density (liquid)	mg L^{-1}	1,009,840	1,004,940	1,024,420	1,010,910
Concentration of H_2O	mg L^{-1}	991,646	991,554	985,137	991,037
Total dissolved solids	mg L^{-1}	18,194	13,386	39,283	19,873
CO_2 (aq)	mg L^{-1}	82	15	2	36
SiO_2 (aq)	mg L^{-1}	–	111	67	–
$B(OH)_3$(aq)	mg L^{-1}	80	528	235	261
Sodium (Na$^+$)	mg L^{-1}	*6,390*	4,036	12,097	7,405
Potassium (K$^+$)	mg L^{-1}	116	62	112	–
Lithium (Li$^+$)	mg L^{-1}	–	2	2	–
Ammonium $\left(NH_4^+\right)$	mg L^{-1}	–	73	46	–
Calcium (Ca^{2+})	mg L^{-1}	381	21	2,293	89
Iron (Fe^{2+})	mg L^{-1}	81	<1	2	–
Magnesium (Mg^{2+})	mg L^{-1}	43	7	158	31
Strontium (Sr^{2+})	mg L^{-1}	55	6	124	–
Barium (Ba^{2+})	mg L^{-1}	18	1	30	–
Chloride (Cl$^-$)	mg L^{-1}	10,865	3,460	*22,607*	10,966
Bromide (Br$^-$)	mg L^{-1}	–	57	98	–
Fluoride (F$^-$)	mg L^{-1}	–	3	1	–
Iodide (I$^-$)	mg L^{-1}	–	14	20	<DL
Bicarbonate $\left(HCO_3^-\right)$	mg L^{-1}	67	760	40	647
Acetate $\left(C_2H_3O_2^-\right)$	mg L^{-1}	–	*4,836*	1,340	–
Carbonate $\left(CO_3^{2-}\right)$	mg L^{-1}	–	5	0.1	2
Sulfate $\left(SO_4^{2-}\right)$	mg L^{-1}	2	36	1	436
Total suspended solids	mg L^{-1}	4	234	191	398
Barium sulfate (barite)	mg L^{-1}	4	6	–	–

(Continued)

TABLE 3.5 (Continued)
Example Brine Compositions from Across the U.S. Using OLI Stream Analyzer V10.0

Formation:		Wattenberg	Reef Ridge	Vedder	Vedder
			(San Joaquin Basin)	(San Joaquin Basin)	(San Joaquin Basin)
Source ID:		Li Thesis (avg) [67]	IDUSGS#3805 [51]	IDUSGS#3813 [51]	IDUSGS#46781 [51]
Strontium carbonate (strontianite)	mg L^{-1}	–	4	–	–
Calcium carbonate (calcite)	mg L^{-1}	–	115	191	398
Silicon dioxide (lechatelierite)	mg L^{-1}	–	109	–	–

Bolded values indicate that this species was the charge balance ion. "–" represents a species not measured by the reference and hence not input into OLI Stream Analyzer. For solid species that are not saturated (i.e. Q/K <1), the Q/K value is listed to show how close this species is to being saturated. Sources are average values if samples were collected over time. Concentrations of dissolved species often increase with time.

[a] Osmotic pressure calculated using OLI's activity of water, not from the OLI osmotic pressure output.

[b] These Marcellus cases are analyzed in more detail in the "Influence of Colloids on Mineralization in Unconventional Oil and Gas Reservoirs and Wellbores: A Case Study with the Marcellus Shale" chapter of this book.

[c] Citrate was added to the produced water during modeling to mimic adding a scale-inhibitor before deep-well injection. The citrate is not in the brine as produced.

A number of different produced waters from across the United States are shown in Table 3.5. However, it should be noted that the compositional information provided is far from complete. We present a few cases studies, rather than ranges of compositions, so that readers can use these compositions in software programs, such as Dow WAVE [27,68], to model treatment processes. (For more details on the full range of compositions, see references [37,46,69].) No two authors measured all of the same species, so the numbers listed here should be considered to be partial, not complete, speciation of produced brine. For example, acetate ions were only sometimes measured. In addition, the organic composition of the brines was rarely listed and has not been included in the table above. All produced brines will have some dissolved hydrocarbons and other organic materials. The presence of dissolved organic carbon can foul membranes, allow microbes to grow, and inhibit kinetics of sulfate solubility, among other adverse impacts. The authors see the general lack of data on the dissolved organic material to be a major concern for modeling. For example, software programs, such as Dow WAVE, allow users to model ion exchange (IX), ultrafiltration (UF), nanofiltration (NF), and reverse osmosis (RO) processes. However, the user is not able to include organic species, such as acetate, in Dow WAVE to predict

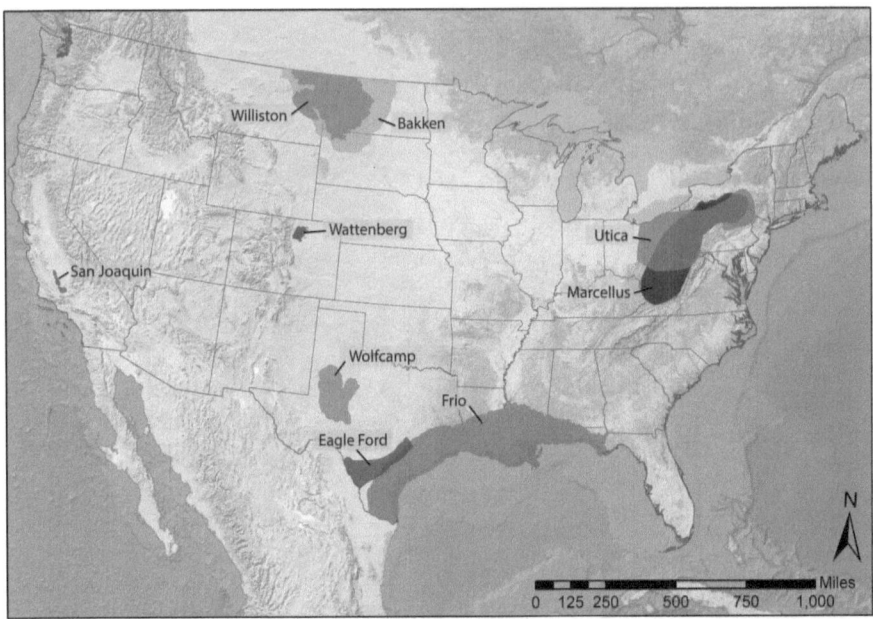

FIGURE 3.2 Geographical locations and extents of the formations in the United States that were examined in this study [52–57].

how much organic species will be absorbed onto IX resins or will permeate through UF/NF/RO membranes. This can be a problem because many downstream applications, including reinjection for water flooding, may have tight specifications on the dissolved organic carbon concentration.

Additionally, nearly all of the brines have suspended solids that will likely have to be removed before reinjection or reuse. Scale-forming ions include carbonate, sulfate, calcium, barium, and strontium. Given that the alkalinity of the brines was often not measured, the list of suspended minerals should not be considered complete due to the lack of information on carbonate speciation, which can lead to the formation of scale minerals such as calcite, aragonite, and siderite. Some brines have naturally occurring species that may be dissolved at high temperatures but which have the capability of precipitating at room temperature. Silica is one of these types of compounds which may be dissolved initially but can begin to polymerize and to precipitate out of solution as the temperature of the produced water decreases.

The ratio of monovalent to divalent ions also varies widely across different produced waters. Some are mainly NaCl brines with only small concentrations of divalent cations and divalent anions (such as brines detailed by Kharaka from the Frio Formation [61]), whereas some have large concentrations of divalent cations (such as the brines from the Marcellus Shale). Depending on whether the produced brines are mixed with sulfate-containing fresh waters or whether the brines are reinjected into different reservoirs with sulfate-rich chemistry, the removal of some scale-forming, divalent ions may be required before mixing or before reinjection into a new reservoir for EOR. In particular, standards for reinjection include limits on several ions [48,70].

The wide variation in brine composition, including differences in the concentrations of scale-forming ions and in the concentrations of organic species, necessitates customizing the pretreatment steps to generate the specific expected brine chemistry for the intended product or end use. A summary of commonly used water treatment processes and technologies, which are or will likely be relevant to produced water treatment, are discussed in the next section.

3.3   STANDARD WASTEWATER TREATMENT TECHNIQUES

Processes to convert brine into reusable products will consist of some combination of (1) pretreatment, (2) brine concentration, (3) salt crystallization, and/or (4) final disposal. Figure 3.3 shows a flow diagram of some potential steps required for produced brine treatment that will be discussed throughout this chapter.

3.3.1   BRINE PRETREATMENT

As seen in Figure 3.3, a number of steps are needed to remove fouling and scaling elements from the brine before reinjection or before resources are recovered from the brine. Typical pretreatment steps include the following: (1) removal of large, suspended particulates; (2) removal of light oil droplets; (3) removal of small-scale suspended particulates; (4) removal of microbes; (5) removal of dissolved organics; and (6) removal of scale-forming dissolved divalent ions, such as Ca^{2+}, Sr^{2+}, Ba^{2+}, and $SO_4{}^{2-}$. The standard methods corresponding to these pretreatment steps are: (1) coagulation/flocculation, (2) flotation, (3) disinfection, (4) media/micro/ultrafiltration, (5) adsorption, and (6) chemical precipitation, ion exchange, and/or nanofiltration. Variations in these treatment steps can be used to produce alternative liquid or solid products, such as barium sulfate (barite) or calcium sulfate (gypsum).

3.3.1.1   Removal of Large Particles via Coagulation and Flocculation

Large heavy suspended particles can be removed through gravity separation. This process can be sped up through the use of coagulants, such as but not limited to ferric chloride ($FeCl_3$), aluminum potassium sulfate (alum), or poly-aluminum chloride (PAC). In typical coagulation methods, chemicals are added to neutralize charges on particle surfaces, which tend to make large crystals form and make small particles clump together. In addition to the chemicals listed above, it can be advantageous to add lime or soda ash, or to add recycle streams rich in divalent ions to help precipitate species early on in the treatment process so that they do not precipitate out at later stages of treatment.

3.3.1.2   Flotation

There are several commercially available options for separating light oil droplets and/or lighter-than-brine particulates from the brine [38,71–75]. Common options include induced air flotation (IAF) and dissolved air flotation (DAF). The introduction of air at this stage could induce precipitation and/or combustion of recoverable hydrocarbons, which could be desired; however, if these oxidation side reactions are undesired, then nitrogen rather than air should be used. In the oil and gas industry,

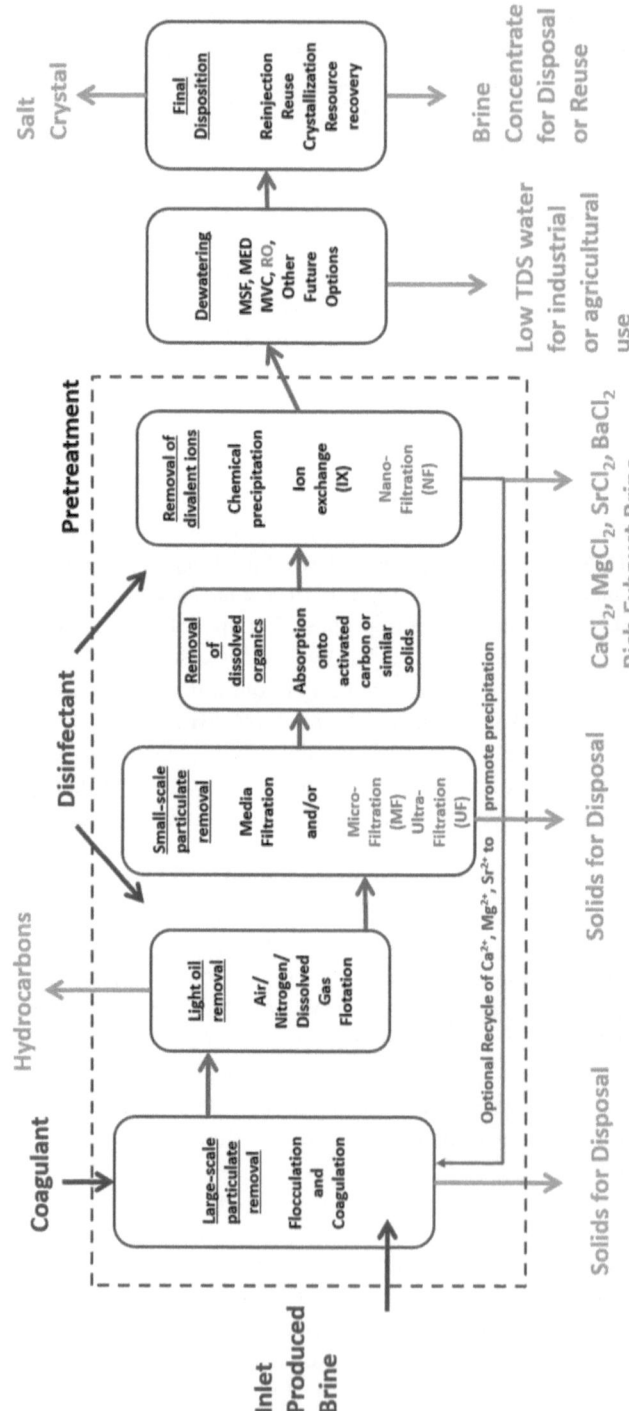

FIGURE 3.3 Pretreatment and dewatering stages for produced brine treatment and end products for use, reuse, or disposal. Membrane processes are highlighted in light blue.

nitrogen gas is often used as a substitute for air in flotation devices to prevent the oxidation of oils and/or the oxidation of reduced species in this stage. In both IAF and DAF, gases help suspended oils and light solids rise to the surface. The main difference between the two processes is that in induced flotation, gases such as pressurized nitrogen are added, whereas in dissolved flotation, the brine is depressurized so that gases evolve from the liquid, much like opening a soda can.

3.3.1.3　Disinfection

In many cases, produced brines will require disinfection to prevent biofouling. This will depend on the brine, its exposure to the atmospheric conditions, and whether the brine is mixed with fresh water before reinjection or reuse. Anaerobic thermophiles can thrive in the water within oil and gas reservoirs [76], and they particularly can thrive at the transition zone between where reinjected produced brine meets the reservoir. This transition zone often has the energetics needed for chemotrophs to survive off the difference in the redox potential of the reinjected brine and the reservoir. The reinjected brine can often be oxygen rich due to any combination of the following: (1) dissolved oxygen if the brine itself comes into contact with the air, (2) oxygenated chemicals added to the fluid, and (3) oxygen from surface waters mixed with the brine before reinjection.

Problems associated with microorganisms include the following: (1) reduced flow due to biofilms, (2) increased corrosion on steel surfaces in piping due to acid and/or sulfide generation, and (3) gas souring due to sulfide generation [77]. For more information on the types of chemicals used in disinfection, see [43,46,49].

3.3.1.4　Media, Micro, and Ultrafiltration

Fine solids as small as $0.1\,\mu m$ can be removed by passing brine through vessels filled with sand or sand-like materials, called media filtration [38,73,78,79]. Media filtration is typically a low upfront cost option but can result in higher O&M costs than equivalent membrane processes due to the cost of replacing the sand bed which fills up with small suspended solids.

The media filtration step additionally can be replaced by a microfiltration (MF) membrane process. MF membrane modules are sold by several companies and can remove particulates of size between 0.1 and $10\,\mu m$. If fine-particulate filtration is required, ultrafiltration membranes typically can remove particulates with diameters as small as $0.01\,\mu m$ ($10\,nm$). Back-washing is a crucial part of MF and UF membrane processes, both in crossflow and dead-end configurations, for the control of membrane fouling.

3.3.1.5　Adsorption

Similar to media filtration but using different packing material, adsorption typically represents the use of a fixed bed of activated carbon or similar materials with large surface area and active sites for selective removal of species of interest. If there is no media filtration or MF/UF steps before the activated carbon adsorption step, then the adsorption step can double as a media filtration step. Due to the higher cost of activated carbon compared with sand, this is typically not ideal, unless the suspended solids concentration is low. Organic constituents and chemical oxygen demand, which includes phenols and hydrocarbons, can be absorbed onto the surface of the activated carbon. For example, Okiel *et al.* found that ~5 g of activated carbon per liter of input brine was effective in reducing the oil concentration in the brine from ~1,000 to ~50 mg L^{-1} [80,81].

Similarly, Karnib *et al.* found that activated carbon can be quite effective in removing metal ions, resulting in 70% to 85% removal of cadmium, lead, nickel, and zinc [82]. Heavy metals absorbed onto the activated carbon can be regenerated using concentrated HCl.

Activated carbon is typically labeled as granular activated carbon (GAC) or powdered activated carbon (PAC). GAC is typically easier to regenerate than PAC. Typically, GAC is regenerated via thermal mechanisms, by which organic matter absorbed onto the pores is thermally oxidized, freeing up the surface. It should be noted that up to 10% of the carbon itself can be oxidized, meaning that there will need to be some make-up of activated carbon even when using regeneration techniques [83]. Thermal regeneration is mostly effective for organic matter absorbed onto the carbon, but care must be taken if the activated carbon also absorbs heavy metals and NORM.

3.3.1.6 Removal of Scale-Forming Ions

The removal of scale-forming ions can potentially generate valuable chemicals and minerals from produced brines, such as barium sulfate (barite), which is a critical material for the drilling industry [18,33]. The removal of scale-forming ions is important in many applications, particularly for dewatering via RO or RO-like processes to prevent membrane fouling and for reinjection applications to prevent mineral scale formation in the well and/or reservoir. For example, sulfate levels may be required to be below 50 mg L^{-1} for reinjection applications [48,70] depending on location. Scaling-forming ions can be removed through any of the following methods:

1. Chemical addition: Na_2CO_3 and/or Na_2SO_4 can be added to precipitate Ba^{2+}/Sr^{2+}/Ca^{2+}. (Note that these chemicals are likely to be added prior to coagulation/flocculation. This could also include the addition of sulfate-rich effluent streams from coal operations.)
2. Ion exchange resins: cation/anion resins, often regenerated using concentrated NaCl, NaOH, and/or HCl
3. NF membranes: H_2O and monovalent ions selectively pass through the membrane, leaving a concentrated stream with a higher TDS than the inlet brine and with a higher ratio of divalent-to-monovalent ions.

Chemical addition for the removal of calcite or barite can be modeled using standard software, such as OLI Flowsheet (OLI Systems Inc., Parsippany, NJ [35]). Ion exchange resins and NF membranes can be modeled using WAVE software (Dow Chemical Company, Midland, MI [68]), as well as software by other companies.

3.3.1.6.1 Chemical Addition

One method used to remove scale-forming cations, such as Ba^{2+}/Sr^{2+}/Ca^{2+}/Mg^{2+}, is to add NaOH, Na_2CO_3, and/or Na_2SO_4 early in the treatment train to help precipitate out these species as solids. Some precipitable solids include gypsum ($CaSO_4 \cdot 2H_2O$), barite ($BaSO_4$), strontianite ($SrCO_3$), calcite ($CaCO_3$), brucite ($MgO·H_2O$), and celestine ($SrSO_4$). As seen previously in Table 3.5, many of these minerals are already present in the brine, which means that the brine might already be saturated or supersaturated with these species. After removing solid carbonates, the pH of the brine would likely be adjusted back to a neutral or lower pH using HCl. One reason to use HCl to lower the pH to below neutral is to drive off carbon dioxide from the brine so as to prevent carbonate scaling downstream of this step. For example, Hayes and Severin report the use of a surge tank with a pH decrease to 4 to prevent carbonate scaling in a downstream brine concentrator [84].

3.3.1.6.2   Ion Exchange Resins

Ion exchange is the process by which problematic cations (or anions) are replaced with a cation (or anion) that is less problematic for downstream processes. There are five main types of IX resins: (1) strong acid cation (SAC), (2) weak acid cation (WAC), (3) strong base anion (SBA), (4) weak base anion (WBA), and (5) chelating ligand exchange resins. For cation exchange resins, a negatively charged functional group is chemically bonded to the backbone support, which is polymeric. For anion exchange resins, a positively charged functional group is chemically bonded to the backbone support. Examples of functional groups are listed below for each resin type.

a. SAC: exchange resins are polymeric materials functionalized with strong acids, such as sulphonate (R—SO_3^-). If the goal is to remove divalent, scale-forming cations, then the SAC would be initially loaded with monovalent Na^+ cations, and then as the produced brine flows through the SAC IX, the Mg^{2+}, Ca^{2+}, Sr^{2+}, and Ba^{2+} cations would exchange with the Na^+ ions initially associated with the resin. SAC resins typically are more selective in the following order: $Ra^{2+} > Ba^{2+} > Sr^{2+} > Ca^{2+} > Mg^{2+}$ [85]. This softening-type process removes divalent ions from brine which might otherwise cause scaling if the brine were sent to a downstream dewatering process, in addition to removing nearly all Ra^{2+} [85], a common component of NORM. It should also be noted that this type of ion exchange has little effect on overall TDS, osmotic pressure, or pH. For example, the osmotic pressure of a 2 molal NaCl solution is roughly 100 atm, which is similar to the osmotic pressure of a brine with 1 molal NaCl and 0.5 molal $CaCl_2$.[2]

b. WAC: exchange resins use functionalized acid groups that are weaker than sulphonate, such as the carboxylate anion functional group (R—CO_2^-), for ion exchange. As such, this type of resin is selective for species that would form precipitable carbonates, such as calcium carbonate. For some species such as calcium, it may be advantageous to use WAC IX resins rather than SAC IX resins because WAC IX resins require less concentrated acid to regenerate than SAC IX resins.

---

[2] We mention osmotic pressure here because the minimum work of separation pure water from a brine solution is proportional to its osmotic pressure. In Section 3.3.2, we discuss water separation technologies.

   c. SBA: exchange resins use functionalized base groups, such as positively charged quaternary ammonium groups, for ion exchange. These resins are typically shipped in the Cl^- form but can be operated in either Cl^- or OH^- form. The Cl^- is useful for replacing SO_4^{2-} ions with 2 Cl^-. As in the Na^+ form of SAC, the Cl^- form of SBA exchange resin has little effect on TDS or pH. However, it could cause a slight increase in osmotic pressure, because the osmotic pressure due to sulfate species is typically half of the osmotic pressure due to chloride species, which take its place in the brine.

   d. WBA: exchange resins use functionalized base groups, such as positively charged primary, secondary, or tertiary amine groups for ion exchange. The groups are less basic than the SBA IX resins and hence require less base to regenerate. WBA resins typically use functionalized amine groups and initially are loaded with Cl^-. When the brine flows through the exchange resin, the Cl^- exchanges with sulfate or other monovalent anions, such as bromide.

   e. Chelating resins: In addition to the four main types of ion exchange resins, there are also specialty resins focused on the selective resin of specific ions (mostly dissolved ionic metals) from water streams. Chelating resins and thiol-functionalized resins have the capability of selectively removing heavy metal cations, particularly Ni^{2+}, Co^{2+}, Cd^{2+}, Hg^{2+}, Fe^{3+}, Cu^{2+}, Zn^{2+}, Cr^{3+}, Pb^{2+}, and Fe^{2+}. Heavy metal species which tend to bond with sulfide can be selectively removed using these types of chelating resins. These resins are regenerated in a similar manner as SAC resins (HCl is added to remove heavy metals and then the resin is regenerated into Na^+ form using NaOH).

For all IX resins, regeneration of the resin is crucial, because to be cost effective these resins must be regenerated >1,000 times. Most resins can be most easily regenerated through the use of HCl and NaOH. For example, in the case of a SAC IX, after a bed of resin is saturated in Mg^{2+}, Ca^{2+}, Sr^{2+}, and Ba^{2+} cations, the produced water would be switched to a second bed, and then HCl would be added into the saturated resin bed. This generates a NaCl-, $MgCl_2$-, $CaCl_2$-, $SrCl_2$-, and $BaCl_2$-containing brine that could potentially be mixed with a sulfate-rich effluent stream to generate gypsum and barite. To fully regenerate the SAC IX, concentrated NaOH would be added to return the resin to the Na^+ form so that it can ion-exchange with produced brine again. It should be noted that the resin could instead be regenerated using concentrated NaCl in one step. This would create a different composition for the divalent-rich waste product. The choice between regeneration with HCl/NaOH or concentrated NaCl will depend on what chemicals are available and whether any of these chemicals can be generated at the treatment facility from the brine.

   To regenerate a SBA IX, the process would be similar, but the order of using NaOH and HCl would be changed. In this case, there would be a waste product of mostly Na_2SO_4. Once again, it is possible to regenerate an SBA using concentrated NaCl in one step rather than a two-step NaOH/HCl process, but the choice will depend on what chemicals are available. Due to the large concentration of Mg^{2+}, Ca^{2+}, Sr^{2+}, and Ba^{2+} cations and the low concentration of sulfate ions in most produced waters, particularly in brine from Appalachia, the SAC IX resins or WAC IX resins appear to be more relevant than the SBA IX resins or WBA IX resins.

Concentrated NaCl could be generated on-site via one of the dewatering techniques to be discussed later. In principle, the HCl and NaOH required to regenerate IX resins could be generated on-site from concentrate NaCl solutions via electrolysis or electrodialysis, but this is difficult because electrolysis and electrodialysis processes typically have extremely strict limitations on the concentration of divalent species to prevent scale formation on the electrodes. In general, both the use of ion exchange resin for produced water treatment and the regeneration of ion exchange resin using NaCl/NaOH/HCl generated from the treated brine on-site are areas that require additional research to prove optimal performance.

The application of IX resins to remove divalent ions from produced brines could benefit from future research and development into the use of continuously looping, countercurrent resin bed designs for both the adsorption onto and desorption from the resin. Current fixed bed geometries are thermodynamically inefficient because of their non-steady-state driving forces, which limit the use of an otherwise quite useful IX resin with applications to remove divalent species from high-salinity brines. The use of continuously looping, countercurrent resin bed designs rather than fixed bed design could increase the usefulness of IX resins to selectively remove species and to be regenerated using high-salinity NaCl brines generated on-site from downstream RO and mechanical vapor recompression (MVR) processes (see Figures 3.5 and 3.6 in Section 3.4 for examples).

3.3.1.6.3 Nanofiltration (NF) Membranes

Nanofiltration is a membrane-based separation process with pore sizes that fall in between UF and RO processes. NF membranes are selective for ions based on size and charge. As such, monovalent ions like Na^+ and Cl^- are able to pass more easily through NF membranes than through RO membranes. Currently, the typical maximum applied pressure for commercially available NF membranes is around 40 bar. NF membranes are ideal for the separation of Mg^{2+}, SO_4^{2-}, Na^+, and Cl^- brines into a concentrated brine comprised mainly of Mg^{2+} and SO_4^{2-}, and a lower salinity permeate comprised mainly of Na^+ and Cl^-. NF membranes use applied hydraulic pressure as the driving force, whereas IX resins use pH adjustment or ion concentration as driving forces. Ultimately, the energy for both processes comes from electricity, either to drive a pump or to drive the splitting of NaCl into HCl and NaOH.

While NF membranes are not nearly as selective for Na^+ over Ca^{2+} as ion exchange resins are, NF membranes can play an important role in produced water treatment because treatment using NF membranes is a steady-state process that does not require the input of chemicals for performing separation. Chemical precipitation as well as currently available IX resin processes require chemical input.

As a case study, in Table 3.6, we show the inlet and outlets of a sample NF treatment process, as calculated by WAVE software. In Table 3.6, we model the treatment of the Morton brine (Frio Formation) using a two-stage NF membrane system [60]. The Morton brine is pressurized to 40 bar then is fed to a NF90 membrane [86]. The concentrate of this membrane is then sent to a NF270 membrane [87]. In this

TABLE 3.6

One-Pass, Two-Stage (NF-270 / NF-90) Process

Stream		Morton Inlet [60]	NF Concentrate	NF Permeate	Concentrate Percentage	Permeate Percentage
Relative volumetric flow rate		1.00	0.34	0.66	34%	66%
Total dissolved solids	mg L^{-1}	68,312	107,412	48,153	53%	47%
SiO$_2$(aq)	mg L^{-1}	74	90	66	41%	59%
CO$_2$(aq)	mg L^{-1}	1	2	1	68%	66%
Sodium (Na$^+$)	mg L^{-1}	21,824	32,261	16,447	50%	50%
Potassium (K$^+$)	mg L^{-1}	265	400	195	51%	49%
Ammonium $\left(NH_4{}^+\right)$	mg L^{-1}	32	47	24	50%	50%
Calcium (Ca^{2+})	mg L^{-1}	3,961	7,904	1,926	68%	32%
Magnesium (Mg^{2+})	mg L^{-1}	151	317	65	71%	29%
Strontium (Sr^{2+})	mg L^{-1}	289	590	134	69%	31%
Chloride (Cl$^-$)	mg L^{-1}	41,594	65,464	29,289	54%	46%
Bicarbonate $\left(HCO_3{}^-\right)$	mg L^{-1}	15	31	6	70%	26%
Sulfate $\left(SO_4{}^{2-}\right)$	mg L^{-1}	49	141	1.5	98%	2%

example, the permeates are mixed. Table 3.6 shows the composition and the relative flows of the inlet and two outlets. (Note that this is just one of many possible configurations of NF and/or RO membranes which could be used to separate fresh water, monovalent ions, and divalent ions.) In this configuration, the volumetric flow rate is roughly 2:1 between the permeate and the concentrate, whereas the Na$^+$/K$^+$/NH$_4{}^+$ are split roughly 1:1 between the permeate and the concentrate and Ca^{2+}/Mg^{2+}/Sr^{2+} are split roughly 1:2 between the permeate and the concentrate. Note that sulfate is almost entirely kept in the concentrate.

In Table 3.7, we show the inlet and outlets, as calculated by WAVE software, for a case in which a Marcellus Shale produced brine is sent to a two-stage NF membrane system. The Phan #1 brine is pressurized to 40 bar and then is fed to a NF270 membrane [66]. The permeate from the NF270 is pressurized to 40 bar and then sent to a NF90 membrane. In this example case, the concentrates from each membrane are mixed in Table 3.7 to simplify the presentation of the results. Real processes might recycle these streams separately depending on the end-product application. The goal of these case studies has been to show how conventional membrane technologies can be used to convert produced brines into a high-salinity concentrate stream high in divalent cations, and a low-salinity concentrate stream comprised mainly of monovalent ions.

3.3.1.7   Other Pretreatment Steps (Silica and Boron)

One area of concern for brine dewatering processes is silica scale. Dissolved silica is likely to be present, especially in brines produced from silicate-rich formations. Due to the increasing solubility of silica in aqueous solutions with increasing

TABLE 3.7
Two-Pass, Two-Stage (NF-270 / NF-90) Process

Stream		Phan#1 Inlet [66]	NF Concentrate	NF Permeate	Concentrate Percentage	Permeate Percentage
Relative volumetric flow rate		1.00	0.73	0.27	73%	27%
Total dissolved solids	mg L^{-1}	74,794	94,196	22,661	92%	8%
SiO$_2$(aq)	mg L^{-1}	28	35	7	91%	7%
Sodium (Na$^+$)	mg L^{-1}	18,995	23,408	7,152	90%	10%
Potassium (K$^+$)	mg L^{-1}	155	190	61	89%	11%
Ammonium (NH$_4^+$)	mg L^{-1}	10	12	6	84%	16%
Calcium (Ca^{2+})	mg L^{-1}	6,339	8,295	1,075	95%	5%
Magnesium (Mg^{2+})	mg L^{-1}	660	867	101	96%	4%
Strontium (Sr^{2+})	mg L^{-1}	1,317	1,726	216	95%	4%
Barium (Ba^{2+})	mg L^{-1}	2,340	3,068	380	95%	4%
Chloride (Cl$^-$)	mg L^{-1}	44,866	56,169	13,663	91%	8%

temperature and due to the fact that brines are produced from reservoirs at significantly higher temperatures than they are at in the treatment process (at or near room temperature), silica is likely to be saturated or supersaturated at the surface. After the brine is sent to a NF membrane and/or a brine dewatering step (such as RO/MVR), the concentration of silica can become supersaturated and precipitation can occur, which could damage membranes or surfaces on other dewatering equipment. This is a crucial ongoing area of research because silica precipitation can be the species whose concentration limits the extent of the overall brine concentration.

In principle, there are other minerals with similar temperature-dependent solubility concerns. Note, however, that the opposite of this occurs for calcite and siderite when reinjecting produced brines underground. Depending on the treatment process, treated produced brines could be nearly saturated with calcite and siderite, which can precipitate out on the walls of the reinjection well when the brines reach the higher temperatures at depth in the well. This precipitation can cause decreased flow rates at a fixed pressure drop while reinjecting.

Another neutral species that can be problematic for produced water treatment is boron, but it is problematic for very different reasons than silica. At pH values below ~9, the dominant form for boron is the neutral species $B(OH)_3$. However, at pH values above this, the dominant form for boron is the anion $B(OH)_4^-$. Since $B(OH)_3$ is neutral and stable, it does not absorb onto most surfaces and will pass right through RO membranes. For some seawater RO applications, there is an ion exchange step after the RO membrane in which the permeate (rich in boron) is first passed through an OH$^-$ anion exchange resin, causing the pH to increase and the $B(OH)_4^-$ anion to be associated with the resin rather than OH$^-$. The addition of N-methyl-D-glucamine (NMDG) functional group on IX resins is effective for the removal of boron from low-salinity streams, particularly from the low-salinity permeate of an RO membrane [88].

3.3.2 WATER RECOVERY VIA BRINE CONCENTRATION

In the pretreatment steps above, fresh water is not generated by any of the processes. Here, we will focus on processes that separate out low-salinity water, leaving behind a concentrated brine. First, we briefly discuss the thermodynamics of water separation and then discuss the technologies that are commercially available to dewater mid- to high-salinity brines.

3.3.2.1 Brine Thermodynamics

Dissolved solutes in an aqueous solution depress the activity of water (a_w). The activity of water is an intrinsic property of a solution, and the activity of water in a solution decreases with an increase in the concentration of dissolved solutes. Pure water naturally flows to brines with lower a_w. To force pure water out of a brine with a_w less than 1 requires the input of useful work, either in the form of hydraulic pressure, electricity, or heat at a different temperature than the environment. The minimum amount of work required to remove pure water from a brine is independent of the separation process (i.e., evaporative, membrane, electrochemical, etc.) [89–91]. The minimum amount of work is equal to the difference of the Gibbs free energy between the output (concentrate plus permeate) and the inlet brine. This calculation can be done using software, such as OLI Stream Analyzer or OLI Flowsheet, as presented in Wenzlick et al. [32].

3.3.2.2 Non-membrane, Thermally Driven Evaporative Processes

Multi-effect distillation (MED) relies on a thermal input in the first step. The water vapor generated in the first effect is then condensed and used to evaporate water in the second stage, and this can be continued for several stages until there is not enough of a temperature gradient to drive further evaporation. Multistage flash (MSF) distillation also relies on both a thermal input and a series of different steps (effects vs. stages), just as in MED. However, the main difference between MSF and MED is that in MED, the heat transfer goes directly into evaporating water from the brine in the MED plant, whereas in MSF, the heat transfer goes to a pressurized liquid, which does not boil/flash until the pressure is dropped in a subsequent stage. So, while MED processes potentially need less surface area for the heat exchange to occur than MSF processes, a possible concern for MED processes is the increased potential for scaling on the heat exchange surface. The increased risk of scaling is due to the evaporation of water directly on the heat exchange surfaces in MED rather than in the bulk in a downstream flash tank as in MSF [89,92]. The key point here is that both MED and MSF rely on thermal input and changes in temperature, and the choice between the two is largely due to whether scale formation on the heat exchange surface will potentially occur. It should also be noted that MED and MSF can be combined in hybrid processes where some heat transfers to a high-pressure liquid (MSF) and some heat transfers to a boiling brine (MED).

3.3.2.3 Non-membrane, Electrically Driven Evaporative Processes

MVR is another commercially utilized brine concentration process, which has broad applicability due to its ability to treat highly saline water. The inlet salinity can range from brackish water (~10 g L^{-1} TDS) to near saturation (~300 g L^{-1} TDS), with varying efficiencies [32]. This process has been used to concentrate high-salinity groundwater to remove water from RO concentrate and to desalinate seawater [93]. There is even some experimental data for MVR operation on produced brines at commercial-scale [84].

MVR units operate by converting the water in the inlet saline brine into steam, then compressing the steam using a mechanical compressor. The high temperature, high-pressure steam then exchanges heat with the inlet saline brine to evaporate water. The incoming saline water is preheated by both the outlet product water and the discharge brine [84,93]. As the latent heat of the steam is used to heat the bulk saline fluids, the efficiency of the process is greater than that of traditional distillation [84]. However, MVR processes are still less efficient than RO membrane processes for desalination applications [94], and MVR processes are also less flexible than RO processes because the equipment needs time to ramp up and reach operating temperatures.

3.3.2.4 Membrane, Electrically Driven, Non-Evaporative Processes

Here, we focus on RO processes that are commercially available and relevant to produced water treatment. As was seen in Tables 3.6 and 3.7 (in the NF portion of Section 3.3.1.6), the permeate from NF membrane system can be quite low in TDS, even if the inlet brine is high in TDS. As such, the permeate from NF membranes could be sent to RO membranes, where low-salinity water can be extracted. The pretreatment steps listed previously will be crucial for RO membranes because these types of membranes can be fouled by any combination of organics, microorganisms, and/or mineral scale. Assuming that the pretreatment steps remove fouling species, then moderate-pressure RO membranes typically can concentrate brines up to an outlet TDS of around $75 \, g \, L^{-1}$, and high-pressure RO membranes can concentrate up to an outlet TDS of around $120 \, g \, L^{-1}$, according to calculations using Dow WAVE (see: Tables 3.8 and 3.9) [68].

Ultimately, when recycle streams are included, combinations of commercially available NF and RO membranes can dewater brine up to around $120 \, g \, L^{-1}$. This is likely around the limit that commercially available NF and RO technologies can achieve. This $\sim 120 \, g \, L^{-1}$ brine could be sent to thermal processes, such as MED, MSF, or MVR. Perhaps in the future, there will be advanced membrane processes which can concentrate brine to $250 \, g \, L^{-1}$ or higher TDS levels.

3.3.2.5 Crystallizers

For applications such as road salt and/or volume reduction for disposal in solid land-fills, crystallizers will be required to generate solid salts after having first concentrated the brine as much as possible using NF/RO/MED/MSF/MVR equipment. It should also be noted that some of the MED/MSF/MVR equipment versions can concentrate brine to the point in which solids form within the equipment.

A hybrid brine concentrator and crystallizer technology has been developed by industry members in which a non-scaling brine, such as $CaCl_2$, is concentrated to high-salinity via MVR/MED/MSF technology [U.S. Patent 9822023] [95,96]. At elevated temperatures, mixtures of $MgCl_2$ and $CaCl_2$ can be concentrated to quite high mass fractions ($\sim 60 \, wt\%$) without forming solid crystals, provided that the

TABLE 3.8

NF90 (40 bar) Pass 1 Followed by SW30-HRLE (68 bar) Pass 2

Stream		Dow WAVE Modeled	Wattenberg Li Thesis (avg) [67]	NF+RO Concentrate	RO Permeate	Concentrate Percentage	Permeate Percentage
Relative volumetric flow rate			1.00	0.17	0.83	17%	83%
Total dissolved solids	mg L^{-1}	17,720	94,196	422		98%	2%
CO$_2$ (aq)	mg L^{-1}	70	73	70		60%[b]	40%[b]
SiO$_2$ (aq)	mg L^{-1}	50[a]	291[c]	1.4		97.8%	2.2%
Sodium (Na$^+$)	mg L^{-1}	6,390	36,456	155.3		98.1%	1.9%
Potassium (K$^+$)	mg L^{-1}	116	660	3.3		97.8%	2.3%
Ammonium (NH$_4^+$)	mg L^{-1}	50[a]	274	3.5		94.3%	5.7%
Calcium (Ca^{2+})	mg L^{-1}	381	2,211	2.2		99.6%	0.4%
Magnesium (Mg^{2+})	mg L^{-1}	43	249	0.3		99.6%	0.4%
Strontium (Sr^{2+})	mg L^{-1}	55	319	0.3		99.6%	0.4%
Barium (Ba^{2+})	mg L^{-1}	18[c]	104[c]	0.1		99.6%	0.4%
Chloride (Cl$^-$)	mg L^{-1}	44,866	62,023	253.6		98.2%	1.8%
Bicarbonate (HCO$_3^-$)	mg L^{-1}	81	460	2.3		60%[b]	40%[b]
Sulfate (SO$_4^{2-}$)	mg L^{-1}	2[c]	12[c]	0		100%	0%

[a] 50 mg/L of silica was added to represent typical silica concentration because silica was not measured in the Wattenberg case study.

[b] The percentage is the fraction for the sum of dissolved CO$_2$ and HCO$_3^-$.

[c] Both silica and barite are supersaturated.

concentration of other chlorides (NaCl, SrCl$_2$, and BaCl$_2$) can be kept low [96]. The osmotic pressure of CaCl$_2$ and MgCl$_2$ brines can be >1000 bar without forming crystals. This is well beyond the osmotic pressure that can be achieved via RO-like processes that rely on applied hydraulic pressure to drive water out of brine solutions. If done carefully (i.e. at the right temperatures and concentrations), the concentrated CaCl$_2$ brine can be added to produced water to induce the precipitate of NaCl. Both the solid NaCl and the concentrated CaCl$_2$ could potentially be sold in various markets, such as for road salt and deicing fluid (with appropriate pretreatment steps beforehand to remove Sr^{2+}, Ba^{2+}, and NORM).

The important point here is that there are variations in MVR/MED/MSF technologies which can be used to create solid salts (such as NaCl) from produced brines but potentially without having the precipitation occur directly on heat exchange surfaces. Variants of these technologies for converting produced brine into valuable products will likely require separation processes, such as the IX and NF technologies discussed previously, to separate produced brine into streams rich in (1) SrCl$_2$ and BaCl$_2$, (2) MgCl$_2$ and CaCl$_2$, and (3) NaCl and KCl. Combinations of standard thermal processes with advanced membrane processes is hopefully a growing area of future research.

TABLE 3.9
SW30-HRLE (65 bar) SEAMAXX (80 bar) Pass 1

Stream		Dow WAVE Modeled	Wattenberg Li Thesis (avg) [67]	HPRO Concentrate	HPRO Permeate	Concentrate Percentage	Permeate Percentage
Relative volumetric flow rate			1.00	0.135	0.865	13.5%	86.5%
Total dissolved solids	$mg\,L^{-1}$	17,720	120,688	1,980	90.4%	9.6%	
CO_2 (aq)	$mg\,L^{-1}$	70	75	71	53%[b]	47%[b]	
SiO_2 (aq)	$mg\,L^{-1}$	50[a]	326[c]	8	86.0%	14.0%	
Sodium (Na^+)	$mg\,L^{-1}$	6,390	42,567	726	90.1%	9.9%	
Potassium (K^+)	$mg\,L^{-1}$	116	762	15	88.9%	11.1%	
Ammonium $\left(NH_4^{\,+}\right)$	$mg\,L^{-1}$	50[a]	299	11	80.8%	19.2%	
Calcium (Ca^{2+})	$mg\,L^{-1}$	381	2,707	17	96.0%	4.0%	
Magnesium (Mg^{2+})	$mg\,L^{-1}$	43	305	2	96.0%	4.0%	
Strontium (Sr^{2+})	$mg\,L^{-1}$	55	391	2	96.1%	4.1%	
Barium (Ba^{2+})	$mg\,L^{-1}$	18[c]	128[c]	1	96.1%	4.1%	
Chloride (Cl^-)	$mg\,L^{-1}$	44,866	72,662	1,187	90.5%	9.5%	
Bicarbonate $\left(HCO_3^{\,-}\right)$	$mg\,L^{-1}$	81	526	11	53%[b]	47%[b]	
Sulfate $\left(SO_4^{\,2-}\right)$	$mg\,L^{-1}$	2[c]	15[c]	0	100%	0%	

[a] 50 mg L^{-1} of silica was added to represent typical silica concentration because silica was not measured in the Wattenberg case study.

[b] The percentage is the fraction for the sum of dissolved CO_2 and $HCO_3^{\,-}$.

[c] Both silica and barite are supersaturated.

3.4 EXAMPLE END-USE APPLICATIONS

Here, we present a few examples and process flow diagrams of possible brine treatment methods to generate several of the industrial products listed in the Introduction, using the brines detailed in Table 3.5. Given the different characteristics of the brines, specific applications may be more or less suitable for each brine and different treatment processes will be necessary.

1. Ten-pound brine
    Brines with moderate-to-high salinity and low concentrations of divalent cations (Mg^{2+}, Ca^{2+}, Sr^{2+}, and Ba^{2+}) would be ideal for generating ten-pound brine or road salt. For example, the Eagle Ford [59], Frio [61], and Wolfcamp Permian [63] brines could be treated to obtain 10-lb. brine using the process flow depicted in Figure 3.4.
2. Water recovery for irrigation
    Brines ideal for beneficial reuse, such as irrigation or livestock watering, are those that require minimal treatment before reuse. These include

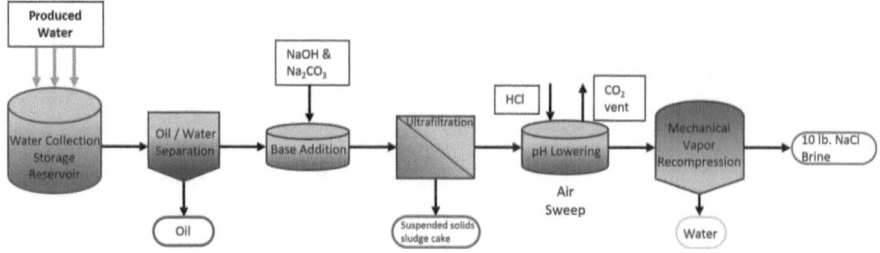

FIGURE 3.4 Example process flow case for creating 10-lb. brine; for more detail see reference [32].

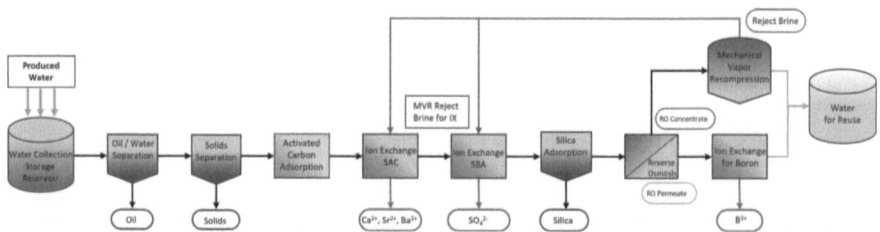

FIGURE 3.5 Example process flow for treating low-salinity brine for beneficial reuse, including irrigation or livestock watering.

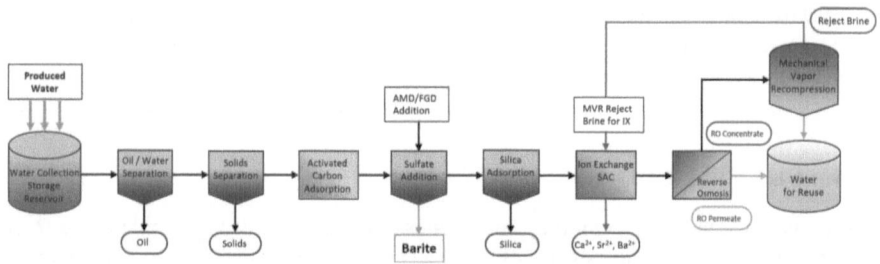

FIGURE 3.6 Example process flow for treating barium-containing brine to generate barite.

low-salinity produced waters such as the Vedder brine (IDUSGS #46781) [51] and the Wattenberg brine [67]. A possible treatment scenario is depicted in the process workflow shown in Figure 3.5 (see also Tables 3.8 and 3.9).

3. Barite

Brines ideal to treat for the generation of barite are those containing high Ba^{2+} and high Sr^{2+} levels relative to other ions. These include the Marcellus Phan #1 and #2 brines [66] and the Marcellus Scenario 2 brines. Possible treatment process flows include the treatment steps detailed in Figure 3.6. Note that this process generates both barite and low-salinity water as products.

3.5 CONCLUSIONS

Managing produced brines continues to grow increasingly important as volumes of brine produced during oil and gas production and other drilling operations continue to grow. Regulatory agencies as well as industry members continue to investigate best practices for produced water management due to the economic and environmental impacts of produced water disposal, treatment, and reuse. Given the growing interest in implementing alternative management techniques to disposal by injection, a wide array of treatment technologies and reuse strategies can be considered for managing produced waters and reducing the volume of brine sent to deep-well injection. The necessary treatment steps for a particular brine depend on the wide-ranging brine composition and TDS levels, as well as the intended output products. Several of the most prominent treatment technologies were discussed in this chapter. Numerous pretreatment and treatment options are commercially available for removing contaminants and minerals from effluent streams. However, many of the technologies discussed here would require novel configurations for treating produced brines. As such, continued research and field testing will be necessary to optimize processes that manage produced brines in an environmentally sustainable manner and that recover valuable resources from the those brines.

REFERENCES

1. Veil, J., *U.S. Produced Water Volumes and Management Practices in 2017.* 2020: Groundwater Research & Education Foundation.
2. Veil, J., *U.S. Produced Water Volumes and Management Practices in 2012.* 2015.
3. Kondash, A.J., N.E. Lauer, and A. Vengosh, The intensification of the water footprint of hydraulic fracturing. *Science Advances*, 2018. **4**(8): p. eaar5982.
4. McDevitt, B., et al., Emerging investigator series: Radium accumulation in carbonate river sediments at oil and gas produced water discharges: Implications for beneficial use as disposal management. *Environmental Science: Processes & Impacts*, 2019. **21**(2): pp. 324–338.
5. McDevitt, B., et al., Isotopic and element ratios fingerprint salinization impact from beneficial use of oil and gas produced water in the Western US. *Science of the Total Environment*, 2020. **716**: p. 137006.
6. McLaughlin, M.C., et al., Mutagenicity assessment downstream of oil and gas produced water discharges intended for agricultural beneficial reuse. *Science of the Total Environment*, 2020. **715**: p. 136944.
7. McLaughlin, M.C., et al., Water quality assessment downstream of oil and gas produced water discharges intended for beneficial reuse in arid regions. *Science of the Total Environment*, 2020. **713**: p. 136607.
8. Veil, J.A., et al., *A white paper describing produced water from production of crude oil, natural gas, and coal bed methane.* 2004, Argonne National Lab., IL (US).
9. Looney, B., *BP Statistical Review of World Energy.* 2020, BP Statistical Review: London.
10. *Underground Injection Control (UIC): Class II Oil and Gas Related Injection Wells.* 2019 [cited 2020 8/12/2020]; Available from: https://www.epa.gov/uic/class-ii-oil-and-gas-related-injection-wells.
11. McCurdy, R., *Underground Injection Wells for Produced Water Disposal.*
12. *Assessment of the potential impacts of hydraulic fracturing for oil and gas on drinking water resources.* 2015, U.S. Environmental Protection Agency, Office of Research and Development.

13. Castillo, C., C. Kervévan, and D. Thiéry, Geochemical and reactive transport modeling of the injection of cooled Triassic brines into the Dogger aquifer (Paris basin, France). *Geothermics*, 2015. **53**: pp. 446–463.

14. Cihan, A., J.T. Birkholzer, and M. Bianchi, Optimal well placement and brine extraction for pressure management during CO_2 sequestration. *The International Journal of Greenhouse Gas Control*, 2015. **42**: pp. 175–187.

15. Su, B., et al., Study on seawater nanofiltration softening technology for offshore oilfield water and polymer flooding. *Desalination*, 2012. **297**: pp. 30–37.

16. Kharaka, Y.K., et al., Deep well injection of brine from Paradox Valley, Colorado: potential major precipitation problems remediated by nanofiltration. *Water Resources Research*, 1997. **33**(5): pp. 1013–1020.

17. Paukert Vankeuren, A.N., et al., Mineral reactions in shale gas reservoirs: barite scale formation from reusing produced water as hydraulic fracturing fluid. *Environmental Science & Technology*, 2017. **51**(16): pp. 9391–9402.

18. McDevitt, B., et al., Maximum removal efficiency of barium, strontium, radium, and sulfate with optimum AMD-Marcellus flowback mixing ratios for beneficial use in the Northern Appalachian Basin. *Environmental Science & Technology*, 2020. **54**(8): pp. 4829–4839.

19. Tasker, T., et al., Environmental and human health impacts of spreading oil and gas wastewater on roads. *Environmental Science & Technology*, 2018. **52**(12): pp. 7081–7091.

20. *Trends in U.S. Oil and Natural Gas Upstream Costs*. 2016, U.S. Energy Information Administration: Washington, DC.

21. *Flowback and Produced Waters: Opportunities and Challenges for Innovation*. 2016, The National Academies of Sciences, Engineering, and Medicine: Washington, DC.

22. Vengosh, A., et al., A critical review of the risks to water resources from unconventional shale gas development and hydraulic fracturing in the United States. *Environmental Science & Technology*, 2014. **48**(15): pp. 8334–8348.

23. Keranen, K.M., et al., Potentially induced earthquakes in Oklahoma, USA: Links between wastewater injection and the 2011 Mw 5.7 earthquake sequence. *Geology*, 2013. **41**(6): pp. 699–702.

24. Horton, S., Disposal of hydrofracking waste fluid by injection into subsurface aquifers triggers earthquake swarm in central Arkansas with potential for damaging earthquake. *Seismological Research Letters*, 2012. **83**(2): pp. 250–260.

25. ch2m, *Oklahoma Water for 2060 Produced Water Reuse and Recycling*. 2017.

26. Shaffer, D.L., et al., Desalination and reuse of high-salinity shale gas produced water: Drivers, technologies, and future directions. *Environmental Science & Technology*, 2013. **47**(17): pp. 9569–9583.

27. Kondash, A.J., et al., The impact of using low-saline oilfield produced water for irrigation on water and soil quality in California. *Science of the Total Environment*, 2020. **733**: p. 139392.

28. Stallworth, A.M., et al., Laboratory method to assess efficacy of dust suppressants for dirt and gravel roads. *Transportation Research Record*, 2020. **2674**(6): pp. 188–199.

29. Guerra, K., K. Dahm, and S. Dundorf, *Oil and gas produced water management and beneficial use in the Western United States*. 2011: US Department of the Interior, Bureau of Reclamation Washington, DC.

30. *10LB Brine with K-Control*. 2017 May 21, 2021; Available from: http://www.intrepid-potash.com/wp-content/uploads/2017/02/INTR_2018-Spec-Sheet_Brine_10-lb-K1-CARLSBAD_TJ_B2.pdf.

31. Acharya, H.R., et al., *Cost Effective Recovery of Low-TDS Frac Flowback Water for Re-use*. 2011, National Energy Technology Laboratory: Morgantown, WV.

32. Wenzlick, M. and N. Siefert, Techno-economic analysis of converting oil & gas produced water into valuable resources. *Desalination*, 2020. **481**: p. 114381.

33. He, C., T. Zhang, and R.D. Vidic, Co-treatment of abandoned mine drainage and Marcellus Shale flowback water for use in hydraulic fracturing. *Water Research*, 2016. **104**: pp. 425–431.
34. Zhang, T., et al., Co-precipitation of radium with barium and strontium sulfate and its impact on the fate of radium during treatment of produced water from unconventional gas extraction. *Environmental Science & Technology*, 2014. **48**(8): pp. 4596–4603.
35. *OLI Flowsheet v.10.0*, OLI Systems. 2019 [cited 8/12/2020]; Available from: https://www.olisystems.com/software/flowsheet-esp/.
36. *Aspen Plus v.11*, AspenTech. 2019 [cited 8/12/2020]; Available from: https://www.aspentech.com/en/products/engineering/aspen-plus.
37. Fakhru'l-Razi, A., et al., Review of technologies for oil and gas produced water treatment. *Journal of Hazardous Materials*, 2009. **170**(2): pp. 530–551.
38. Igunnu, E.T. and G.Z. Chen, Produced water treatment technologies. *International Journal of Low-Carbon Technologies*, 2012. **9**: p. 1–21.
39. Duraisamy, R.T., A.H. Beni, and A. Henni, State of the art treatment of produced water. In W. Elshorbagy (ed.) *Water Treatment*, 2013: pp. 199–222. doi: 10.5772/53478.
40. Munirasu, S., M.A. Haija, and F. Banat, Use of membrane technology for oil field and refinery produced water treatment—A review. *Process Safety and Environmental Protection*, 2016. **100**: pp. 183–202.
41. Giwa, A., et al., Brine management methods: Recent innovations and current status. *Desalination*, 2017. **407**: pp. 1–23.
42. Pramanik, B.K., L. Shu, and V. Jegatheesan, A review of the management and treatment of brine solutions. *Environmental Science: Water Research & Technology*, 2017. **3**(4): pp. 625–658.
43. Arena, J.T., et al., Management and dewatering of brines extracted from geologic carbon storage sites. *International Journal of Greenhouse Gas Control*, 2017. **63**: pp. 194–214.
44. Kaplan, R., et al., Assessment of desalination technologies for treatment of a highly saline brine from a potential CO_2 storage site. *Deslination*, 2017. **404**: pp. 87–101.
45. Semblante, G.U., et al., Brine pre-treatment technologies for zero liquid discharge systems. *Desalination*, 2018. **441**: pp. 96–111.
46. Jiménez, S., et al., State of the art of produced water treatment. *Chemosphere*, 2018. **192**: pp. 186–208.
47. Chang, H., et al., Potential and implemented membrane-based technologies for the treatment and reuse of flowback and produced water from shale gas and oil plays: A review. *Desalination*, 2019. **455**: pp. 34–57.
48. Al-Ghouti, M.A., et al., Produced water characteristics, treatment and reuse: A review. *Journal of Water Process Engineering*, 2019. **28**: pp. 222–239.
49. Conrad, C.L., et al., Fit-for-purpose treatment goals for produced waters in shale oil and gas fields. *Water Research*, 2020. **173**: p. 115467.
50. *OLI Studio v10.0*, OLI Systems. 2019 [cited 8/12/2020]; Available from: https://www.olisystems.com/oli-studio-stream-analyzer.
51. Blondes, M., et al., *US Geological Survey National Produced Waters Geochemical Database, v2. 3*. US Geological Survey data release, 2018: p. F7J964W8.
52. *U.S. shale oil and natural gas maps, in Maps: Oil and Gas Exploration, Resources, and Production*. EIA.
53. *COGCC - Wattenberg Gas Field: Undifferentiated Field Boundaries*. 2011, Colorado Oil and Gas Conservation Commission.
54. Gaswirth, S.B., et al., *Assessment of Undiscovered Oil Resources in the Bakken and Three Forks Formations, Williston Basin Province, Montana, North Dakota, and South Dakota, 2013 in USGS Fact Sheet*. 2013, U.S. Geological Survey, Central Energy Resources Team: Denver, Colorado.

55. Gaswirth, S.B., *USGS National and Global Oil and Gas Assessment Project - Permian Basin Province, Midland Basin, Wolfcamp Shale Assessment Unit Boundaries and Assessment Input Data Forms, Version 2.0: U.S. Geological Survey data release.* 2019, USGS.

56. Tennyson, M., *USGS National and Global Oil and Gas Assessment Project-San Joaquin Basin Province, Monterey Formation Assessment Unit Boundaries: U.S. Geological Survey data release,* USGS, Editor. 2019.

57. Warwick, P.D., et al., *National Assessment of Oil and Gas - Paleogene System and Cretaceous-Tertiary Coalbed Assessment Units of the Gulf Coast (Provinces 047, 048 and 049).* 2007, U. S. Geological Survey, Energy Resources Program.

58. Engle, M.A., et al., Origin and geochemistry of formation waters from the lower Eagle Ford Group, Gulf Coast Basin, south central Texas. *Chemical Geology,* 2020. **550**: p. 119754.

59. Slutz, J., et al., *Key Shale Gas Water Management Strategies: An Economic Assessment Tool.* 2012, Society of Petroleum Engineers.

60. Morton, R.A., et al., *Salinity variations and chemical compositions of waters in the Frio Formation, Texas Gulf Coast.* 1981, U.S. Department of Energy Division of Geothermal Energy.

61. Kharaka, Y.K., E. Callender, and W.W. Carothers. Geochemistry of geopressured geo-thermal waters from the Texas Gulf Coast. In *Third Geopressured-Geothermal Energy Conference.* 1977. Lafayette.

62. Knauss, K.G., J.W. Johnson, and C.I. Steefel, Evaluation of the impact of CO_2, co-contaminant gas, aqueous fluid and reservoir rock interactions on the geologic seques-tration of CO_2. *Chemical Geology,* 2005. **217**: pp. 339–350.

63. Stuckman, M., N. Siefert, Editor. Personal communication, 2020.

64. Kurz, M., J. Hamling, and J. Hurley, *Williston Basin Water Treatment Technology Test Bed.* 2019, University of North Dakota Energy & Environmental Research Center.

65. Spencer, M., R. Garlapalli, and J.P. Trembly, Geochemical phenomena between Utica-Point Pleasant shale and hydraulic fracturing fluid. *AIChE Journal,* 2019. **66**(4). doi:10.1002/aic.16887.

66. Phan, T.T., J.A. Hakala, and S. Sharma, Application of isotopic and geochemical sig-nals in unconventional oil and gas reservoir produced waters toward characterizing in situ geochemical fluid-shale reactions. *Science of the Total Environment,* 2020. **714**: p. 136867.

67. Li, H., *Produced water quality characterization and prediction for Wattenberg field.* 2000–2019-CSU Theses and Dissertations, 2013.

68. *WAVE, v.1.77a,* Dow Chemical Company. 2019 [cited 8/12/2020]; Available from: https://www.dupont.com/water/resources/design-software.html.

69. Benko, K.L. and J. Drewes, Produced water in the Western United States: Geographical distribution, occurrence, and composition. *Environmental Engineering Science,* 2008. **25**: pp. 239–246.

70. Carrero-Parreño, A., et al., Optimal pretreatment system of flowback water from shale gas production. *Industrial & Engineering Chemistry Research,* 2017. **56**(15): pp. 4386–4398.

71. AQWATEC, *Technical Assessment of Produced Water Technologies.* 1st ed. 2009, Golden, CO: Colorado School of Mines.

72. Rubio, J., M. Souza, and R. Smith, Overview of flotation as a wastewater treatment technique. *Minerals Engineering,* 2002. **15**(3): pp. 139–155.

73. Valavala, R., et al., Pretreatment in reverse osmosis seawater desalination: A short review. *Environmental Engineering Research,* 2011. **16**(4): pp. 205–212.

74. Edzwald, J.K., Dissolved air flotation and me. *Water Research,* 2010. **44**(7): pp. 2077–2106.

75. Al-Maamari, R.S., et al., Flotation, filtration, and adsorption: Pilot trials for oilfield produced-water treatment. *Oil and Gas Facilities*, 2014. **3**(02): pp. 56–66.
76. Canganella, F. and J. Wiegel, Anaerobic thermophiles. *Life*, 2014. **4**(1): pp. 77–104.
77. Lipus, D., et al., Predominance and metabolic potential of Halanaerobium spp. in produced water from hydraulically fractured Marcellus shale wells. *Applied and Environmental Microbiology*, 2017. **83**(8): p. e02659.
78. Greenlee, L.F., et al., Reverse osmosis desalination: Water sources, technology, and today's challenges. *Water Research*, 2009. **43**(9): pp. 2317–2348.
79. Çakmakce, M., N. Kayaalp, and I. Koyuncu, Desalination of produced water from oil production fields by membrane processes. *Desalination*, 2008. **222**(1): pp. 176–186.
80. Okiel, K., M. El-Sayed, and M.Y. El-Kady, Treatment of oil–water emulsions by adsorption onto activated carbon, bentonite and deposited carbon. *Egyptian Journal of Petroleum*, 2011. **20**(2): pp. 9–15.
81. Dastgheib, S.A., et al., Treatment of produced water from an oilfield and selected coal mines in the Illinois Basin. *International Journal of Greenhouse Gas Control*, 2016. **54**: pp. 513–523.
82. Karnib, M., et al., Heavy metals removal using activated carbon, silica and silica activated carbon composite. *Energy Procedia*, 2014. **50**: pp. 113–120.
83. Brooks, D., R. Roll, and W. Naylor, Wastewater technology fact sheet granular activated carbon adsorption and regeneration. *Environmental Protection Agency, USA*, 2000. **832**: pp. 1–7.
84. Hayes, T. and B.F. Severin, *Barnett and Appalachian Shale Water Management and Reuse Technologies*. 2012.
85. Bi, Y., et al., Removal of radium from synthetic shale gas brines by ion exchange resin. *Environmental Engineering Science*, 2016. **33**(10): pp. 791–798.
86. *FilmTec™ NF90-4040*. 2020 [cited 2020 8/12/2020]; Available from: https://www.dupont.com/products/filmtecnf904040.html.
87. *FilmTec™ NF270-4040*. 2020 [cited 2020 8/12/2020]; Available from: https://www.dupont.com/products/filmtecnf2704040.html.
88. Guan, Z., et al., Boron removal from aqueous solutions by adsorption—A review. *Desalination*, 2016. **383**: pp. 29–37.
89. Mistry, K.H. and J.H. Lienhard, Generalized least energy of separation for desalination and other chemical separation processes. *Entropy*, 2013. **15**(6): pp. 2046–2080.
90. Mistry, K.H., et al., Entropy generation analysis of desalination technologies. *Entropy*, 2011. **13**(10): pp. 1829–1864.
91. Thiel, G.P., et al., Energy consumption in desalinating produced water from shale oil and gas extraction. *Desalination*, 2015. **366**: pp. 94–112.
92. Ghaffour, N., T.M. Missimer, and G.L. Amy, Technical review and evaluation of the economics of water desalination: Current and future challenges for better water supply sustainability. *Desalination*, 2013. **309**: pp. 197–207.
93. Arena, J.T., et al., Management and dewatering of brines extracted from geologic carbon storage sites. *International Journal of Greenhouse Gas Control*, 2017. **63**: pp. 194–214.
94. Bartholomew, T.V., et al., Osmotically assisted reverse osmosis for high salinity brine treatment. *Desalination*, 2017. **421**: pp. 3–11.
95. *CLEAN WATER LLC: Turning Production Water into Pure Water*. 2017 [cited 2020 8/21]; Available from: https://www.cleanwaterllc.net/.
96. Booth, D.W. and T.L. Anderson, *Method and system for water treatment, U.S.P. Office*, Editor. 2017, Clean Water LLC: United States.

4 Transport of Major Elements in Produced Water through Reactive Porous Media

Zi Ye and Valentina Prigiobbe
Stevens Institute of Technology

CONTENTS

4.1   INTRODUCTION

Fracking is used to maximize the extraction of underground resources (oil, natural gas, geothermal energy, and water) from low permeability formations [1, 2]. It is considered to be an important approach for future fossil fuel extraction, particularly within the period of transition to renewable energy systems. In the United States, projections of future oil and gas extraction using hydraulic fracturing forecast up to 16 trillion bbl by 2050 [3]. In addition to oil and gas, operations generate a large amount of wastewater, which is technically called flowback and produced water [4]. Since the beginning of fracking, millions of cubic meters of wastewater have been brought to the surface. This water comprises only 4%–8% of returned injection fluids, while 92%–96% is made up of the saline or brackish brine from the shale formation [5].

DOI: 10.1201/9781003091011-4

Flowback and produced waters naturally contain heavy metals and radionuclides, such as radium [6–10]. The average composition of produced water from fracking in the Marcellus shale (PA), which is the largest and the best documented shale gas resource in the United States, is reported in Table 4.1 in conjunction with acidic mine drainage and seawater, for comparison. The table lists the concentrations of the inorganic compounds observed, although organics are also present, primarily due to the composition of the fracturing fluids, which is reflected as total carbon (c_T). However, their concentration is within the micromolar range, and their effect on inorganic ion speciation and interaction is negligible [11,12]. Organics that have been observed in this range of concentration include [13,14,9] 7,12-dimethylbenz(a) anthracene, aldrin, dieldrin, heptachlor epoxide, benzo(a)pyrene, heptachlor, dibenz(a,h)anthracene, betahexachlorocyclohexane benzene, toluene, ethylbenzene, xylene, phenols, bis (2-chloroethyl) ether, and 2-butanone.

Table 4.1 shows that the produced water salinity, given as total dissolved solids (TDS), can be as high as four to ten times the salinity of the seawater [15], resulting in an ionic strength between 1 and 4 molar. Barium (Ba^{2+}), calcium (Ca^{2+}), sodium (Na^+), strontium (Sr^{2+}), and radium (^{228}Ra and ^{226}Ra) are among the alkaline earth elements and radionuclides with the largest concentrations. More infrequently, toxic trace elements may be present at elevated concentrations as well, making the water a potential hazard in the case of a spill [16, 17].

TDS, total dissolved solids.

TABLE 4.1
Average Composition of Brine from Fracking [18–24] and Mining, and Seawater [25–28]

	Produced Water	Mine Drainage	Seawater
Ca^{2+} (mg L^{-1})	7,220	27–503	460
Ba^{2+} (mg L^{-1})	2,224	$6-10^{-3}$	
Sr^{2+} (mg L^{-1})	1,695		16
Mg^{2+} (mg L^{-1})	632	23–465	1,421
Na^+ (mg L^{-1})	24,123	1–204	11,638
Fe dissolved (mg L^{-1})	40.8	2–302	
^{228}Ra (pCi L^{-1})	~30–300		
^{226}Ra (piC L^{-1})	~100–1,000		
Mn (mg L^{-1})	2.5		0.01
Li (mg L^{-1})	95	30–2,700	0.17
Co (mg L^{-1})	10	192–289	0.10
Cl^- (mg L^{-1})	57,447	1–147	22,089
SO^{2-} (mg L^{-1})	71	780–4,110	2,928
Br^- (mg L^{-1})	511	3–772	66
As (mg L^{-1})	<77		
c_T (mg L^{-1})	48		
TDS (mg L^{-1})	144,748	1,091–4,550	34,483
pH	5.5–7	2.6–5.6	8–8.27
Temperature (°C)	75–85		18–30

Typically, the wastewater is temporarily stored in tanks to be then either recycled, reused upon treatment, or disposed of [6,20,29]. Spills during these activities may occur with potential negative impact on the environment [30,16,31,32]. The leakage may reach water resources through either natural pathways, lining failure at wells and storage tanks, or along pipeline [33,34,35,36], and the transport of the solutes within the water may change with the mineralogy of the soil [37]. Inorganic and organic pollution associated to spills has been found in soil and river sediments as reported in previous works [16,17,38]. Of particular concern are elements such as barium and radium that have harmful effect on human health [39, 40].[226]Ra ([226]Ra/[228]Ra ~ 0.1) may even reach concentrations up to 15,000 pCi/L; however, the variability among sites is large and it has been observed to be around three orders of magnitude [41]. Given the potential negative impact of produced water spills, there is an urgent need to identify a practice that can mitigate the consequences of accidental leaks and options are either a wastewater treatment, which may be expensive, or a more cost-effective approach based on the inclusion of a material within fractures and around storage pools and wells that can adsorb toxic elements avoiding their migration toward the environment [42, 43].

In this chapter, we present an experimental and modeling work on the transport of alkaline earth elements in produced water, which are congeners of radium, namely Ba^{2+}, Sr^{2+}, Ca^{2+}, and magnesium (Mg^{2+}) in addition to sodium Na^+ [44]. Column-flood tests were conducted using produced water from a shale gas site (Marcellus shale) and reactive porous media made of ubiquitous minerals such as sand, goethite, hydrous ferric oxide (HFO), activated alumina, and manganese oxide. These minerals were chosen because of their known selective adsorption of alkaline earth elements [45–51]. Conventional proppant materials were also tested for reference. A reactive transport model was developed for the transport of Ba^{2+}, Sr^{2+}, Ca^{2+}, Mg^{2+}, and Na^+ through the porous medium coated with manganese oxide, which gave the greatest attenuation of the investigated elements. The model was implemented in the open-source geochemical software PHREEQC v.3 [52], also used in previous works [53–58], and compared with experimental results.

4.2 MATERIALS

4.2.1 Produced Water

Produced water was obtained from MSEEL (Marcellus Shale Energy and Environmental Laboratory, West Virginia). The water was taken from a waste-storage tank, unfiltered and not acidified. The concentration of major alkaline earth elements, i.e., barium, calcium, magnesium, sodium, and strontium in, in the received water was measured using ICP-OES (Model 5100 SVDV, Agilent, U.S.A.). Upon characterization, the pH was adjusted to 8 and filtered using a 0.22 µm filter (GE Healthcare, U.S.A.) to remove precipitates, which consist mostly of iron oxides. Then, the water was analyzed again and the composition upon filtration determined (Table 4.2). The pH was adjusted to 8 as it represents an average value for produced water at the production site [59,16].

TABLE 4.2

Composition of the Solutions of Produced Water Used in this Work; Symbols Diluted Synthetic Produced Water (dSPW)$_{1/100}$ and dSPW$_{1/1000}$ Indicate Synthetic Produced Water Diluted at 1:100 and 1:1000, Correspondingly, with Respect to Corrected Synthetic Produced Water (cSPW)

	NPW	cNPW	cSPW	dSPW1/100	dSPW1/1000
Ba^{2+} (mg L^{-1})	3,864.00	3,692.00	3,621.00	31.23	3.88
Ca^{2+} (mg L^{-1})	7,266.00	5,914.00	6,295.00	45.00	7.01
Co^{2+} (mg L^{-1})	2.00	1.48	2.00	0.02	0.002
Fe^{3+} (mg L^{-1})	240.00	11.00	1.50	0.78	0.34
K^{+} (mg L^{-1})	334.70	122.40	495.00	2.81	0.35
Li^{+} (mg L^{-1})	61.32	47.41	59.51	0.12	0.01
Mg^{2+} (mg L^{-1})	975.80	740.50	790.00	8.09	0.93
Mn^{2+} (mg L^{-1})	8.51	4.05	10.00	0.10	0.01
Na^{+} (mg L^{-1})	24,387.00	24,259.00	26,275.00	247.61	26.74
Sr^{2+} (mg L^{-1})	1,763.00	1,719.00	1,709.00	15.08	1.79
Cl^{-} (mg L^{-1})	Nd	nd	5,802.35*	517.98*	60.66*
TDS (mg L^{-1})	42,900	61,500	55,500	904	137
I_m (mol L^{-1})	1.10	0.95	1.34	0.02	0.003
I (mol L^{-1})	-	-	1.03	0.01	0.001
pH SO$_4{}^{2-}$ (mg L^{-1})	3–5 < 60	~8.03 nd	~8.02 -	~8.01 -	~8.01 -

nd, not determined; TDS, total dissolved solids; *, calculated from solution composition; I_m, is the ionic strength calculated using Eq. 4.1; and I is the ionic strength determined using Eq. 4.2.

Synthetic produced water was prepared, and it included only the major alkaline earth elements and chloride ion. It was created as an analog to understand the role of organics and other minor species within the produced water on the transport of the elements of interest. The preparation of synthetic produced water was carried out using chloride salts, namely BaCl$_2$, CaCl$_2$, CoCl$_2$, FeCl$_3$, KCl, LiCl, MgCl$_2$, MnCl$_2$, NaCl, and SrCl$_2$ (Fisher chemical, U.S.A.). The pH was then adjusted to 8, which is generally established in storage tanks at extraction sites. Solutions of the synthetic produced water were also created with ratios of 1:100 and 1:1000 to study the effect of dilution of the brine on cation adsorption upon mixing of the brine with fresh water [60, 38].

The produced water composition was determined on various types of created solutions and the results of the analyses are reported in Table 4.2.

In the table, it is possible to read that the composition of the natural produced water (NPW) as received from MSEEL is comparable to the average composition of the produced water from other oil- and gas-field sites in the Marcellus shale as reported in Table 4.1. The largest ion concentrations were determined for Na$^+$, Ca^{2+}, Ba^{2+}, Sr^{2+}, Mg^{2+}, and Cl$^-$, which therefore control the ionic strength. Upon pH adjustment and filtration, the corrected natural produced water (cNPW) showed a major reduction in iron concentration, suggesting that precipitates of iron were formed.

The corrected synthetic produced water (cSPW) composition compared well with that of cNPW, and upon dilution the concentrations dropped to two and three orders of magnitudes, as expected. The ionic strength of natural and produced water was determined indirectly from the measured TDS as [61],

$$I_m = 2.5 \cdot 10^{-5} \text{TDS}, \tag{4.1}$$

upon filtration and it is reported in Table 4.2. In addition, for the synthetic produced water I was also calculated as,

$$I = \frac{1}{2_i} c_i z_i^2, \tag{4.2}$$

where c_i and z_i are the concentration and the charge of the ith-species, respectively. In the calculations of I, all the major ions used for preparation as listed in Table 4.2 were considered.

4.2.2 GRANULAR MATERIALS

Natural sand (SiO_2, Sakrete, U.S.A.) with average diameter within 0.595 and 0.841 mm, determined by sieving the material, was used in this work. The measured BET-specific surface area for the sand was equal to $0.230 \, m^2 g^{-1}$, which is comparable to values measured in earlier works [62,56]. The material was used uncoated as well as coated with iron and manganese oxides to create hydrous ferric oxide-coated sand (FCS) and manganese oxide-coated sand (MCS). For the synthesis of material coated with goethite, glass beads were used instead (goethite-coated beads, GCB). Additional granular materials were used in this work, namely activated alumina (AA, Buzick Lumber, U.S.A) and resin-coated sand (RCS, Carbo, U.S.A). The average size of RCS was 0.420 mm and the measured BET-specific surface area for the material was equal to $0.005 \, m^2 g^{-1}$. Methods for coating and characterization are reported in our previous works [63,44].

All the granular materials were characterized using X-ray powder diffraction (Ultima IV, Rigaku, U.S.A.), scanning electron microscopy-energy dispersive X-ray spectroscopy (SEM-EDS, Zeiss, U.S.A.), and Brunauer–Emmett–Teller (BET) surface area (Quantachrome Instrument Version 11.0, U.S.A.).

The X-ray diffraction analysis of the MCS is reported in Figure 4.1. The figure shows that the synthesized mineral is made of α-MnO_2.

Figure 4.2 reports the results from the characterization of the all granular materials used in this work. The photos along the upper row of this figure show the characteristic colors of the minerals, while the images in the following rows are from SEM-EDS analyses. All the materials were supercoated with 5 nm of gold to increase the surface conduction. As expected, the dominant species on the sand surface were silicon (Si) and oxygen (O) with iron (Fe), aluminum (Al), and manganese (Mn) on FCS, AA, and MCS, respectively. More than 99.99% purity was obtained for the RCS and AA materials.

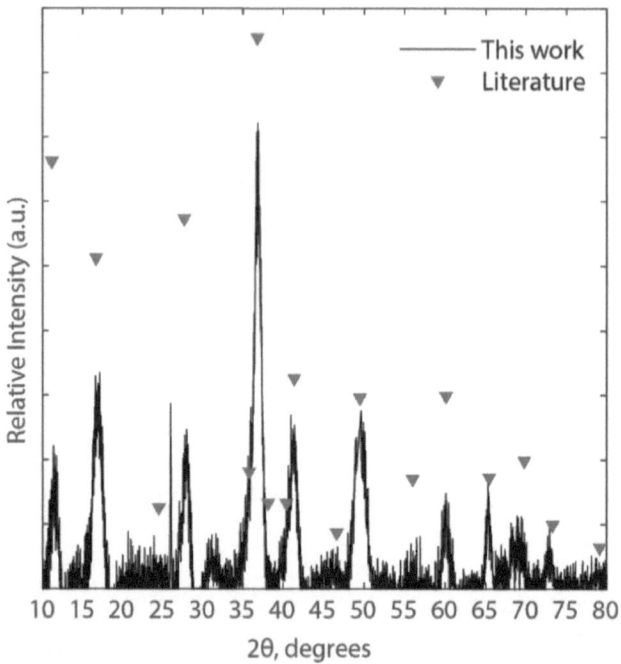

FIGURE 4.1 X-ray diffraction analysis of the synthesized manganese oxide used to coat sand compared with literature data [64].

4.3 SET-UP AND EXPERIMENTAL PROCEDURE

Transport experiments were performed by supplying a brine to a column of 15 mm of diameter and 200 mm of length (DWK Life Sciences, Germany) through a peristaltic pump (Masterflex L/S, U.S.A.). The brine was injected from the bottom to the top to avoid fingering due to density-driven flow. The column was filled with granular materials, i.e., sand, RCS, GCB, FCS, AA, and MCS. Sensors of pH and temperature (Liquiline CM442, Endress Hauser, U.S.A.) were installed inline at the inlet and at the outlet of the column to measure continuously the flow streams. Grab samples were taken using a fraction collector (Frac-920, GE Healthcare Life Sciences, U.S.A.) and analyzed off-line using the ICP-OES. Further details about the experimental procedure are reported in our previous work [44].

During the tests, the column was initially flooded with a solution of pH 5 adjusted by using HCl, representing a condition of a slightly acidic uncontaminated shallow aquifer [65]. After the column stabilized at that pH, another solution containing the produced water was supplied to the column with a flow rate equal to 1 mL min^{-1}. The type of solution injected into the column is specified for each test in Table 4.3.

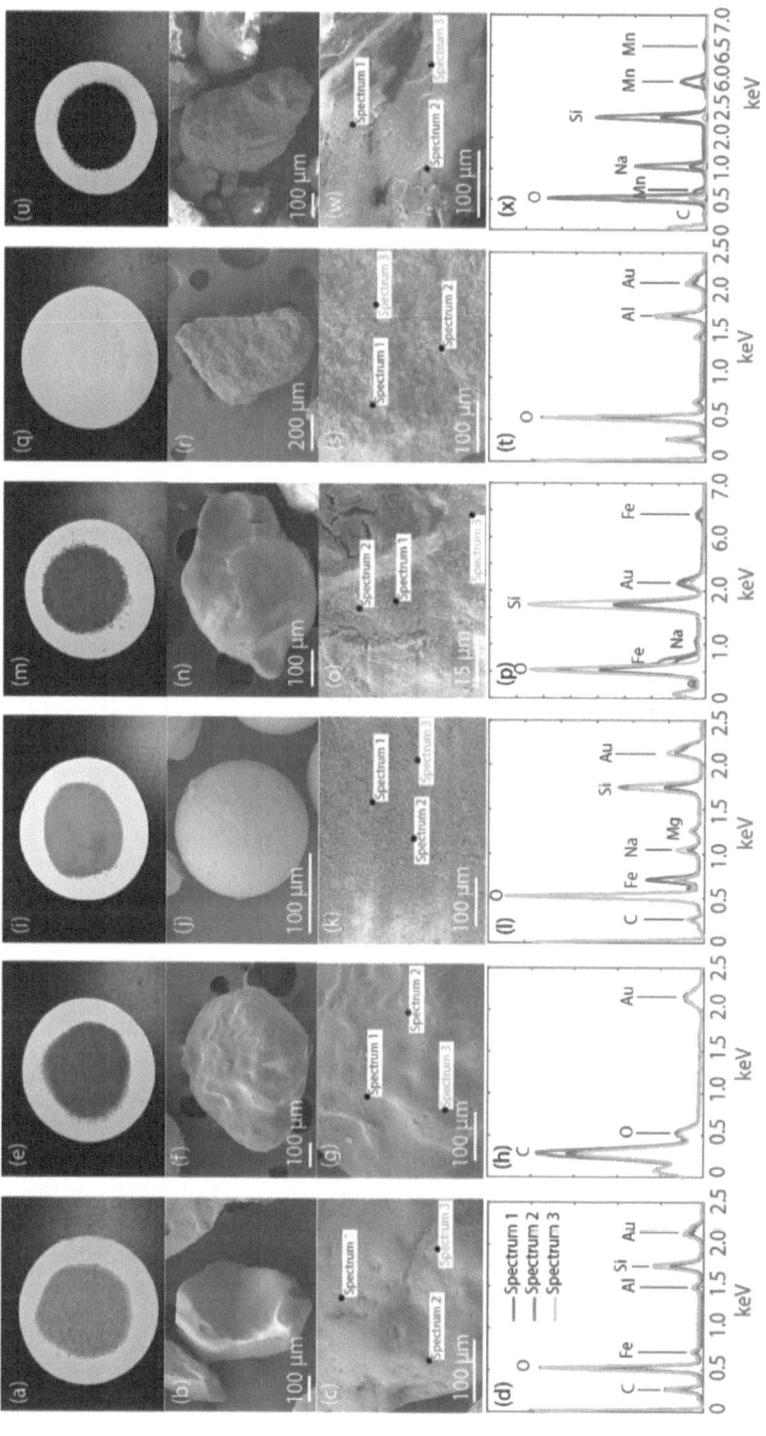

FIGURE 4.2 Photos and SEM-EDS images and analyses of granular materials: (a–d) sand, (e–h) RCS, (i–l) GCB, (m–p) FCS, (q–t) AA, and (u–x) MCS.

TABLE 4.3

Summary of the Experimental Conditions Applied During all Transport Tests Including the Injected Solution into the Column Upon Stabilization with a Solution of pH 5

Experiments	Material	Solution	pHini	pHinj	n, (–)	Q, mL/min	D 10^{-5}, m^2/s	$v \cdot 10^{-4}$, m/s
1	Sand	cNPW	5.12	8.01	0.21	1.01	1.810	4.54
2	RCS	cNPW	4.81	8.00	0.41	0.91	0.235	2.09
3	GCB	cNPW	5.05	8.01	0.40	0.66	0.041	1.55
4	GCB	dSPW 1/100	5.01	8.02	0.40	0.67	0.159	1.58
5	FCS	cNPW	5.03	8.04	0.54	0.99	0.289	1.73
6	AA	cNPW	5.12	8.05	0.65	0.80	0.035	1.16
7	AA	cSPW	4.85	8.02	0.65	0.76	0.025	1.11
8	AA	dSPW 1/100	5.00	8.01	0.65	0.80	0.035	1.16
9	AA	dSPW 1/1000	5.02	8.01	0.52	0.54	0.042	0.98
10	MCS	NPW	5.00	3.21	0.48	1.00	0.125	1.98
11	MCS	cNPW	4.84	8.05	0.48	0.48	0.056	0.97
12	MCS	cNPW	4.96	8.03	0.48	0.57	0.064	1.11
13	MCS	cSPW	5.07	8.02	0.48	0.60	0.054	1.19
14	MCS	cSPW	4.85	8.01	0.48	0.67	0.133	1.33
15	MCS	dSPW 1/100	5.00	8.01	0.48	0.58	0.079	1.15
16	MCS	dSPW 1/1000	5.03	8.01	0.48	0.45	0.202	0.86
17	MCS	dSPW 1/1000	4.89	8.05	0.48	0.27	0.183	0.53

AA, activated alumina; cNPW, corrected natural produced water; FCS, ferric oxide-coated sand; GCB, goethite-coated beads; MCS, manganese oxide-coated sand; RCS, resin-coated sand

4.4   MODELING

Selected experiments were described with a reactive transport model implemented in the open-source geochemical software PHREEQC v.3 [52]. The model couples chemical reactions in solution and at the solid–liquid interface with transport equations.

4.4.1   GEOCHEMICAL MODEL

A geochemical model to describe the adsorption of the alkaline earth elements onto MnO_2 was developed. The model is based on the CD-MUSIC (Charge Distribution MUltiSIte Complexation) triple-layer model [66, 67], and it is coupled with the Pitzer activity coefficient method [68] to account for the effect of the ionic strength.

Following previous works [69–77], the generalized formulation of the chemical reactions at the solid–liquid interface and the associated mass action equations can be written as,

$$SOH + H^+ \leftrightarrow SOH_2^+, K_1 = \frac{\{SOH_2^+\}}{\{SOH\}a_{H^+}} e^{(\Psi_0 F / RT)}, \tag{4.3}$$

$$SOH \leftrightarrow SO^-, H^+, K_2 = \frac{\{SO^-\}a_{H^+}}{\{SOH\}} e^{(-\Psi_0 F/RT)} \qquad (4.4)$$

$$SOH + M \leftrightarrow SOM + H, K_{M,1} = \frac{\{SOM^+\}a_{H^+}}{\{SOH\}a_{M^{2+}}} {}^{[(-\Psi+2\Psi)F/RT]} \qquad (4.5)$$

$$SOH + M^{2+} + H_2O \leftrightarrow SOMOH + 2H^+, K_{M,2} = \frac{\{SOMOH\}a^{2+}}{\{SOH\}a_{M^{2+}}a_{H_2O}} {}^{[(-\Psi+\Psi)F/RT]} \qquad (4.6)$$

$$SOH + Na^+ \leftrightarrow SONa + H^+, K_{Na} = \frac{\{SONa\}a_{H^+}}{\{SOH\}a_{Na^+}} e^{[(-\Psi_0+\Psi_1)F/RT]} \qquad (4.7)$$

$$SOH_2^+ + Cl^- \leftrightarrow SOH_2Cl + K_{Cl} = \frac{SOH_2Cl}{\{SOH\}a_{Cl^-}} e^{[(-\Psi_0-\Psi_1)F/RT]} \qquad (4.8)$$

where the symbol M^{2+} represents a divalent cation, i.e., either Ba^{2+}, Ca^{2+}, Mg^{2+}, or Sr^{2+}; a_i is the activity of the subscript species; $\{SO^-\}$ is the reactive surface site concentration, mol kg^{-1}; the symbols within curly brackets ($\{\cdot\}$) indicate the concentration of a surface complex, mol kg^{-1}; Ψ_0 and Ψ_1 are the potentials of the solid surface and of the inner Helmholtz plane (IHP), V; F is the Faraday constant; R is the molar gas constant; and T is the temperature, K. The mass action equations of these chemical reactions were coupled with the equation for the total surface site,

$$Z_t = \{SOH\} + \{SOH_2^+\} + \{SO^-\} + \{SOM^+\} + \{SOMOH\} + \{SONa\} + \{SOH_2Cl\}, \quad (4.9)$$

and the equations for surface charge,

$$\sigma_0 = \frac{F}{AS}\left(\{SOH_2^+\} + \{SOH_2Cl\} - \{SO^-\} - \{SOM^+\} - \{SOMOH\} - \{SONa\}\right) \quad (4.10)$$

$$\sigma_1 = \frac{F}{AS}\left(2\{SOM^+\} + \{SOMOH\} + \{SONa\} - \{SOH_2Cl\}\right), \qquad (4.11)$$

$$\sigma_2 = 8 \cdot 10RTIEE_0C \quad \sinh\left(\Psi_2 F/2RT\right), \qquad (4.12)$$

$$\sigma_0 + \sigma_1 + \sigma_2 = 0, \qquad (4.13)$$

$$\sigma_0 = \left(\Psi_0 - \Psi_1\right)C_1, \qquad (4.14)$$

$$\sigma_2 = \left(\Psi_1 - \Psi_2\right)C_2, \qquad (4.15)$$

TABLE 4.4

Model Parameters Related to the Surface Complexation Model from Literature and Determined in Our Earlier Work [44]

	Literature	This Work
$\log K_1$	1.8[a]	1.8
$\log K_2$	-8.2[a], -4.2[b]	-4.2
$\log K_{Ba,1}$		-4.6
$\log K_{Ba,2}$		-9.44
$\log K_{Ca,1}$	$-3.8 - -4.0$[b]	-6.06
$\log K_{Ca,2}$		-11
$\log K_{Mg,1}$	-3.3[b]	-5.4
$\log K_{Mg,2}$		-14.2
$\log K_{Sr,1}$	$-4.45 - -4.65$[c]	-6.27
$\log K_{Sr,2}$	$-13.78 - -13.88$[c]	-10.5
$\log K_{Na}$	$-2.7 - -3.3$[b], -3.7[d]	$-4.4 - -6$
N_s (site nm^{-2})	$16-171$[d,e,f]	85
C_1 (F m^{-2})	$1.1 - 1.57$[a]	1
C_2 (F m^{-2})	0.2[d]	0.2
A (m^{2}g^{-1})	359[e]	300
Mass (g)	-	50

Reference considered for the values are [a]Sverjensky et al. [75], [b] Balistrieri et al. [95], [c] Karaseva et al. [96], [d] Sahai et al. [72], [e] Trivedi et al. [97], and [f] Pourret et al. [98]

where A is the specific surface area, m^{2}g^{-1}; S is the solid concentration, g L^{-1}; Ψ_2 is the potential of the outer Helmholtz plane (OHP); E is the dielectric constant (i.e., relative permittivity) of the solvent; and E_0 is the permittivity of vacuum. The intrinsic equilibrium constants of the chemical reactions (Eqs. 4.3–4.7) at the solid–liquid interface, namely, K_1, K_2, $K_{M,1, KM,2}$, and K_{Na}, and other parameters are listed in Table 4.4.

In this model, the adsorption of negatively charged species, such as Cl$^-$, was not accounted for as under the applied conditions their adsorption is negligible. At those conditions, the mineral surfaces are negatively charged (Figure 4.3) and predominantly contain SOH$^-$ sites, where cations adsorption is favored, instead.

4.4.2 Transport Model

One-dimensional (1D) reactive transport model was developed to describe the migration of the major cations in the produced water through the selected granular materials. The model combines the geochemical model reported in Section 4.4.1 for the derivation of the adsorption isotherms and the mass conservation equations for transport. It was implemented in PHREEQC [52] and solved using the Pitzer activity coefficient method [68] to account for the effect of the large ionic strength (between 0.001 and 1.1 mol kg^{-1}). The domain was discretized in 40 cells with a time step of 0.1 second. Further details about the model are reported in our previous work [44].

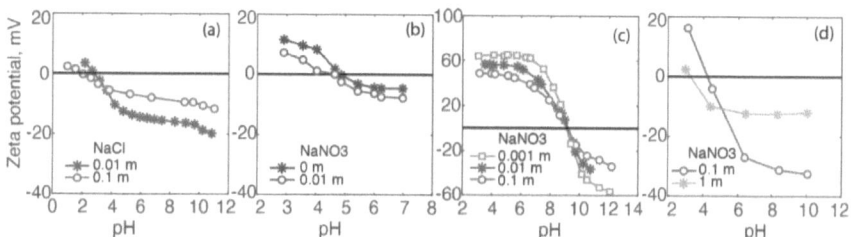

FIGURE 4.3   Literature data of zeta potential (Ψ_d, mV) of (a) sand at 0.01 and 0.1 m of NaCl [78]; (b) HFO at 0 and 0.01 m of NaNO$_3$ [79]; (c) AA at 0.001, 0.01 and 0.1 m of NaNO$_3$ [80]; and (d) MnO$_2$ at 0.1 and 1 m of NAO$_3$ [81,82].

4.5   RESULTS FROM TRANSPORT TESTS

The summary of the experimental conditions applied during the transport tests is given in Table 4.3. Here, the granular materials used in this work are listed together with the types of produced water solutions injected into the column, the pH of the initial (pH$_{ini}$) and injected (pH$_{inj}$) solutions, the porosity (n, –), the estimated flow rate (Q, mL min^{-1}), the longitudinal hydrodynamic dispersion (D, m^{2}s^{-1}), and the interstitial fluid velocity (v, m s^{-1}). The values Q, v, and D were determined by fitting a front of Na$^+$ with the analytic solution of the 1D advection-dispersion equation [83] under conditions where no Na$^+$ adsorption onto the porous medium surface was expected [63].

4.5.1   TESTS USING NATURAL PRODUCED WATER

The breakthrough curves of Ba^{2+}, Ca^{2+}, Mg^{2+}, Na$^+$, and Sr^{2+} and pH are shown in Figure 4.4. The concentration front of each cation is normalized by the corresponding injected concentration and represented as a function of the pore volume injected (i.e., PV $= Qt/(V \cdot n)$, with V the total volume of the granular material within the column). As it is possible to see in this figure, the cations in the effluent of experiments 1 and 2, where sand and RCS were used, respectively, break through the column at 1 PV and, simultaneously, the pH reaches the value of the injected solution. This indicates that the transport of these elements through these traditional proppants [36] occurs without retardation and, therefore, they travel at the interstitial flow velocity.

   In experiments 3, 5, and 6, where GCB, HFO-coated sand (FCS), and AA were used, respectively, the cation front appears at 1 PV, but the pH does not reach the value of the injected solution, simultaneously. The pH front stabilizes around a constant value of approximately 6.2. This suggests that the adsorption of the cations onto the GCB, FCS, and AA surfaces indeed occurred and if a larger specific surface area would have been provided, a more significant adsorption would have been possible determining greater retardation. It is interesting to notice in parts s through x of Figure 4.4, a small peak in the concentration front appears. It resembles the fast concentration wave or pulse observed on our earlier works, which is due to pH-dependent adsorption and hydrodynamic dispersion [84,85].

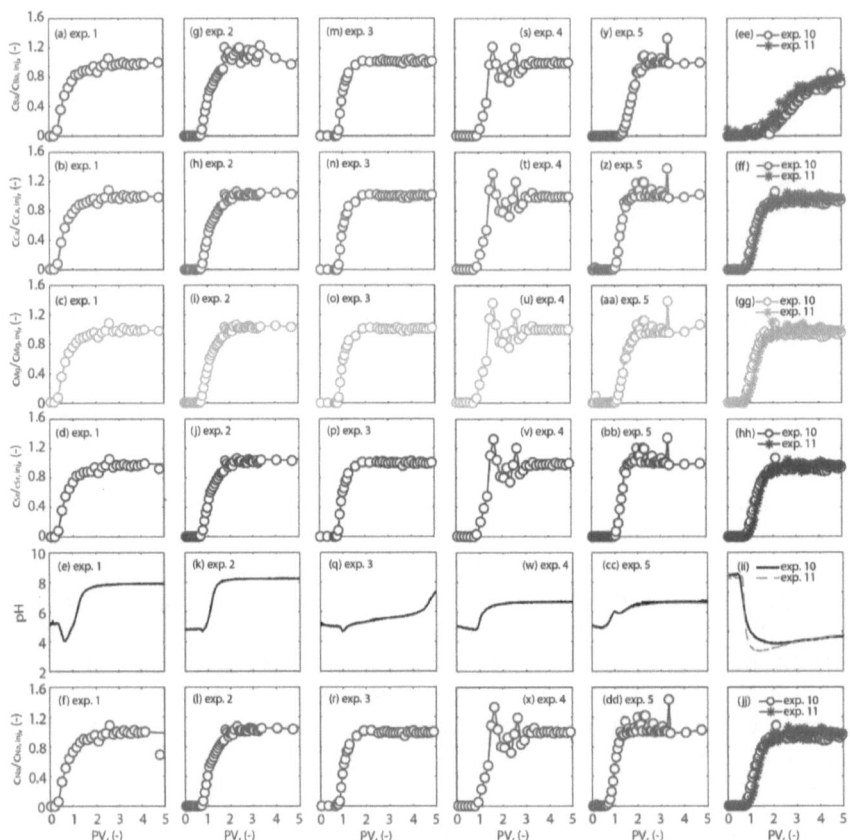

FIGURE 4.4 Measured breakthrough curves of the major cations and pH during transport experiments 1, 2, 3, 5, 6, 11, and 12 of NPW through granular materials such as (a–f) sand, (g–l) RCS, (m–r) GCB, (s–x) AA, and (y–dd) FCS, and (ee–jj) MCS.

In experiments 11 and 12, where MnO_2-coated sand (MCS) was used, all the cation fronts appear at 1 PV, except for barium, which is characterized by a broader front centered around 3 PVs. In these experiments, the pH decreases and stabilizes around 2, suggesting that the alkaline earth elements adsorbed onto MnO_2-coated sand, with the largest extent in the case of Ba^{2+}. Similar trends were observed during experiments 10 (Figure 4.5), where the pH of the NPW was not adjusted to 8 but was left at 3. In this test, all the measured cations except for barium breakthrough at 1 PV and the pH stabilizes around 2.5 indicating that adsorption was negligible under these conditions due to the almost neutral surface charge of the MCS at that pH and salinity (Figure 4.3).

The greatest reactivity of MnO_2-coated sand with respect to the other investigated materials is due to the small value of the point of zero charge (pzc) of MnO_2 mineral and the large negative surface potential even at the ionic strength of 1 m (Figure 4.3).

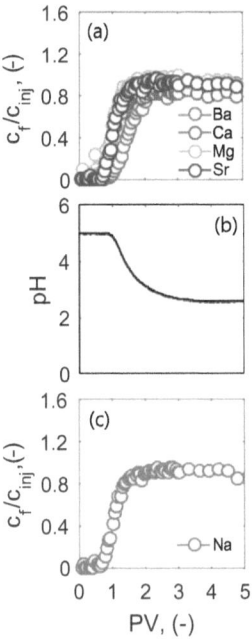

FIGURE 4.5 Measured breakthrough curves of the major cations and pH during experiment 10 where NPW, without pH correction, was injected into a column filled with MCS.

Our results confirm earlier observations, which indicate that the affinity of metals for MnO_x surface follow the order:

$Mg < Ca < Sr < Ba$ [86]. The work by Moon et al. [87] also reports that Ra is was favored over other alkaline earth elements. As a matter of fact, Mn oxides have been employed for the removal of Ra from seawater [88] and radioactive wastewater [89–91].

4.5.2 TESTS USING SYNTHETIC PRODUCED WATER

On the basis of the results obtained using NPW, three granular materials were selected, namely GCB, AA, and MCS, for further investigations using synthetic produced water.

Figure 4.6 reports the effluent composition during the tests run using GCB and synthetic produced water ($dSPW_{1/100}$). However, the measured cations breakthrough the column at 1 PV and the pH reaches approximately 7 after 5 PVs. Therefore, this material was not further investigated.

Figure 4.7 reports the effluent composition during the tests run using AA and synthetic produced water. Comparing the concentration fronts measured in experiment 7 and those in experiment 6 (Figure 4.4), where NPW was used, it is possible to see that the transport behavior of the cations is similar. In both cases, the pH

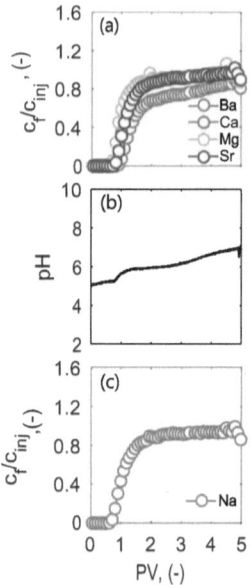

FIGURE 4.6 Measured breakthrough curves of the major cations and pH during experiment 4 where dSPW$_{1/100}$ was injected into a column filled with GCB.

fronts stabilized around 6.5, indicating that the adsorption of the elements was not controlled by minor species dissolved into the NPW.

In the case of the diluted synthetic solutions (experiments 8 and 9), a different transport behavior was observed. Broader concentration fronts formed and the cations broke through the column between 1 PV (for Na$^+$) and 4 PVs (for Ba^{2+}) stabilizing around pH. These results suggest that adsorption indeed occurred. Even if adsorption of the alkaline earth elements might have not been important, adsorption for anions such as Cl$^-$ must have occurred. The Coulombic force of interaction between the negatively charged ions and the positively charged surface of AA favors adsorption. The point of zero charge (pzc) of α-Al$_2$O$_3$, which is a metastable solid form in AA, is between pH 7.6 and 8.6 [92], and it is not significantly affected by the ionic strength [93]. Therefore, adsorption of anions in produced water onto AA may be expected at these conditions.

Larger specific surface area of the synthesized materials, such as GCB and FCS, might have increased the retardation of the cations. The larger the surface, the larger the adsorption capacity of the material. But, the specific surface area is not the only factor influencing adsorption. The reactivity of the surface due to its mineralogy is critical. Our results confirm earlier observations that manganese oxide has significant affinity toward alkaline earth elements. Therefore, MCS was further investigated using synthetic produced water.

Figure 4.8 reports the results of the experiments. Here, it is possible to see a different transport behavior of the divalent cations when either cNPW or cSPW were used under the same operating conditions (experiments 11 and 12 in Figure 4.4 and

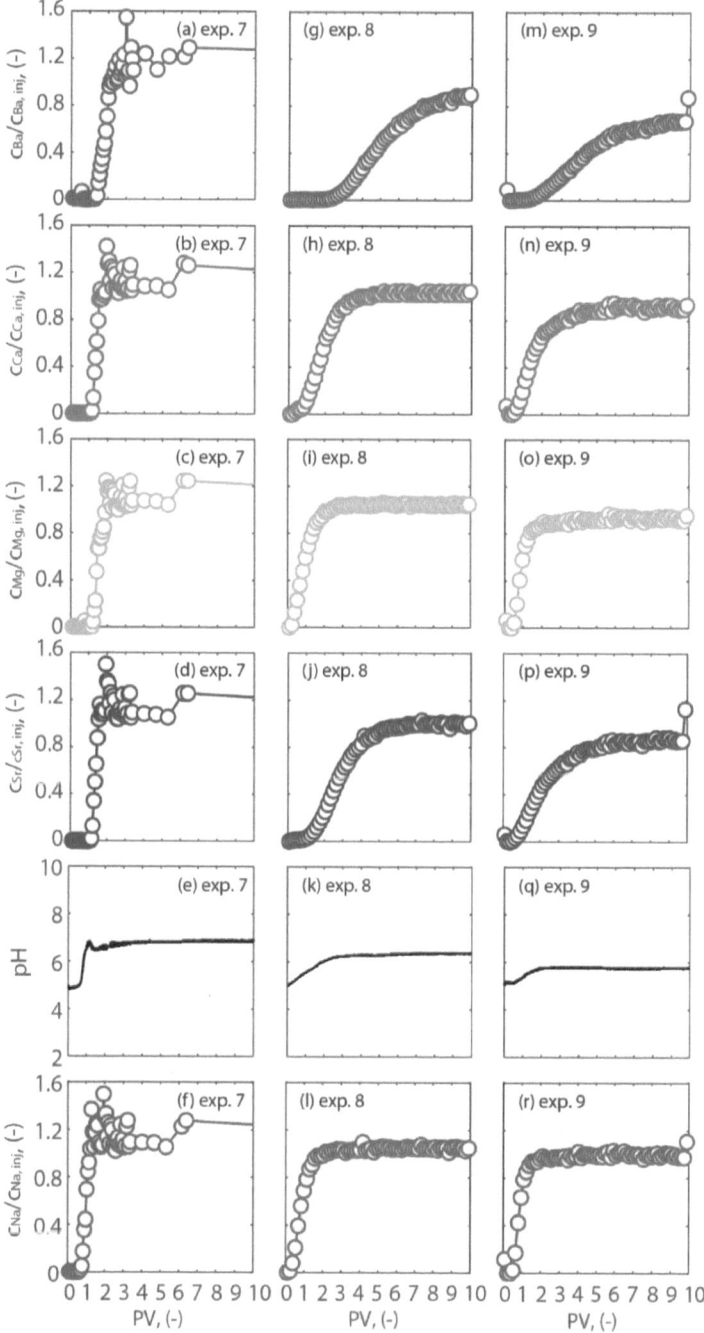

FIGURE 4.7 Measured breakthrough curves of the alkaline earth elements and pH during the transport experiments (namely 7, 8, and 9) using AA and various solutions of synthetic produced water, i.e., (a–f) cSPW, (g–l) $dSPW_{1/100}$, and (m–r) $dSPW_{1/1000}$.

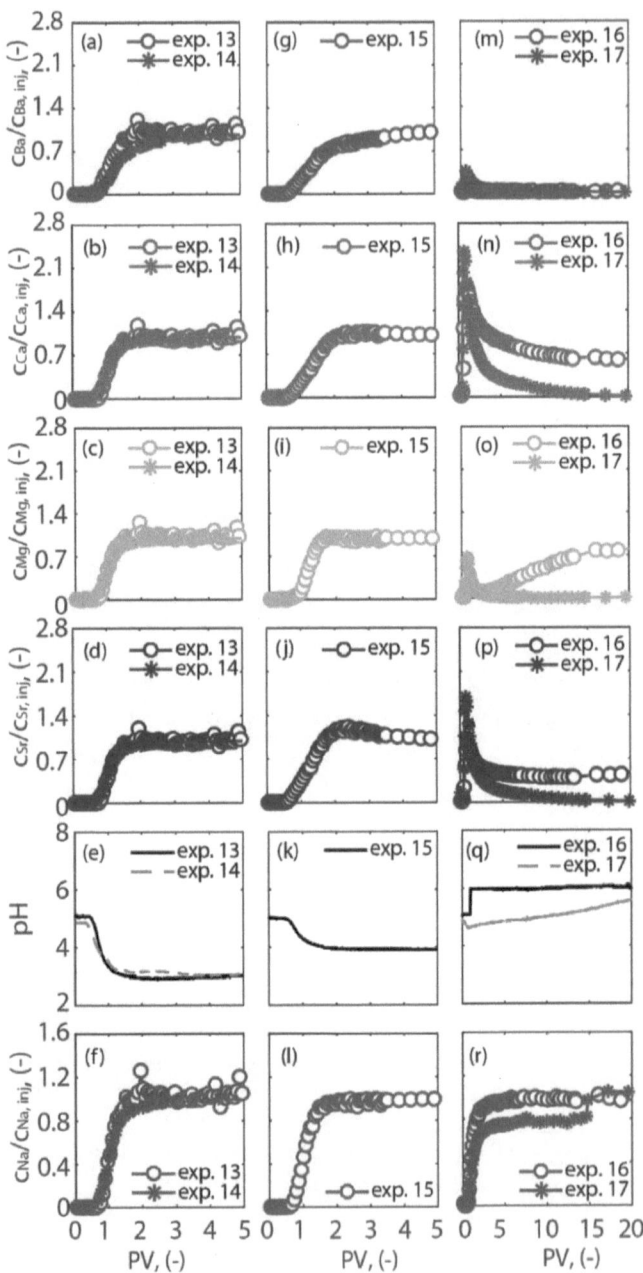

FIGURE 4.8 Measured breakthrough curves of the major cations and pH during transport experiments (namely 13, 14, 15, 16, and 17) through MnO_2-coated sand for various solutions of synthetic produced water, (a–c) cSPW, (d–f) $dSPW_{1/100}$, and (g–i) $dSPW_{1/1000}$. Markers with symbol and gray dash line (in parts a through c, g through i) are the results of experiments 14 and 17 ran as duplicate of experiments 13 and 16, respectively.

experiments 13 and 14 in Figure 4.8, respectively). In the case of the tests using cNPW, there is a slight delay of barium with respect to the other cations; whereas, when cSPW was used, all cations broke through, simultaneously. This different behavior could have been due to different surface complexation reactions occurring at the solid–liquid interface in the presence of cNPW, where minor organic species were present.

When the diluted synthetic produced water was used (experiments 15, 16, and 17 in Figure 4.8), the concentration fronts broaden and a peak, centered between 1 and 2 PVs, arose. The concentration curves stabilize around 150 PVs for all divalent cations, except for barium where the retarded front starts to appear at 500 PVs [44]. In all cases, retardation is due to adsorption which is favored at the pH of injection (ca. 8). The concentration peaks resemble the fast pulse observed in previous works and they are due to pH-dependent solute transport and hydrodynamic dispersion [94,85]. In these earlier works, it was observed that hydrodynamic dispersion favors the formation of a fast concentration pulse by creating a mixing zone between the initial and the injected solutions within the pore space. Upon mixing, the pH is acidic and adsorption of the cations is negligible. Therefore, a solute pulse forms and moves at the interstitial flow velocity. Hydrodynamic dispersion is generally regarded as second-order effect as it only smooths the concentration fronts, whereas, adsorption controls the retardation of the fronts. The larger the adsorption at given chemical conditions, the slower the fronts. However, even in the presence of significant adsorption, mixing between the initial and the injected solutions in the pore space, due to hydrodynamic dispersion, can create chemical conditions which do not favor the adsorption of the solutes and therefore the solutes travel like a conservative tracer. This is shown in our previous paper where we observed the pulse of strontium in a reactive porous medium containing iron oxide [94]. In the work reported in this chapter, both ionic strength and pH control the adsorption process. As ionic strength decreases, pH-dependent adsorption dominates the reactive transport behavior, favoring the formation of the fast wave upon mixing. Overall, results from experiments 16 and 17 confirm the notable reactivity of MnO_2 mineral. In Figure 4.8, it is possible to see that the maximum normalized concentration of the fast pulse reaches for calcium and magnesium values larger than 1. This is because of competitive adsorption of the cations on the same reactive surface sites. Initially, fast adsorption occurs onto the mineral surface then it is followed by the desorption of those cations which have less affinity with the surface.

Warner et al. [21] and Lauer et al. [38] observed significant retention of alkaline earth elements by river sediments. The authors suggest that upon dilution of the high-salinity wastewater, discharged from a treatment plant, low-salinity river water rapid adsorption onto river sediments could occur because of the drop of ionic strength. This indeed may also be the case of a spill of produced water from a storage tank or a well. Mixing of the brine with fresh groundwater could quickly reduce the ionic strength favoring the adsorption of alkaline earth elements onto soil. However, our work also shows that a pulse due to pH-dependent solute adsorption and hydrodynamic dispersion may arise, traveling at the average flow velocity and enhancing the transport of the contaminants which move at a speed larger than expected.

4.6 MODELING AND SIMULATIONS OF TRANSPORT THROUGH MCS

Transport experiments 13–17 were described using the reactive transport model reported above. The aim was to understand if a reactive transport model based on geochemical reactions present in the literature could describe the transport of alkaline earth elements in produced water. The estimated values of the model parameters, i.e., the intrinsic reaction constants (K), the density of sites (N_s), the capacitances (C_1 and C_2), and the specific surface area (A), are listed in Table 4.4 and further details including a PHREEQC file used to simulate the experiments are available in our earlier work [44]. Large discrepancy between our work and the literature was found for the equilibrium constants related to the adsorption/desorption reactions of Mg^{2+} and Na^+.

Figures 4.9–4.11 show the measurements with the model upon fitting. As it is possible to notice in these figures, the model agrees well with the data, except for the pH. The model can predict the emergence of the fast wave for the various investigated species as well as the retarded front of barium in experiment 16. This suggests that the model developed in this work is a valuable tool to describe the transport of alkaline earth elements in produced water and identify optimal operating conditions and mineral properties for alkaline earth element removal from produced water.

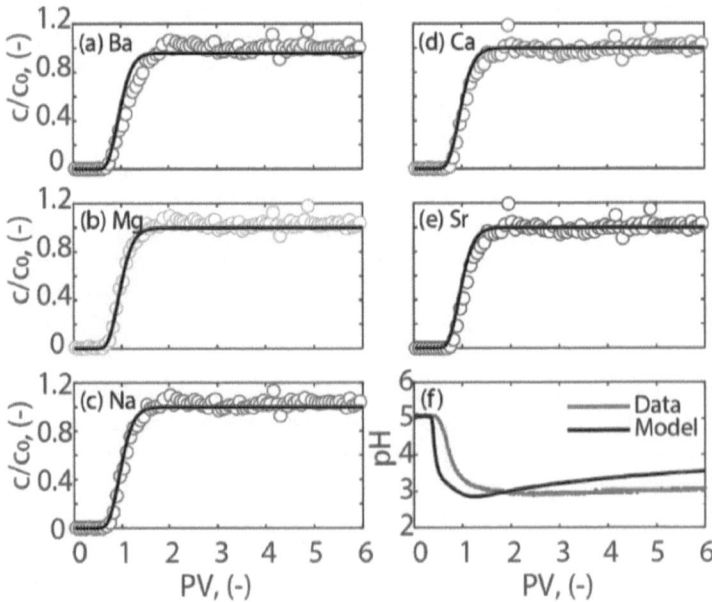

FIGURE 4.9 Measurements (dots and dash line) and modeling (black line) of experiment 13 using cSPW (Table 4.3). Parts (a)–(f) report the concentration of Ba^{2+}, Mg^{2+}, Na^+, Ca^{2+}, Sr^{2+}, and pH as a function of PV injected.

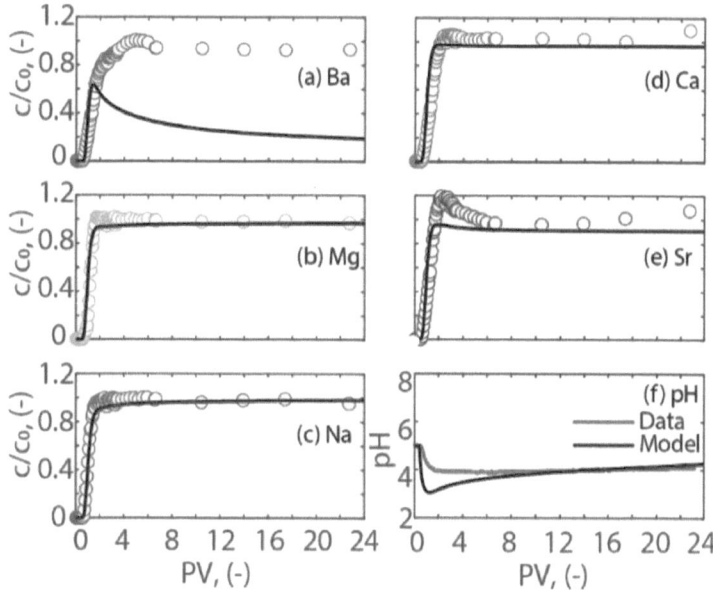

FIGURE 4.10 Measurements (dots and dash line) and modeling (black line) of experiment 15 using dSPW$_{1/100}$ (Table 4.3). Parts (a)–(f) report the concentration of Ba^{2+}, Mg^{2+}, Na$^+$, Ca^{2+}, Sr^{2+}, and pH as a function of PV injected.

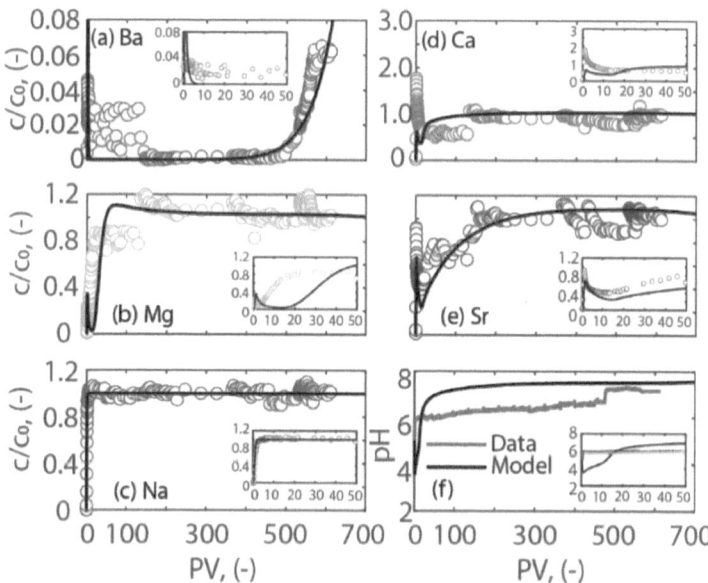

FIGURE 4.11 Measurements (dots and dash line) and modeling (black line) of experiment 16 using dSPW$_{1/1000}$ (Table 4.3). Parts (a) through (f) report the concentration of Ba^{2+}, Mg^{2+}, Na+, Ca^{2+}, and S$_r^{2+}$, and pH as a function of PV injected.

4.7 CONCLUSIONS

In this chapter, an experimental and modeling work was presented where proppants and natural minerals were tested to study the transport behavior of alkaline earth elements in produced water. In all cases, no retardation of the ions was observed at the salinity conditions of the produced water, but strong retardation in the pH front was measured, indicating that adsorption indeed occurred. When using manganese oxide and upon dilution of produced water, the concentration fronts of all major divalent cations were retarded. However, a fast traveling pulse was formed due to the combined effect of pH-dependent solute transport and hydrodynamic dispersion. This phenomenon confirmed that significant adsorption occurred under those conditions. But, pH-dependent adsorption and hydrodynamic dispersion can favor fast solute transport. A reactive transport model was developed and implemented in PHREEQC to describe these results. It agrees well with the data. This opens the possibility for the application of the model to design a reactive MnO_2-based material to mitigate the potential negative impact of produced water spills on potable shallow aquifers.

Overall, these results suggest that manganese oxide could be used as a reactive material in proppants, in the lining of temporary storage tanks and in the well cases to retard the migration of the major toxic elements in produced water. However, mixing must be controlled to avoid the emergence of an instability at the concentration fronts favoring the formation of fast waves.

BIBLIOGRAPHY

1. Robert B. Jackson, Ella R. Lowry, Amy Pickle, Mary Kang, Dominic DiGiulio, and Kaiguang Zhao. The depths of hydraulic fracturing and accompanying water use across the United States. *Environ. Sci. Technol.*, 49(15): 8969–8976, 2015.
2. J. Quinn Norris, Donald L. Turcotte, Eldridge M. Moores, Emily E. Brodsky, and John B. Rundle. Fracking in tight shales: What is it, what does it accomplish, and what are its consequences? In Jeanloz, R. and Freeman, K.H., editor, *Annual Review of Earth and Planetary Sciences*, volume 44, pages 321–351. 2016. doi:10.1146/annurev-earth-060115-012537.
3. EIA. Annual energy outlook 2019 with projections to 2050. https://www.eia.gov/outlooks/aeo/pdf/aeo2019.pdf, 2019. Online; accessed 12 November 2019.
4. Devin L. Shaffer, Laura H. Arias Chavez, Moshe Ben-Sasson, Santiago Romero-Vargas Castrillón, Ngai Yin Yip, and Menachem Elimelech. Desalination and reuse of high-salinity shale gas produced water: Drivers, technologies, and future directions. *Environ. Sci. Technol.*, 47(17): 9569–9583, 2013.
5. Andrew J. Kondash, Elizabeth Albright, and Avner Vengosh. Quantity of flowback and produced waters from unconventional oil and gas exploration. *Sci. Total Environ.*, 574: 314–321, 2017.
6. Devin L. Shaffer, Laura H. Arias Chavez, Moshe Ben-Sasson, Santiago Romero-Vargas Castrillón, Ngai Yin Yip, and Menachem Elimelech. desalination and reuse of high-salinity shale gas produced water: Drivers, technologies, and future directions. *Environ. Sci. Technol.*, 47(17): 9569–9583, 2013.
7. Sam Almond, Sarah. A. Clancy, Richard. J. Davies, and Fred Worrall. The flux of radionuclides in flowback fluid from shale gas exploitation. *Environ. Sci. Pollut. Res. Int.*, 21(21): 12316–12324, 2014.

8. Genevieve A. Kahrilas, Jens Blotevogel, Edward R. Corrin, and Thomas Borch. Downhole transformation of the hydraulic fracturing fluid biocide glutaraldehyde: Implications for flowback and produced water quality. *Environ. Sci. Technol.*, 50(20): 11414–11423, 2016.

9. Erin E. Yost, John Stanek, Robert S. DeWoskin, and Lyle D. Burgoon. Overview of chronic oral toxicity values for chemicals present in hydraulic fracturing fluids, flowback, and produced waters. *Environ. Sci. Technol.*, 50(9): 4788–4797, 2016.

10. Chuanxia Tong, Huazhou Huang, Huan He, and Bo Wang. Chemical characteristics and development significance of trace elements in produced water with coalbed methane in tiefa basin. *ACS Omega*, 4(17): 17561–17568, 2019.

11. Gregory P. Thiel and John H. Lienhard. Treating produced water from hydraulic fracturing: Composition effects on scale formation and desalination system selection. *Desalination*, 346: 54–69, 2014.

12. Werner Stumm and James J. Morgan. *Aquatic Chemistry: Chemical Equilibria and Rates in Natural Waters*, 3rd ed. John Wiley & Sons, New York, 1996.

13. ALLConsulting. Technical summary of oil & gas produced water treatment technologies. U.S. Department of Energy, National Petroleum Technology Office: Washington, D.C., 2005.

14. Katie Guerra, Katharine Dahm, and Steve Dundorf. Oil and gas produced water management and beneficial use in the western United States, science and technology program report no. 157. U.S. Department of the Interior Bureau of Reclamation, 2011.

15. GWPC. Modern shale gas development in the United States: A primer - ground water protection council. Prepared for U.S. Department of Energy, the National Energy Technology Laboratory. 2009.

16. Nancy E. Lauer, Jennifer S. Harkness, and Avner Vengosh. Brine spills associated with unconventional oil development in North Dakota. *Environ. Sci. Technol.*, 50(10): 5389–5397, 2016.

17. William D. Burgos, Sa Castillo-Meza, Travis L. Tasker, Thomas J. Geeza, Patrick J. Drohan, Xiaofeng Liu, Joshua D. Landis, Jens Blotevogel, Molly McLaughlin, Thomas Borch, and Nathaniel R. Warner. Watershed-scale impacts from surface water disposal of oil and gas wastewater in western Pennsylvania. *Environ. Sci. Technol.*, 51(15): 8851–8860, 2017.

18. Elise Barbot, Natasa S. Vidic, Kelvin B. Gregory, and Radisav D. Vidic. Spatial and temporal correlation of water quality parameters of produced waters from devonian-age shale following hydraulic fracturing. *Environ. Sci. Technol.*, 47(6): 2562–2569, 2013.

19. Lara O. Haluszczak, Arthur W. Rose, and Lee R. Kump. Geochemical evaluation of flowback brine from Marcellus gas wells in Pennsylvania, USA. *Appl. Geochem.*, 28: 55–61, 2013.

20. Kyle J. Ferrar, Drew R. Michanowicz, Charles L. Christen, Ned Mulcahy, Samantha L. Malone, and Ravi K. Sharma. Assessment of effluent contaminants from three facilities discharging marcellus shale wastewater to surface waters in pennsylvania. *Environ. Sci. Technol.*, 47(7): 3472–3481, 2013.

21. Nathaniel R. Warner, Cidney A. Christie, Robert B. Jackson, and Avner Vengosh. Impacts of shale gas wastewater disposal on water quality in western Pennsylvania. *Environ. Sci. Technol.*, 47(20): 11849–11857, 2013.

22. Tieyuan Zhang, Richard W. Hammack, and Radisav D. Vidic. Fate of radium in marcellus shale flowback water impoundments and assessment of associated health risks. *Environ. Sci. Technol.*, 49(15): 9347–9354, 2015.

23. Can He, Meng Li, Wenshi Liu, Elise Barbot, and Radisav D. Vidic. Kinetics and equilibrium of barium and strontium sulfate formation in Marcellus shale flowback water. *J. Environ. Eng.*, 140(5): B4014001, 2014.

24. Eunyoung Jang, Yunjai Jang, and Eunhyea Chung. Lithium recovery from shale gas produced water using solvent extraction. *Appl. Geochem.*, 78: 343–350, 2017.

25. Bernd G. Lottermoser. *Mine Wastes*, 3rd ed. Springer-Verlag, Berlin, 2010.

26. Ramesh Chitrakar, Hirofumi Kanoh, Yoshitaka Miyai, and Kenta Ooi. Recovery of lithium from seawater using manganese oxide adsorbent ($H_{1.6}Mn_{1.6O4}$) derived from Li1.6Mn1.6O4. *Ind. Eng. Chem. Res.*, 40(9): 2054–2058, 2001.

27. Shaked Stein, Amos Russak, Orit Sivan, Yoseph Yechieli, Eyal Rahav, Yoram Oren, and Roni Kasher. Saline groundwater from coastal aquifers as a source for desalination. *Environ. Sci. Technol.*, 50(4): 1955–1963, 2016.

28. LithiumMine. LithiumMine.com. http://www.lithiummine.com/, 2017.

29. Brian D. Lutz, Aurana N. Lewis, and Martin W. Doyle. Generation, transport, and disposal of wastewater associated with marcellus shale gas development. *Water Resour. Res.*, 49(2): 647–656, 2013.

30. Avner Vengosh, Robert B. Jackson, Nathaniel Warner, Thomas H. Darrah, and Andrew Kondash. A critical review of the risks to water resources from unconventional shale gas development and hydraulic fracturing in the united states. *Environ. Sci. Technol.*, 48(15): 8334–8348, 2014.

31. Lauren A. Patterson, Katherine E. Konschnik, Hannah Wiseman, Joseph Fargione, Kelly O. Maloney, Joseph Kiesecker, Jean-Philippe Nicot, Sharon Baruch-Mordo, Sally Entrekin, Anne Trainor, and James E. Saiers. Unconventional oil and gas spills: Risks, mitigation priorities, and state reporting requirements. *Environ. Sci. Technol.*, 51(5): 2563–2573, 2017.

32. Namita Shrestha, Govinda Chilkoor, Joseph Wilder, Venkataramana Gadhamshetty, and James J. Stone. Potential water resource impacts of hydraulic fracturing from unconventional oil production in the Bakken shale. *Water Res.*, 108: 1–24, 2017.

33. Nathaniel R. Warner, Robert B. Jackson, Thomas H. Darrah, Stephen G. Osborn, Adrian Down, Kaiguang Zhao, Alissa White, and Avner Vengosh. Geochemical evidence for possible natural migration of marcellus formation brine to shallow aquifers in pennsylvania. *Proc. Natl. Acad. Sci. U.S.A.*, 109(30): 11961–11966, 2012.

34. Garth T. Llewellyn, Frank Dorman, J. L. Westland, Dave Yoxtheimer, Paul Grieve, Todd Sowers, Elizabeth Humston-Fulmer, and Susan L. Brantley. Evaluating a groundwater supply contamination incident attributed to marcellus shale gas development. *Proc. Natl. Acad. Sci. U.S.A.*, 112(20): 6325–6330, 2015.

35. Dominic C. DiGiulio and Robert B. Jackson. Impact to underground sources of drinking water and domestic wells from production well stimulation and completion practices in the pavillion, Wyoming, field. *Environ. Sci. Technol.*, 50(8): 4524–4536, 2016.

36. Feng Liang, Mohammed Sayed, Ghaithan A. Al-Muntasheri, Frank F. Chang, and Leiming Li. A comprehensive review on proppant technologies. *Petroleum*, 2(1): 26–39, 2016.

37. Zhang Cai, Hang Wen, Sridhar Komarneni, and Li Li. Mineralogy controls on reactive transport of marcellus shale waters. *Sci. Total Environ.*, 630: 1573–1582, 2018.

38. Nancy E. Lauer, Nathaniel R. Warner, and Avner Vengosh. Sources of radium accumulation in stream sediments near disposal sites in Pennsylvania: Implications for disposal of conventional oil and gas wastewater. *Environ. Sci. Technol.*, 52(3): 955–962, 2018.

39. Julia Kravchenko, Thomas H. Darrah, Richard K. Miller, H. Kim Lyerly, and Avner Vengosh. A review of the health impacts of barium from natural and anthropogenic exposure. *Environ. Geochem. Health*, 36(4): 797–814, 2014.

40. U.S.EPA. Radionuclide basics: Radium. https://www.epa.gov/radiation/radionuclide-basics-radium#radiumenvironment, 2019.

41. Wenjia Fan, Kim F. Hayes, and Brian R. Ellis. Estimating radium activity in shale gas produced brine. *Environ. Sci. Technol.*, 52(18): 10839–10847, 2018.

42. Valentina Prigiobbe and X. Meng. Reactive propping agent to immobilize heavy metals and radionuclides in the subsurface during hydraulic fracturing, 2015.

43. Alen V. Gusa and Radisav D. Vidic. Development of functionalized proppant for the control of norm in marcellus shale produced water. *Environ. Sci. Technol.*, 53(1): 373–382, 2019.

44. Zi Ye and Valentina Prigiobbe. Transport of produced water through reactive porous media. *Water Res.*, 185: 116258, 2020.

45. Henry V. Mott, Sarabjit Singh, and Venkateshwer R. Kondapally. Factors affecting radium removal using mixed iron-manganese oxides. *J. Am. Water Works Assoc.*, 85(10): 114–121, 1993.

46. Shuddhodan P. Mishra and Dhanesh Tiwary. Ion exchangers in radioactive waste management. Part VIII: Radiotracer studies on adsorption of strontium ions on hydrous manganese oxide. *Radiochimica Acta*, 69(2): 121–126, 1995.

47. Shuddhodan P. Mishra and Dhanesh Tiwary. Ion exchangers in radioactive waste management. Part XI. Removal of barium and strontium ions from aqueous solutions by hydrous ferric oxide. *Int. J. Appl. Radiat. Isot.*, 51(4): 359–366, 1999.

48. Deepak Garg and Dennis A. Clifford. Removing radium from water by plain and treated activated alumina. *U.S. Environmental Protection Agency, Washington, D.C., EPA/600/R-92/048*, 2002.

49. Meagan Eagle Gonneea, Paul J. Morris, Henrieta Dulaiova, and Matthew A. Charette. New perspectives on radium behavior within a subterranean estuary. *Mar. Chem.*, 109(3–4): 250–267, 2008.

50. Nayereh Ghaeni, Mojtaba S. Taleshi, and Fatemeh Elmi. Removal and recovery of strontium (Sr(II)) from seawater by Fe_3O_4/MnO_2/fulvic acid nanocomposite. *Mar. Chem.*, 213: 33–39, 2019.

51. Hiroki Tamura, Noriaki Katayama, and Ryusaburo Furuichi. Modeling of ion-exchange reactions on metal oxides with the frumkin isotherm. 1. acid–base and charge characteristics of MnO_2, TiO_2, Fe_3O_4, and Al_2O_3 surfaces and adsorption affinity of alkali metal ions. *Environ. Sci. Technol.*, 30(4): 1198–1204, 1996.

52. USGS. Phreeqc (version 3)–a computer program for speciation, batch-reaction, one-dimensional transport, and inverse geochemical calculations. http://wwwbrr.cr.usgs.gov/projects/GWC_coupled/phreeqc/, 2016.

53. C. Anthony, J. Appelo, and Paul Wersin. Multicomponent diffusion modeling in clay systems with application to the diffusion of tritium, iodide, and sodium in opalinus clay. *Environ. Sci. Technol.*, 41(14): 5002–5007, 2007.

54. Muhammad Muniruzzaman and Massimo Rolle. Modeling multicomponent ionic transport in groundwater with IPhreeqc coupling: Electrostatic interactions and geochemical reactions in homogeneous and heterogeneous domains. *Adv. Water Resour.*, 98: 1–15, 2016.

55. Riccardo Sprocati, Matteo Masi, Muhammad Muniruzzaman, and Massimo Rolle. Modeling electrokinetic transport and biogeochemical reactions in porous media: A multidimensional Nernst-Planck-Poisson approach with PHREEQC coupling. *Adv. Water Resour.*, 127: 134–147, 2019.

56. Colin J. McNeece and Marc A. Hesse. Challenges in coupling acidity and salinity transport in porous media. *Environ. Sci. Technol.*, 51(20): 11799–11808, 2017.

57. Muhammad Muniruzzaman and Massimo Rolle. Experimental investigation of the impact of compound-specific dispersion and electrostatic interactions on transient transport and solute breakthrough. *Water Resour. Res.*, 53(2): 1189–1209, 2017.

58. Lucien Stolze, Jakob B. Wagner, Christian D. Damsgaard, and Massimo Rolle. Impact of surface complexation and electrostatic interactions on ph front propagation in silica porous media. *Geochimica et Cosmochimica Acta*, 277: 132–149, 2020.

59. Lara O. Haluszczak, Arthur W. Rose, and Lee R. Kump. Geochemical evaluation of flowback brine from Marcellus gas wells in Pennsylvania, USA. *Appl. Geochem.*, 28: 55–61, 2013.

60. Michael A. Chen and Benjamin D. Kocar. Radium sorption to Iron (Hydr)oxides, Pyrite, and Montmorillonite: Implications for mobility. *Environ. Sci. Technol.*, 52(7): 4023–4030, 2018.

61. Klaus Knödel, Gerhard Lange, and Hans-Jürgen Voigt. *Environmental Geology: Handbook of Field Methods and Case Studies.* Springer, Heidelberg, 2007.

62. Florian Huber and Johannes Luetzenkirchen. Uranyl retention on quartz-new experimental data and blind prediction using an existing surface complexation model. *Aquat. Geochem.*, 15(3): 443–456, 2009.

63. Zi Ye and Valentina Prigiobbe. Effect of ionic strength on barium transport in porous media. *J. Contam. Hydrol.*, 209: 24–32, 2018.

64. Marijan Gotić, Tanja Jurkin, Svetozar Musić, Klaus Unfried, Ulrich Sydlik, and Anamarija Bauer-Šĕgvić. Microstructural characterizations of different Mn-oxide nanoparticles used as models in toxicity studies. *J. Mol.*, 1044: 248–254, 2013.

65. C. Anthony, J. Appelo, and Dieke Postma. *Geochemistry, Groundwater, and Pollution*, II Edition. CRC Press, London, 2005.

66. Tjisse Hiemstra, Willem H. Van Riemsdijk, and Gerard (Jerry) Hendrik Bolt. Multisite proton adsorption modeling at the solid-solution interface of (hydr)oxides - a new approach.1. model description and evaluation of intrinsic reaction constants. *J. Colloid Interface Sci.*, 133(1): 91–104, 1989.

67. Tjisse Hiemstra, Peter Venema, and Willem H. Van Riemsdijk. Intrinsic proton affinity of reactive surface groups of metal (hydr)oxides: The bond valence principle. *J. Colloid Interface Sci.*, 184(2): 680–692, 1996.

68. Kenneth S. Pitzer. Thermodynamics of electrolytes. I. theoretical basis and general equations. *J. Phys. Chem.*, 77(2): 268–277, 1973.

69. Kim F. Hayes, George Redden, Wendell Ela, and James O. Leckie. Surface complexation models: An evaluation of model parameter estimation using FITEQL and oxide mineral titration data. *J. Colloid Interface Sci.*, 142(2): 448–469, 1991.

70. Chin-Pao Huang and Werner Stumm. Specific adsorption of cations on hydrous γ-Al_2O_3. *J. Colloid Interface Sci.*, 43(2): 409–420, 1973.

71. Sabine Goldberg. Use of surface complexation models in soil chemical systems. *Adv. Agron.*, 47: 233–329. 1992. Academic Press.

72. Nita Sahai and Dimitri A. Sverjensky. Evaluation of internally consistent parameters for the triple-layer model by the systematic analysis of oxide surface titration data. *Geochim. Cosmochim. Acta*, 61(14): 2801–2826, 1997.

73. Dimitri A. Sverjensky. Physical surface-complexation models for sorption at the mineral–ater interface. *Nature*, 364: 776–780, 1993.

74. Dimitri A. Sverjensky. Interpretation and prediction of triple-layer model capacitances and the structure of the oxide-electrolyte-water interface. *Geochim. Cosmochim. Acta*, 65(21): 3643–3655, 2001.

75. Dimitri A. Sverjensky. Prediction of surface charge on oxides in salt solutions: Revisions for 1:1 (M^+L^-) electrolytes. *Geochim. Cosmochim. Acta*, 69(2): 225–257, 2005.

76. Christophe Tournassat, Sylvain Grangeon, Philippe Leroy, and Eric Giffaut. Modeling specific ph dependent sorption of divalent metals on montmorillonite surfaces. A review of pitfalls, recent achievements and current challenges. *Am. J. Sci.*, 313(5): 395–451, 2013.

77. Colin J. McNeece and Marc A. Hesse. Reactive transport of aqueous protons in porous media. *Adv. Water Resour.*, 97: 314–325, 2016.

78. Mohamed Zuki Fathiah and Robert G. J. Edyvean. The role of ionic strength and mineral size to zeta potential for the adhesion of *p.putida* to mineral surfaces. 2015.

79. Tanjina Nur, Paripurnanda Loganathan, T. Chung Nguyen, Saravanamuthu Vigneswaran, Gurmeet Singh, and Jeevetha Kandasamy. Batch and column adsorption and desorption of fluoride using hydrous ferric oxide: Solution chemistry and modeling. *Chem. Eng. J.*, 247: 93–102, 2014.

80. Juan Luis Reyes Bahena, Aurora Robledo Cabrera, Alejandro Valdivieso, and Ronaldo Herrera Urbina. Fluoride adsorption onto α-al$_2$o$_3$ and its effect on the zeta potential at the alumina-aqueous electrolyte interface. *Sep. Sci. Technol.*, 37(8): 1973–1987, 2002.

81. Stanisław Chibowski, Elżbieta Grzadka, and Jacek Patkowski. Comparison of the influence of a kind of electrolyte and its ionic strength on the adsorption and electrokinetic properties of the interface: Polyacrylic acid/MnO$_2$/electrolyte solution. *Colloids Surf. A*, 326(3): 191–203, 2008.

82. Elżbieta Grzadka and Stanis-law Chibowski. Influence of a kind of electrolyte and its ionic strength on the conformation changes of polyacrylic acid during its coming from the bulk solution to the surface of MNO$_2$. 2008.

83. Charles Willard Fetter. *Contaminant Hydrogeology/C.W. Fetter.* Macmillan Pub. Co.; Maxwell Macmillan Canada; Maxwell Macmillan International New York: Toronto; New York, 1993.

84. Valentina Prigiobbe, Marc A. Hesse, and Steven L. Bryant. Hyperbolic theory for flow in permeable media with pH-dependent adsorption. *SIAM J. Appl. Math.*, 73(5): 1941–1957, 2013.

85. Valentina Prigiobbe and Steven L. Bryant. ph-dependent transport of metal cations in porous media. *Environ. Sci. Technol.*, 48(7): 3752–3759, 2014.

86. James W. Murray. The interaction of metal ions at the manganese dioxide-solution interface. *Geochimica et Cosmochimica Acta*, 39(4): 505–519, 1975.

87. Do Sik Moon, William C. Burnett, Samaa Nour, Philip Horwitz, and Andrew Bond. Preconcentration of radium isotopes from natural waters using MnO$_2$ Resin. *Appl. Radiat. Isot.*, 59(4): 255–262, 2003.

88. Philip H. Towler, J. David Smith, and David R. Dixon. Magnetic recovery of radium, lead and polonium from seawater samples after preconcentration on a magnetic adsorbent of manganese dioxide coated magnetite. *Anal. Chim. Acta*, 328(1): 53–59, 1996.

89. Alfred Edward (Ted) Ringwood, Sue E. Kesson, N.G. Ware, William O. Hibberson, and Anthony S. Major. Immobilization of high-level nuclear-reactor wastes in synroc. *Nature*, 278(5701): 219–223, 1979.

90. Edward R. Landa and David F. Reid. Sorption of radium-226 from oil-production brine by sediments and soils. *Environ. Geol.*, 5(1): 1–8, 1983.

91. Naeem Qureshi and Steven Nelson. Radium removal by HMO and manganese greensand. *J. Am. Water Works Assoc.*, 95(3): 101–108, 2003.

92. Marek Kosmulski. The ph-dependent surface charging and points of zero charge: V. update. *J. Colloid Interface Sci.*, 353(1): 1–15, 2011.

93. Natalia Mayordomo, Harald Foerstendorf, Johannes Lützenkirchen, Karsten Heim, Stephan Weiss, Ursula Alonso, Tiziana Missana, Katja Schmeide, and Norbert Jordan. Selenium(IV) sorption onto γ-Al$_2$O$_3$: A consistent description of the surface speciation by spectroscopy and thermodynamic modeling. *Environ. Sci. Technol.*, 52(2): 581–588, 2018.

94. Valentina Prigiobbe, Marc A. Hesse, and Steven L. Bryant. Fast strontium transport induced by hydrodynamic dispersion and pH-dependent sorption. *Geophys. Res. Lett.*, 39(18), 2012. L18401.

95. Laurie S Balistrieri and James W. Murray. The surface chemistry of δMnO$_2$ in major ion sea water. *Geochim. Cosmochim. Acta*, 46(6): 1041–1052, 1982.

96. Olga N. Karasyova, Lyudmila I. Ivanova, and Leonid Z. Lakshtanov. Strontium adsorption on manganese oxide (δ-MnO$_2$) at elevated temperatures: Experiment and modeling. *Geochem. Int.*, 57(10): 1107–1119, 2019.

97. Paras Trivedi and Lisa Axe. A comparison of strontium sorption to hydrous aluminum, iron, and manganese oxides. *J. Colloid Interface Sci.*, 218(2): 554–563, 1999.

98. Olivier Pourret and Málanie Davranche. Rare earth element sorption onto hydrous manganese oxide: A modeling study. *J. Colloid Interface Sci.*, 395: 18–23, 2013.

5 Prediction of Barium Sulfate Deposition in Petroleum and Hydrothermal Systems

Derek M. Hall and Serguei N. Lvov
National Energy Technology Laboratory
and
The Pennsylvania State University

Isaac K. Gamwo
National Energy Technology Laboratory

CONTENTS

5.1 MINERAL DEPOSITION IN PETROLEUM AND HYDROTHERMAL SYSTEMS

Mineral deposition is a critical problem faced by engineers working with petroleum and hydrothermal fluids. These problems are most apparent for petroleum and natural gas (PNG) industry and geothermal systems. If left unchecked, hydrothermal fluids can quickly become supersaturated and form mineral deposits that damage the productivity of a production field. As recent explorations have dramatically expanded the range of temperatures and pressures observed in PNG and geothermal reservoirs, we need to adapt our current modeling practices to capture these new conditions. However, the most sophisticated modeling tools [1] are limited with a range of temperatures and pressures.

DOI: 10.1201/9781003091011-5

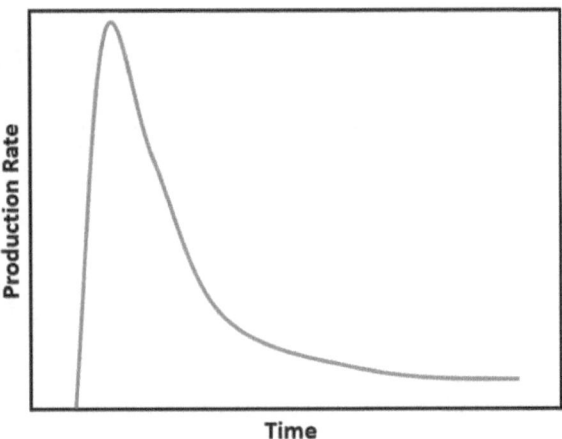

FIGURE 5.1   Typical sketch production profile from unconventional well.

The process of forming these deposits is referred to as scaling, which affects the productivity of many geoengineered systems. One extreme example of a detrimental scaling incident occurred in the Miller oilfield in the North Sea. When an oil well went from 30,000 barrels per day of production to zero within 24 hours [2]. The cause of the production shutdown was a mineral plug of barium sulfate (barite) several hundred feet long which had quickly formed within the well [2]. The Miller oilfield scaling incident is a classic example of how the scaling phenomena can adversely impact production from conventional oil fields, but this is just one instance as several other examples can be found throughout literature [3–7].

Scaling is also a possible cause of the rapid production drop-off observed in unconventional oil and gas fields. Though the use of hydraulic fracturing has played a critical role in the resurgence of oil and gas exploration in shale reservoirs, these unconventional fields are known for their rapid decrease in production over time [8,9]. Here is a typical plot of the production decline from unconventional reservoirs showing the hyperbolic decline and exponential tail decline phases (Figure 5.1) [10].

Their use of acidic and often complex hydraulic fracturing fluids (HFFs) aims to enhance the transport properties, increasing reservoir productivity. However, these reservoirs are subject to injection sequences which introduce the HFFs into the reservoir and then undergo extended shut-in periods which neutralize the incoming fluids through their contact with minerals and fluids contained within the formation [8]. Recent studies have shown that mineral precipitation within these reservoirs, as well as deposits inside fractures, near pore throats and at the surface of proppant materials, leads to reduced porosities. [11]. Minerals found within the fracture network consist of species such as Fe(III)-containing solids, calcite, and barite [8,9,11].

Scaling occurs in environments that share some core commonalities. In a hydrothermal system, scaling occurs if an aqueous solution becomes supersaturated with ions that contain the elements needed to form a mineral such as barite or calcite. In principle, a solution can either be saturated, undersaturated, or supersaturated with a mineral depending on the activity of mineral constituents within the aqueous phase.

In these circumstances, specific outcomes are expected. Supersaturated solutions deposit minerals, undersaturated solutions dissolve minerals, and saturated solutions will neither dissolve nor deposit a mineral as they are in equilibrium with these solid mineral phases. The solubility limit, which determines each of these states, is a thermodynamic property. Kinetics of deposition/dissolution could impact the rate of these processes but their thermodynamics provides the principle possibility of these processes to occur.

The extent of mineral saturation in any solution can be determined through a thermodynamic analysis of the solid and aqueous phase. Through such an analysis, concepts such as scaling tendency, saturation index (sometimes called the scaling index), and solubility products are used to assess if deposition is thermodynamically favorable. One common mineral known to cause scaling issues is barite. Here, barite precipitation and dissolution are described using the International Union of Pure and Applied Chemistry (IUPAC) conventions [12]:

$$BaSO_4(s) \leftrightarrow Ba^{2+}(aq) + SO_4{}^{2-}(aq). \tag{5.1}$$

$BaSO_4(s)$ represents the mineral barite in its solid phase "(s)", and $Ba^{2+}(aq)$ and SO_4^{2-} (aq) are the barium and sulfate charged species (ions) present in an aqueous phase "(aq)". In addition to the ionic species, ion pairs such as $BaSO_4^0$ (aq) can also form within the aqueous phase and are critical to understanding barite solubility in superheated waters as the attraction between ions increases due to decrease of relative permittivity (dielectric constant) of water. Formation of the ion pairs and other complexes is a common phenomenon for any strong electrolyte when temperature of solution increases and its density decreases. This formation of ion pairs and complexes needs to be considered in modeling and predicting mineral deposition in petroleum and hydrothermal systems at high temperatures and low pressures.

5.2 SOLUBILITY PRODUCT CONSTANTS AND SCALING TENDENCIES

Knowing the saturation limit of barite in a solution is a useful tool for predicting its likelihood of deposition. The solubility product constant (K_{sp}), also called the solubility product, is an equilibrium constant (K_{eq}) for a mineral dissolution reaction such as the one shown in Eq. 5.1 and provides a means of quantifying this limit. Like all equilibrium constants, they can be determined from standard Gibbs energy of reaction ($\Delta_r G^0$) values as follows:

$$\ln(K_{eq}) = -\Delta_r G^0 (RT)^{-1}, \tag{5.2}$$

where R is the molar gas constant (in J mol^{-1} K^{-1}) and T is the thermodynamic temperature (in K). Alternatively, K_{eq} can be expressed in terms of activities. For barite dissolution, K_{sp} is as follows:

$$K_{sp} = a_{Ba^{2+}(aq)} a_{SO_4^{2-}(aq)} / a_{BaSO_4(s)}, \tag{5.3}$$

where $a_{Ba^{2+}(aq)}$ is the activity of barium ions, $a_{SO_4^{2-}(aq)}$ is the activity of sulfate ions, and $a_{BaSO_4(s)}$ is the activity of barite which is equal to 1 for pure mineral.

The ion activity product (IAP), also known as Q in some texts, $a_{Ba^{2+}(aq)} \times a_{SO_4^{2-}(aq)}$, of a solution is key indicator of whether a solution is capable of precipitating barite. The ratio of IAP and K_{sp} is also used in predicting scaling tendencies:

$$\Omega = IAP\, K_{sp}^{-1} \tag{5.4}$$

where Ω, the ratio of IAP and K_{sp}, is referred to as the scaling tendency and is used to assess the favorability of barite deposition from a solution [13]. As such, if a solution with an IAP of barium $a_{SO_4^{2-}(aq)}$ions is ever greater than its K_{sp}, then that solution is supersaturated with respect to barite. Conversely, solutions with an IAP lower than K_{sp} are undersaturated and will dissolve the mineral. However, we should note that thermodynamic values only tell us if precipitation and dissolution can happen. Kinetics of these reactions should be considered to determine if they happen in any meaningful timeframe.

Therefore, by calculating the $\Delta_r G^0$ using Eq. 5.1, we can assess the favorability of barite deposition in a given solution. The $\Delta_r G^0$can be calculated for Eq. 5.1 as follows:

$$\Delta_r G^0 = \Delta_f G^0_{Ba^{2+}(aq)} + \Delta_f G^0_{SO^{2-}(aq)} - \Delta_f G^0_{BaSO_4(s)}, \tag{5.5}$$

where $\Delta_r G_i^0$ is the standard Gibbs energy of formation of the i-th chemical in the reaction (eq. 5.5). The standard Gibbs energy of formation does not depend on concentration, but it does depend on the concentration scale. The concentration scales used with aqueous systems are molality (b), molarity (c), and mole fraction (x). Briefly, molality is defined as mol of solute per 1 kg of water, whereas molarity is mol of solute per 1 L of solution. Mole fraction is a dimensionless value that is the ratio of the number of moles of one component of a solution (solute or solvent) to the total number of moles representing all of the solution components. Comparing standard values in literature can be confusing because these concentration scales are rarely reported but often lead to ~10 kJ mol^{-1} differences between sources. For reliable calculations, the origins of the standard Gibbs energy values should be carefully considered before use. The SUPCRT92 database [14] is a great starting point for many systems as it is internally consistent among its many species and valid for a sizable range of temperatures and pressures. As an example, we will calculate $\Delta_r G^0$ for barite precipitation at 25°C and 1 bar as

$$\Delta_r G^0 = \Delta_f G^0_{Ba^{2+}(aq)} + \Delta_f G^0_{SO^{2-}(aq)} - \Delta_f G^0_{BaSO_4(s)}$$

$$= \left(-560.8\,kJ\,mol^{-1}\right) + \left(-744.5\,kJ\,mol^{-1}\right) - \left(1,362.1\,kJ\,mol^{-1}\right) = 56.92\,kJ\,mol^{-1}.$$

With the $\Delta_r G^0$ known, the natural logarithm of the solubility product for barite at 25°C and 1 bar can be readily calculated as:

$$\ln\left(K_{sp}\right) = -\Delta_r G^0 \left(RT\right)^{-1}$$

$$= -\left(56{,}916 \text{ J mol}^{-1}\right)\Big/\left[\left(8.3145 \text{ J mol}^{-1}\text{K}^{-1}\right)(298.15\text{K})\right]$$

$$= -22.96,$$

resulting in a K_{sp} of 1.069×10^{-10}. If the solution only contains barium ions, sulfate ions and water, then the square root of K_{sp} provides a means to estimate the solubility limit of barite in water $[(1.069\times10^{-10})^{0.5} = 1.034\times10^{-5}\text{mol kg}^{-1}$ at 25°C and 1 bar]. The solubility limit we calculated here is consistent with the 1.11×10^{-5} ($\pm$ 8%) mol kg^{-1} obtained experimentally [15].

Note that in the above calculations, we assumed that the activity coefficients of aqueous species equal 1, so activity of the i-th species, a_i, is numerically the same as molal concentration, b_i, while precisely $a_i = b_i / b^0$, where b^0 is the standard molality equals 1 mol kg^{-1}, so the activity is dimensionless.

The introduction of activity and activity coefficients bring up some important concepts about the relationship between concentration and activity. As the concentration of ions in a solution increases, interactions between these ions alter the effective concentration of each species, which we call activity. The deviation due to these interactions are accounted for through activity coefficients, which are obtained using a wide variety of equations that are thoroughly explained elsewhere [13]. As many geochemical fluids contain a plethora of ions, these deviations can be dramatic depending on the conditions. Likewise, minerals can form solid solutions which also require careful treatment. Though we acknowledge these factors and their significance, this chapter will focus on how composition (excluding activity coefficient effects) and temperature impact barite precipitation predictions. Excluding the activity coefficients, considering very dilute solutions is a reasonable approximation.

In many instances, Ω can vary by several orders of magnitude, so another metric is used to quantify the likelihood of barite and other minerals to precipitate from a solution. The saturation index (SI), which is defined as $\text{SI} = \log_{10}(\Omega)$, can provide a similar assessment of the likelihood whether a mineral will dissolve or precipitate in the given solution [13]. An SI equal to zero means the solution is at equilibrium with the mineral in question, while a negative SI indicates the mineral is predicted to dissolve and a positive SI means that the mineral is predicted to precipitate. These perditions are based on thermodynamics and the precipitation/dissolution kinetics should also be taken into account in some cases.

5.3 SPECIATION MODELS AND PREDOMINANCE DIAGRAMS

The solubility product can make the prediction of barite scaling seem deceptively simple. However, the presence of additional species complicates scaling thermodynamics as they can influence the maximum solubility of sulfate and/or barium-containing species. Such complications are frequent in real-world applications as engineers regularly work with natural and produced waters with a diverse range of components. The injection of produced waters [4] or seawater [2,5,6] into the reservoir to enhance oil recovery can increase the concentration of ionic species within the reservoir that can promote the formation of mineral scale. For example, pH, temperature, and fluid density variations can favor the formation of additional barium-containing species such as the barium hydroxide complex, e.g., $BaOH^+$(aq), or the ion pair, e.g., $BaSO_4^0$(aq) [16]. Problems such as these can be solved by developing a more realistic speciation model.

Speciation analysis is an approach that quantifies the thermodynamically favorable phases and their components (molecules, ions, ion pairs, etc.) for a particular system at particular state variables such as temperature and pressure. With respect to barite mineral scaling, speciation analysis can provide insights into how the chemical composition of produced waters can influence solubility of barite. The effects of dissolved salts, gases, and pH as well as temperature and pressure can be analyzed through this approach. As produced waters have a wide range of chemical and physical properties [4,17], speciation analysis is often needed to discern the fate of any mineral components within the production lines.

Two methods are commonly used to examine the speciation of a petroleum and hydrothermal system which shows how solution composition and the solubility limit of minerals can change in response to additional chemicals. The first relies on solving a system of equilibrium constants along with equations of electroneutrality and mass balance and the second relies on minimizing the Gibbs energy of the system [13,18]. The equilibrium constant method relies on identifying chemicals reactions that represent a system of interest, while the minimization method requires all Gibbs energies of formation of all species are available. In addition to the barite dissolution reaction shown in Eq. 5.1, barium hydroxide species is another well-known barium-containing aqueous species. Changing pH of the solution using acids and bases can invoke a wide range of responses. The chemicals used to change the system pH will strongly influence the outcome. Adjustments to pH invoke sulfate reactions involving protons as well as sulfate ions and hydrogen sulfate which can influence the solubility of barium sulfate significantly. Furthermore, a barium sulfate ion pair, $BaSO_4^0$(aq), can impact solubility at higher temperatures particularly in the presence of sulfate-containing electrolytes [16,19,20]. These relations can be expressed using additional chemical reactions:

$$BaOH^+(aq) \leftrightarrow OH^-(aq) + Ba^{2+}(aq), \tag{5.6}$$

$$HSO_4^-(aq) \leftrightarrow SO_4^-(aq) + H^+(aq), \tag{5.7}$$

$$H_2O(l) \leftrightarrow H^+(aq) + OH^-(aq), \tag{5.8}$$

$$BaSO_4^0(aq) \leftrightarrow Ba^{2+}(aq) + SO_4^{2-}(aq). \tag{5.9}$$

Using Eq. 5.2 and the SUPCRT92 database as well as some published ion pair data [16,19,20], K_{eq} values for each of these reactions can be calculated for a wide range of temperatures and pressures. As shown in Table 5.1, at 25°C and 1 bar, the K_{eq} values for each of these reactions can be different by more than 30 orders of magnitude.

If combined with mass balance and charge balance constraints, these equations form a system of equations that provide a means to determine the molal concentration

TABLE 5.1
Overview of Chemical Reactions and Equilibrium Constants
for Eqs. 5.1 and 5.6–5.9 at 25°C and 1 Bar

Chemical Reactions	$\ln(K_{eq})$
K_1 : $BaSO_4(s) \leftrightarrow Ba^{2+}(aq) + SO_4{}^{2-}(aq)$	−22.96
K_2 : $BaOH^+(aq) \leftrightarrow OH^-(aq) + Ba^{2+}(aq)$	−1.14
K_3 : $HSO_4{}^-(aq) \leftrightarrow SO_4{}^{2-}(aq) + H^+(aq)$	−4.56
K_4 : $H_2O(l) \leftrightarrow H^+(aq) + OH^-(aq)$	−32.22
K_5 : $BaSO_4{}^0(aq) \leftrightarrow Ba^{2+}(aq) + SO_4{}^{2-}(aq)$	−6.14

(b_i) of all species within an aqueous phase. Here is a demonstration of how these can be solved assuming the activity coefficients (γ_i) are close to 1 (i.e. $a_i = b_i\gamma_i \approx b_i$), which is only applicable in very dilute solutions [21]. Otherwise, an iterative approach is required to account for how activity coefficients change with the ionic strength of the solution, which is an approach presented elsewhere [13].

The K_{eq} equations for K_1–K_5 can be expressed in terms of activities (using square brackets) according to Eqs. 5.10–5.14:

$$K_1 = \left[b_{Ba^{2+}} \right]\left[b_{SO_4^{2-}} \right], \tag{5.10}$$

$$K_2 = \left[b_{Ba^{2+}} \right]\left[b_{OH^-} \right]/\left[b_{BaOH^+} \right], \tag{5.11}$$

$$K_3 = \left[b_{SO_4^{2-}} \right]\left[b_{H^+} \right]\left[b_{HSO_4^-} \right], \tag{5.12}$$

$$K_4 = \left[b_{OH^-} \right]\left[b_{H^+} \right], \tag{5.13}$$

$$K_5 = \left[b_{Ba^{2+}} \right]\left[b_{SO_4^{2-}} \right]/\left[b_{BaSO_4^0} \right]. \tag{5.14}$$

By selecting a pH titrant concentration and the total amount of moles of barium per kg of water for a system $(b_{total,Ba})$, we can solve the system of equations to determine solubility of barium sulfate at a given pH by including the following mass balance and charge balance equations, Eqs. 5.15 and 5.16, if we assume the counterions do not require additional reactions to be considered:

$$b_{total,Ba} = b_{Ba^{2+}} + b_{BaSO_4^0} + b_{BaSO_4(s)} + b_{BaOH^+}, \tag{5.15}$$

$$0 = 2b_{Ba^{2+}} + b_{BaOH^+} + b_{H^+} - b_{OH^-} - b_{HSO_4^-} - 2b_{SO_4^{2-}} + b_{Na^+}. \tag{5.16}$$

We set the total barium sulfate in the system to be 0.1 mol per kg of water, and we set the pH of the system and the counter ion concentration. If the titrants are NaOH and H_2SO_4, a parametric sweep of pH shows an increase in the solubility of barium sulfate from acidic to neutral conditions and a plateau at high pH (Figure 5.2a). The lower solubility at low pH is due to the higher concentrations of sulfate ions from the addition of the sulfuric acid to decrease the pH. (Figure 5.2b)

The approach presented in this section demonstrates how speciation models and predominance diagrams can be used to visualize solubility and scaling trends for barite as a function of the solution chemistry change. Similar diagrams can be developed for a number of compositional changes such as variations in sulfate and chloride concentrations. These diagrams can also be configured for a range of temperatures and pressures to study their influence on barite solubility. However, the equilibrium constant method becomes increasingly difficult as the number of species and possible reactions increase. For example, if we used HCl instead of H_2SO_4 to control the pH, then reactions with barium chloride are needed; otherwise erroneous results will be obtained. For more complex systems with multiple phases and multiple components such as examples given in Refs. [22–24], Gibbs energy minimization using a commercial software is a more convenient approach.

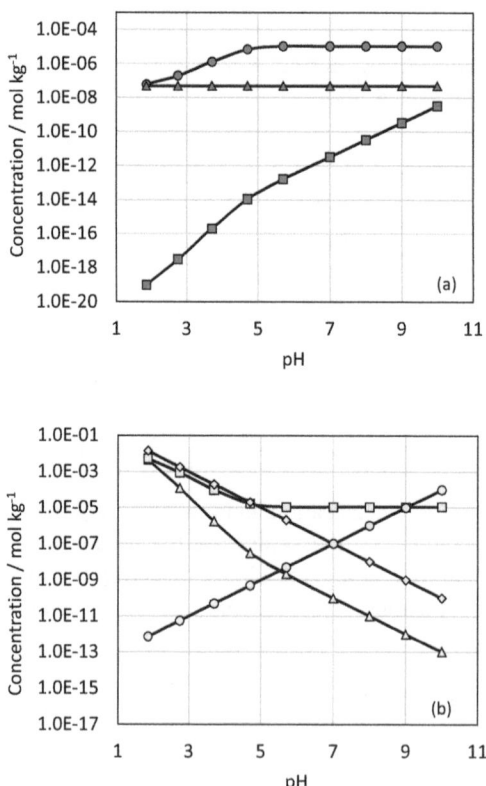

FIGURE 5.2 Predominance diagrams for (a) barium-containing species and (b) non-barium-containing species at 25°C and 1 bar for a $BaSO_4$-H_2O system. The titrants were NaOH and H_2SO_4. For pH 7 and higher, $b_{Na+(aq)} = b_{OH^-(aq)}$. ($b_{Ba^{2+}} = \bullet$, $b_{BaOH^+} = \blacksquare$, $b_{BaSO_4^0} = \blacktriangle$, $b_{H^+} = \blacklozenge$, $b_{OH^-} = \bullet$, $b_{HSO_4^-} = \blacktriangle$, $b_{SO_4^{2-}} = \blacksquare$)

5.4 BARIUM SULFATE SOLUBILITY MEASUREMENTS AT ELEVATED TEMPERATURES AND PRESSURES

Solubility of barite is generally quite low which often results in scaling. At saturated pressures of water, P_{sat}, the solubility increases to around 100°C, then it decreases in a temperature range between 100°C and 300°C. From 25°C to 300°C, increasing pressure from 10 to 100 MPa increases the barite solubility [16]. A few techniques were used for measuring barite solubility such as liquid phase sampling in autoclave and flow systems, weight loss of crystal, conductivity, spectroscopy, liquid scintillation, X-ray fluorescence, inductively coupled plasma-atomic emission spectroscopy, sulfur radioactive tracer, and barium radioactive tracer.

In Ref. [15], the solubility data of barite in pure water for a temperature range between 0°C and 300°C at P_{sat} was collected from 29 publications (Figure 5.3).

In Ref. [15], the data were carefully evaluated and fitted to the following polynomial equation:

$$b = \alpha_0 + \alpha_1 t + \alpha_2 t^2 + \alpha_3 t^3, \tag{5.17}$$

where b is barite solubility in mol per 1 kg of water (molality), t is temperature in degrees Celsius, and the parameters a_i were estimated as:

$$\alpha_0 = 5.176 10^{-6} \, \text{mol} / \text{kg},$$

$$\alpha_1 = 2.852 10^{-7} \, \text{mol}/(\text{kg}°\text{C}),$$

$$\alpha_2 = -2.024 10^{-9} \, \text{mol}/\text{kg}°\text{C}^2,$$

$$\alpha_3 = 3.473 10^{-12} \, \text{mol} / \left(\text{kg}°\text{C}^3\right).$$

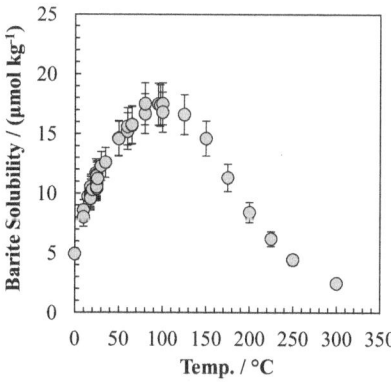

FIGURE 5.3 Experimental solubility data for barite in water at P_{sat} as a function of temperature. (Based on Krumgalz et al. [15].)

Using 47 experimental points with $R^2 = 0.983$, it was observed that discrepancies between the experimental solubilities and those calculated by this polynomial are below 4.34%.

Solubility data for barite in pure water at temperatures above 300°C were presented and discussed in a reference book [25]. Morey and Hesselgesser [26] measured $BaSO_4(s)$ solubility at 500°C and 1,000 bar using a flow method to pass supercritical water at a constant pressure over the mineral with the fluid exiting the system being analyzed to assess the solubility. Their measurement error is uncertain as it is not reported or listed. Strübel studied barite solubility in water at three conditions: temperature range of 200°C–600°C and pressures of 15.6–2100 bar; isobaric conditions and densities of 0.326–0.925 g cm^{-3}; and isochoric conditions using a weight loss of crystal method [27]. The authors provided an uncertainty of the solubility measurements of $\pm 20\%$–25%. Gundlach et al. [28] applied a method of liquid phase sampling taken from a high temperature autoclave with following analysis of the sample. The covered temperature and pressure ranges were 200°C–350°C and 15.9–168.6 bar. Their measurement error is uncertain as it is not reported or listed. Blount carried out the $BaSO_4$ solubility measurements [29] using hydrothermal equipment previously described in Ref. [30]. A method for sampling the liquid phase was developed to avoid corrosion and achieve equilibrium in the aqueous phase. The studies were done at temperatures 189°C–279°C and pressures 92–1010 bar. Their measurement error is also uncertain as it is not reported.

Therefore, while a few barite solubility studies were carried out at temperatures above 300°C, precision and reliability of the generated data were not well-defined. Given that Blount and Strübel contradict each other by a factor of 2–5 in regions they overlap, more solubility measurements are needed to confirm the Strübel dataset.

5.5 THERMODYNAMIC MODEL FOR AQUEOUS SPECIES AT HIGH TEMPERATURES AND PRESSURES

Here, we present how a new aqueous thermodynamic model can be used to predict solubilities for a wide range of temperatures and pressures. The widely accepted EOS that is used to predict the solubility of minerals from room temperature up to elevated temperatures is not suitable in many high-temperature and high-pressure (HTHP) conditions. Computational software and databases, such as the SUPCRT92 database, rely on the Helgeson–Kirkham–Flowers (HKF) model. The SUPCRT database and HKF model can reliably predict the solubility of barite up to 300°C. However, this model does not include data related to the ion pair $BaSO_4$, which is a major contributor to the solubility predictions at higher temperatures and pressures. In their seminal works, the authors of the HKF model indicate that their model fails at high temperatures (> 300°C) and low fluid densities (< 0.8 g cm^3) [14,31,32]. Though oil and gas applications do reach these temperatures, geothermal systems more frequently operate in this region [33]. As such, even if we were to fit the ion pair, the HKF model is unable to reliably predict the solubility limits of barite for some hydrothermal systems even though it could in principle work at high temperatures if the fluid density is large enough. A demonstration in the inaccuracy of SUPCRT-based approaches at temperatures > 300°C is shown below through comparisons between SUPCRT modeling results and experimental values of $BaSO_4$ solubility.

Recent advancements in molecular statistical thermodynamics (MST) relating to molecular interactions can provide a more reliable description of how the standard Gibbs energy of solute molecules vary with temperature and pressure [31,34,35]. Previous works have shown that dipole–dipole and ion–dipole statistical thermodynamic expressions improve the predictive capabilities of standard thermodynamic models to a broader range of conditions [31,32]. Mineral solubility and ion pair association reactions ($NaOH^0$, HCl^0, KCl^0) with available experimental data were used to confirm the viability of this new approach for extending the range of solubility predictions beyond those set by the landmark HKF model [18,31,32,36].

Like the HKF model, this approach predicts the apparent standard molar Gibbs energy of formation for aqueous species. Note that the apparent Gibbs energy of formation differs from "typical" standard Gibbs energy of formation in that the standard Gibbs energy of species in their elemental form are only zero at 25°C and 1 bar. For example, molecules such as $H_2(g)$ will have non-zero apparent standard Gibbs energy values at elevated temperatures and pressures. Like the HKF model, the proposed MST-supported model uses the following equation:

$$\Delta G_j^0(T,P) = \Delta_f G_j^0(T_r,P_r) + \left[G_j^0(T,P) - G_j^0(T_r,P_r) \right], \qquad (5.18)$$

to determine the apparent standard Gibbs energy, $\Delta G_j^0(T,P)$, of the j-th species at given temperature (T) and pressure (P). Here, $\Delta_f G_j^0(T_r,P_r)$ is the standard Gibbs energy of the j-th species at the reference temperature of 25°C and pressure of 1 bar, and $\left[G_j^0(T,P) - G_j^0(T_r,P_r) \right]$ is the change in the standard partial molar Gibbs energy of the j-th species from changes in T and P. The second term is determined using a series of expressions that depend on temperature, pressure, properties of pure water, and five fitting parameters:

$$G_j^0(T,P) - G_j^0(T_r,P_r) = \left[\sum_k S_j^k(T_r,P_r) - S_j^0(T_r,P_r) \right](T - T_r) + A_j(P - P_r)$$

$$- C_j \left(T\ln\frac{T}{T_r} + T_r - T \right) + \left(\sum_k G_j^k(T,P) - \sum_k G_j^k(T_r,P_r) \right) \qquad (5.19)$$

The G_j^k terms are the MST-based k-type partial molar Gibbs energy contributions which include the hard sphere (HS) contributions, the electrostatic ion–dipole (ID) and dipole–dipole (DD) contributions, reference state contributions for changes in the standard state density (SS), and Gibbs energy changes for switching from unit mol fraction to unit molality reference states (MS). The model includes entropy terms for each of these contributions $\left(S_j^k \right)$, which can be determined by taking the derivative of each contribution with respect to temperature, and the standard partial molar entropy at the reference point $\left(S_j^0 \right)$. These were reported previously [31]. Lastly, A_j and C_j are two empirical constants used to account for the short-range interactions. The remaining three of the five fitting parameters are within the MST expressions used to determine the G_j^k terms.

Three MST expressions represent the bulk of the temperature and pressure dependencies. The remaining fitting parameters within these expressions are σ_i, σ_w, and p_j which are the ion–ion pair diameter, water molecule diameter, and ion pair dipole moment, respectively. The HS contribution, G_j^{HS}, is determined as [37]:

$$\frac{G_j^{HS}}{RT} = -\ln(1-\eta) + 3D\frac{\eta}{1-\eta} + 3D^2\left(\frac{\eta}{(1-\eta)^2} + \frac{\eta}{(1-\eta)} + \ln(1-\eta)^\circ\right)$$

$$-D^3\left(\frac{3\eta^3 - 6\eta^2 + \beta\eta}{(1-\eta)^3} + 2\ln(1-\eta)\right),$$

(5.20)

where R is the molar gas constant = 8.3145 J K^{-1}mol^{-1}, $\eta = \pi N_A \rho \sigma_w^3/6$, ρ is the molecular density, $D = \sigma_i/\sigma_w$, N_A is the Avogadro number, and $\beta = 1/(kT)$, where k is the Boltzmann constant = 1.3806 10^{-23}J K^{-1}. The mean spherical approximation (MSA) for an ion–dipole interaction was determined as [38]:

$$\frac{G_j^{ID}}{RT} = -N_A e^2 z_j^2 \frac{(1-1/\varepsilon)}{\sigma_j + \sigma_w(\beta_6/\beta_3)},$$

(5.21)

where z_i is the charge number of the ionic species, e is the elementary charge = 1.602 10^{-19}C, and ε is the permittivity of the pure solvent. ε is related to β_6 and β_3 through the well-known Wertheim equation given as [39]:

$$\varepsilon = \frac{\beta_{12}^4 \beta_3^2}{\beta_6^6} = \frac{(1 + b_2/12)^4 (1 + b_2/3)^2}{(1 - b_2/6)^6},$$

(5.22)

where b_2 is a parameter from MSA theory. The dipole–dipole electrostatic term, G_j^{DD}, is calculated as [34,40]:

$$\frac{G_j^{DD}}{RT} = \frac{-8N_A p_j^2(\varepsilon - 1)}{2\sigma_w^3\left(1 - \frac{\beta_{12}}{\beta_3}\right)\left(\frac{\beta_{12}}{\beta_6}\right)^3 + 2\varepsilon\left(\sigma_j + \sigma_w\frac{\beta_6}{\beta_3}\right)^3 + \left(\sigma_j + \sigma_w\frac{\beta_{12}}{\beta_6}\right)^3},$$

(5.23)

where p_j is the last fitting parameter. The Gibbs energy contribution due to the change in the standard state density is given as [35]:

$$\frac{G_j^{SS}}{RT} = -RT\ln(\rho RT/P^*),$$

(5.24)

where $P^* = 1$ bar is the pressure of the ideal gas reference state. Lastly, the Gibbs energy that counts on the difference between unit mol fraction and unit MS is given as [35]:

$$\frac{G_j^{MS}}{RT} = -RT\ln(M_s/b^0),$$

(5.25)

where M_s is the molar mass of the solvent with units of kg mol^{-1}. Equations 5.20–5.25 are MST-based equations and reference state conversion factors contained within the summation terms in Eq. 5.19. With values for the five fitted parameters as well as the standard partial molar Gibbs and entropy values at the reference state for an aqueous species, Eq. 5.19 can be used to reliably predict the apparent Gibbs energy of an aqueous species in HTHP conditions. If coupled with solid-phase standard Gibbs energy values, such as Eq. 5.5, these values can be used to determine solubility limits for HTHP conditions. Model parameters and thermodynamic properties at the reference temperature and pressure have been previously obtained for ten aqueous species and are presented in Tables 5.2 and 5.3 [23,24].

HTHP solubility and association constant predictions were compared to experimental data for minerals that had experimental data available in the range of conditions where the HKF model was known to fail. Quartz solubility comparisons were made between experimental and theoretical values ranging from 150°C to 500°C for pressures of 1–1,500 bar [31,32]. Agreement within a few percent was obtained across this range with solubility values varying from 1×10^{-3} to 30×10^{-3} mol kg^{-1} [26]. Ammonia, hydrochloric acid, potassium chloride, and sodium hydroxide association constants were within 10% of available experimental data up to 500°C and 2,000 bar [31]. Similarly, aluminum oxide and hydroxides (corundum and boehmite) systems were analyzed for a range of temperatures from 25°C to 600°C with pressures from 1 to 1,000 bar [36]. Agreement in HTHP conditions was obtained between experimental values [33] and these predictions [28] for temperatures from 300°C to 600°C at a pressure of 1,000 bar that were within the experimental uncertainty of $\pm 25\%$. The solubility values obtained for these HTHP conditions ranged from 10 to 100 μmol kg^{-1} depending on temperature and pressure [36,41].

TABLE 5.2
Fitted Parameters for the Molecular Statistical Thermodynamics (MST)-Based Thermodynamic Model

Species	A_j / a	σ_j / b	σ_w / b	p_j / c	C_j / d
Cl$^-$(aq)	2.336	2.999	1.372	0.000	−73.78
HCl^0(aq)	0.918	0.807	0.155	1.080	−326.81
K$^+$(aq)	0.615	5.284	1.122	0.000	−0.19
KCl^0(aq)	−3.499	2.999	1.372	35.637	−93.91
Na$^+$(aq)	−0.317	5.053	1.235	0.000	35.65
NaOH^0(aq)	−2.336	5.649	2.415	22.820	−74.99
NH$_3$^{0}(aq)	2.421	2.412	1.423	1.470	75.55
NH$_4$$^+$(aq)	1.449	5.316	1.022	0.000	53.05
OH$^-$(aq)	0.299	2.693	2.179	0.000	−89.08
SiO$_2$^{0}(aq)	2.188	1.184	2.624	1.337	−28.68

Units: a = J / (mol bar), b = Å, c = Debye, d = J / (mol K).

TABLE 5.3

Reference State Values and Charge Number for the Species Presented in Table 5.2

Species	$G_j^0(T_r,P_r)/e$	$S_j^0(T_r,P_r)/d$	z_j
Cl⁻(aq)	−131290	56.74	−1
HCl⁰(aq)	−134359	123.56	0
K⁺(aq)	−282462	101.04	1
KCl⁰ (aq)	−416338	94.31	0
Na⁺(aq)	−261881	58.41	1
NaOH⁰ (aq)	−421915	14.55	0
NH₃⁰(aq)	−26706	107.82	0
NH₄⁺(aq)	−79454	111.17	1
OH⁻(aq)	−157297	-10.71	−1
SiO₂⁰(aq)	−833411	56.48	0

Units: $d = $ J / (mol K), $e = $ J mol⁻¹

5.6 BARITE SOLUBILITY MODELS FOR HIGH TEMPERATURES AND PRESSURES

The MST-based model presented here was used to predict the solubility of barite in water for a wide range of temperatures and pressures. The use of this approach aims to address two issues: (1) a lack of ion pair Gibbs energy values that reliably predicts barite solubility for available experimental data and (2) a thermodynamic model valid for petroleum and hydrothermal systems not coverable by the HKF model which are relevant to energy applications. The needed model parameters were obtained by fitting known apparent Gibbs energy values for Ba^{2+}(aq) and SO_4^{2-}(aq). Following our previous approach for adapting available HKF model species to our MST-based model [31], values from 0°C to 300°C with pressures from 1 to 250 bar were used to determine the model parameters. These range of conditions were used because the ions covered by the HKF model were clearly defined in this region [31]. The limited amount of data available to fit $BaSO_4^0$(aq) required us to use the same approach we used to fit other species with limited influence at ambient conditions such as HCl⁰(aq), KCl⁰(aq), and NaOH⁰(aq) [31]. Briefly, all the experimental solubility data available at high temperatures [28,29], where the effect of the ion pair is expected to be most pronounced, were used to fit the standard Gibbs energy values of $BaSO_4^0$(aq). The values provided by Djamali et al., 2016 and others [16,19,20] dramatically overestimated the solubility of barite if combined with accepted Gibbs energy values from the SUPCRT92 database, so the ion pair dataset from the OLI Systems [1] database was used as it best fit experimental data while using ions from the internally consistent SUPCRT database. These values were then fit to obtain the necessary model parameters. The obtained model parameters for these species are presented in Table 5.4 and the reference thermodynamic properties are given in Table 5.5.

TABLE 5.4
Fitted Parameters Needed to Predict Barite Solubility Using the Molecular Statistical Thermodynamics (MST)-Based Thermodynamic Model

Species	A_j / a	σ_j / b	σ_w / b	p_j / c	C_j / d
$Ba^{2+}(aq)$	−1.941	7.782	0.924	0.000	−63.85
$SO_4^{2-}(aq)$	2.113	5.206	1.728	0.000	−167.53
$BaSO_4^0(aq)$	−16.887	1.670	3.478	1.040	−75.40

Units: a = J/(mol bar), b = Å, c = Debye, d = J/(mol K).

TABLE 5.5
Reference State Values and Charge Number for the Species Presented in Table 5.4

Species	$G_j^0(T_r, P_r)$ / e	$S_j^0(T_r, P_r)/d$	z_j
$Ba^{2+}(aq)$	−560782	9.62	2
SO_4^{2-}	−744459	18.83	−2
$BaSO_4^0(aq)$	−1290320	33.30	0

Units: d = J/(mol K), e = J mol^{-1}

Significantly better agreement was obtained between the experimental data and the new model predictions relative to model predictions by Ref. [16], Ref. [15], and OLI Systems. OLI Systems predictions were shown as their predictions were the second closest to capturing the solubility trends. Barite solubility values from experimental measurements and model predictions are presented in Figures 5.4 and 5.5.

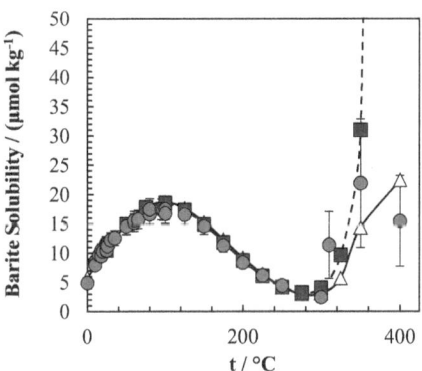

FIGURE 5.4 Barite solubility comparisons for P_{sat} to 350°C and 500 bar at 400°C (●, experimental data [15,28]; △, this work; and ■, OLI Systems).

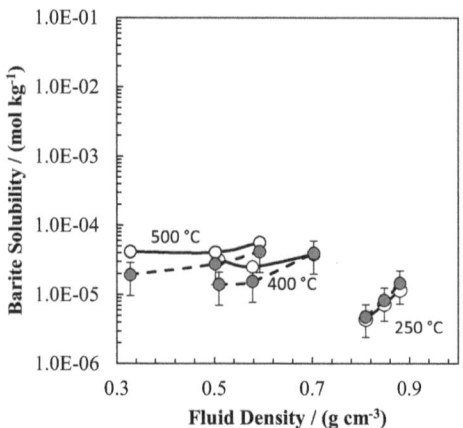

FIGURE 5.5   Barite solubility comparisons between the MST model and experimental data for HTHP conditions (●, experimental data [28,29] and ○, this work.)

As temperatures increased above 300°C, the dominant barium-containing species switched from Ba^{2+}(aq) to $BaSO_4^0$(aq). In general, though the trend in solubility varies considerably with temperature, the new model was able to follow the available data with an accuracy within experimental uncertainty. This is particularly true because the available data sets contained large uncertainties (>20%), in addition to some sizable disagreements between sets (>200%). Tables 5.6–5.8 provide data from the MST model for the conditions examined.

The considerable disagreements between measurements suggest that there are likely errors within one or more of the studies for this system. As such, the reliability of our predictions can be further verified if more experimental data were to become available to help us discern which sets best represent this system.

5.7   SECTION SUMMARY

Scaling and mineral precipitation in produced waters is a problem that plagues the oil, natural gas, and high enthalpy geothermal community. A thermodynamic analysis of the barite-water system provides a means to predict if a mineral will be dissolved or precipitated in a wide range of conditions. Predominance diagrams and speciation models provide a means to understand how changes in solution composition impact mineral behavior for a given system. At ambient conditions, increases in pH above 2 increases the solubility of barite from ~1 to 10 μmol kg^{-1} but further increases to pH beyond 5 had little effect. As temperature increases, the solubility of barite in pure water increases to about 17 μmol kg^{-1} but then decreases to about 5 μmol kg^{-1} by 300°C at saturated pressures. Available HTHP data indicate that the solubility increases at higher temperatures and pressures again to 10s of μmol kg^{-1} at temperatures >300°C. Data available indicate that the $BaSO_4^0$(aq) ion pair plays a key role in governing the solubility limit at HTHP conditions. Therefore, a new MST-based model was extended to provide reliable predictions of barite solubility for a wide range of conditions by providing $BaSO_4^0$(aq), Ba^{2+}(aq), and SO_4^{2-}(aq) Gibbs energy values in HTHP conditions.

TABLE 5.6

Standard Apparent Partial Molar Gibbs Energy Values from the Molecular Statistical Thermodynamics (MST)-Based Models

t / °C	P / Bar	$BaSO_4^0(aq)$ / kJ mol⁻¹	$Ba^{2+}(aq)$ / kJ mol⁻¹	SO_4^{2-} (aq) / kJ mol⁻¹	$BaSO_4(s)$ / kJ mol⁻¹
0	1.0	−1290.59	−560.49	−743.70	−1358.89
25	1.0	−1290.31	−560.78	−744.46	−1362.09
50	1.0	−1292.06	−561.01	−744.76	−1365.50
75	1.0	−1295.16	−561.16	−744.65	−1369.12
100	1.0	−1299.30	−561.24	−744.14	−1372.94
125	2.3	−1304.17	−561.24	−743.24	−1376.95
150	4.8	−1309.57	−561.18	−741.92	−1381.14
175	8.9	−1315.35	−561.03	−740.18	−1385.50
200	15.5	−1321.37	−560.78	−737.93	−1390.02
225	25.5	−1327.57	−560.40	−735.08	−1394.68
250	39.7	−1333.94	−559.82	−731.47	−1399.49
275	59.4	−1340.48	−558.94	−726.80	−1404.42
300	85.8	−1347.22	−557.48	−720.49	−1409.46
325	120.0	−1354.29	−554.88	−711.33	−1414.61
350	163.4	−1362.03	−549.22	−695.48	−1419.83
400	500	−1369.50	−546.03	−687.07	−1429.40

TABLE 5.7

Standard Apparent Partial Molar Gibbs Energy Values from the Molecular Statistical Thermodynamics (MST)-Based Model for High-Temperature and High-Pressure (HTHP) Conditions

t / °C	P / Bar	$BaSO_4^0(aq)$ / kJ mol⁻¹	$Ba^{2+}(aq)$ / kJ mol⁻¹	SO_4^{2-} (aq) / kJ mol⁻¹	$BaSO_4(s)$ / kJ mol⁻¹
189	404	−1321.38	−561.25	−738.93	−1385.97
190	904	−1323.33	−561.83	−738.73	−1383.54
247	92	−1334.14	−559.74	−731.73	−1398.62
248	522	−1335.37	−560.32	−732.89	−1396.42
249	1010	−1337.18	−560.91	−733.37	−1394.23
279	996	−1344.38	−560.31	−729.90	−1400.36
310	99	−1349.95	−555.99	−716.18	−1411.51
350	171	−1361.27	−548.28	−694.26	−1419.80
400	382	−1371.52	−537.70	−669.59	−1430.02
400	500	−1369.50	−546.03	−687.07	−1429.40

(Continued)

TABLE 5.7 (*Continued*)
**Standard Apparent Partial Molar Gibbs Energy Values from the Molecular
Statistical Thermodynamics (MST)-Based Model for High-Temperature and
High-Pressure (HTHP) Conditions**

t / °C	P / Bar	$BaSO_4^0(aq)$ / kJ mol^{-1}	$Ba^{2+}(aq)$ / kJ mol^{-1}	SO_4^{2-} (aq) / kJ mol^{-1}	$BaSO_4(s)$ / kJ mol^{-1}
400	1070	−1368.95	−554.38	−706.93	−1426.43
400	2100	−1373.55	−557.91	−714.74	−1421.07
500	586	−1388.76	−477.61	−561.64	−1453.04
500	914	−1385.72	−531.00	−651.62	−1451.33
500	1276	−1385.79	−543.39	−676.13	−1449.44
600	874	−1398.50	−467.58	−539.75	−1477.38
600	1084	−1398.17	−499.85	−591.16	−1476.28

TABLE 5.8
**Comparison of Molecular Statistical Thermodynamics (MST)-Based Model
Predictions to Available Experimental Data [15,28]**

t / °C	P / bar	$b_{s,model}$ / μmol kg^{-1}	$b_{s, exp}$ / μmol kg^{-1}
0	1	5.88	5.18
25	1	10.47	11.10
50	1	14.89	14.81
75	1	17.82	16.65
100	1.0	18.68	16.93
125	2.3	17.63	15.98
150	4.8	15.25	14.14
175	8.9	12.24	11.71
200	15.5	9.14	9.04
225	25.5	6.38	6.44
250	39.7	4.25	4.24
275	59.4	3.02	2.77
300	85.8	3.14	2.35
325	120.0	5.73	3.30
350	163.4	14.33	21.90
400	500	22.48	15.45

5.8 NOMENCLATURE

Symbol	Units	Name
a_j	Unitless	Species activity
A_j	J mol^{-1} Pa^{-1}	Empirical short-range interaction parameter
b_j	mol kg^{-1}	Species concentration, molality
b^0	mol kg^{-1}	Standard molality
b_2	Unitless	Dipole–dipole interaction parameter from MSA theory
C_j	J mol^{-1}·K^{-1}	Empirical short-range interaction parameter
D	Unitless	(σ_j/σ_w), ratio of diameters
e	C	Elementary charge of an electron
$\Delta_f G^0$	J mol^{-1}	Standard Gibbs energy of formation
$\Delta_r G^0$	J mol^{-1}	Standard Gibbs energy of reaction
$\Delta G_j^0 (T, P)$	J mol^{-1}	Apparent standard partial molar Gibbs energy
$\Delta_f G_j^0 (T_r, P_r)$	J mol^{-1}	Reference standard partial molar Gibbs energy of formation
$G_j^0 (T, P)$	J mol^{-1}	Standard partial molar Gibbs energy
$G_j^0 (T_r, P_r)$	J mol^{-1}	Reference standard partial molar Gibbs energy
$G_j^k (T_r, P_r)$	J mol^{-1}	Reference molecular statistical Gibbs energy contribution
$G_j^k (T, P)$	J mol^{-1}	Molecular statistical Gibbs energy contribution
G_j^{HS}	J mol^{-1}	Hard sphere Gibbs energy contribution
G_j^{ID}	J mol^{-1}	Ion–dipole Gibbs energy contribution
G_j^{DD}	J mol^{-1}	Dipole–dipole Gibbs energy contribution
G_j^{SS}	J mol^{-1}	Standard state solution density Gibbs energy contribution
G_j^{MS}	J mol^{-1}	Standard state unit molality Gibbs energy contribution
K	J K^{-1}	Boltzmann's constant
IAP	Unitless	Ion activity product
K_{eq}	Unitless	Equilibrium constant
K_{sp}	Unitless	Solubility product constant
M_s	kg mol^{-1}	Molar mass of solvent
N_A	mol^{-1}	Avogadro's number
P_r	Pa	Reference pressure
$P*$	Pa	Ideal gas reference state pressure
P	Pa	Pressure
p_j	C m	MST model parameter, solute dipole moment
R	J mol^{-1} K^{-1}	Molar gas constant
SI	Unitless	Scaling index
$S_j^0 (T_r, P_r)$	J mol^{-1} K^{-1}	Reference standard partial molar entropy
$S_j^k (T_r, P_r)$	J mol^{-1} K^{-1}	Reference molecular statistical entropy contribution

(Continued)

Symbol	Units	Name
T_r	K	Reference temperature
T	K	Thermodynamic temperature
T	°C	Temperature
z_i	Unitless	Charge number of an ion
α_0	mol kg^{-1}	Empirical parameter, barite solubility equation
α_1	mol kg^{-1}°C^{-1}	Empirical parameter, barite solubility equation
α_2	mol kg^{-1}°C^{-2}	Empirical parameter, barite solubility equation
α_3	mol kg^{-1}°C^{-3}	Empirical parameter, barite solubility equation
$\beta = 1/kT$	J^{-1}	Thermodynamic beta, MST theory
β_3	Unitless	$(1 + b_2/3)$, Wertheim equation parameter
β_6	Unitless	$(1 - b_2/6)$, Wertheim equation parameter
β_{12}	Unitless	$(1 + b_2/12)$, Wertheim equation parameter
γ_i	Unitless	Species activity coefficient
E	Unitless	Permittivity of water
H	Unitless	$\left(\pi\sigma_w^3 n/6\right)$, hard sphere equation parameter
P	kg m^{-3}	Density of water
σ_j	m	MST model parameter, solute diameter
σ_w	m	MST model parameter, solvent diameter
Ω	Unitless	Scaling tendency

REFERENCES

1. OLI Systems, OLI Studio Stream Analyzer User Guide. V 9.5, 2017.
2. M. Brown, Full scale attack, *Review* 30 (1998) 30–32.
3. M. Amiri, J. Moghadasi, The effect of temperature on calcium carbonate scale formation in Iranian oil reservoirs using OLI ScaleChem software, *Pet. Sci. Technol.* 30 (2012). doi:10.1080/10916461003735145.
4. A.N.P. Vankeuren, J.A. Hakala, K. Jarvis, J.E. Moore, Mineral reactions in shale gas reservoirs: Barite scale formation from reusing produced water as hydraulic fracturing fluid, *Environ. Sci. Technol.* 51 (2017) 9391–9402. doi:10.1021/acs.est.7b01979.
5. M.S. Kamal, I. Hussein, M. Mahmoud, A.S. Sultan, M.A.S. Saad, Oilfield scale formation and chemical removal: A review, *J. Pet. Sci. Eng.* 171 (2018) 127–139. doi:10.1016/j.petrol.2018.07.037.
6. A.A. Olajire, A review of oilfield scale management technology for oil and gas production, *J. Pet. Sci. Eng.* 135 (2015) 723–737. doi:10.1016/j.petrol.2015.09.011.
7. A.T. Kan, J.Z. Dai, G. Deng, G. Ruan, W. Li, K. Harouaka, Y.T. Lu, X. Wang, Y. Zhao, M.B. Tomson, Recent advances in scale prediction, approach, and limitations, *Soc. Pet. Eng. SPE Int. Oilf. Scale Conf. Exhib.* 2018 (2018) 20–21. doi:10.2118/190754-pa.
8. A.D. Jew, M.K. Dustin, A.L. Harrison, C.M. Joe-wong, D.L. Thomas, K. Maher, G.E. Brown, J.R. Bargar, Impact of organics and carbonates on the oxidation and precipitation of iron during hydraulic fracturing of shale, *Energy and Fuels.* 31 (2017). doi:10.1021/acs.energyfuels.6b03220.
9. M.K. Dustin, J.R. Bargar, A.D. Jew, A.L. Harrison, C. Joe-wong, D.L. Thomas, G.E. Brown, K. Maher, Shale Kerogen: Hydraulic fracturing fluid interactions and contaminant release, *Energy and Fuels.* 32 (2018) 8966–8977. doi:10.1021/acs.energyfuels.8b01037.

10. S. Blumsack, Production Decline for Shale Gas Wells, EME 801 Energy Mark. Policy Regul. (2021). https://www.e-education.psu.edu/eme801/node/521.

11. Q. Li, A.D. Jew, A. Kohli, K. Maher, G.E. Brown, J.R. Bargar, Thicknesses of chemically altered zones in shale matrices resulting from interactions with hydraulic fracturing fluid, *Energy and Fuels.* 33 (2019) 6878–6889. doi:10.1021/acs.energyfuels.8b04527.

12. E.R. Cohen, T. Cvitas, J.G. Frey, B. Holmström, K. Kuchitsu, R. Marquardt, I. Mills, F. Pavese, Quantities, units and symbols in physcial chemistry, in: *IUPAC Green B.*, 3rd Edition, IUPAC & RSC Publishing, Cambridge, 2008: pp. 1–236.

13. G. Anderson, Rock-water systems, in: *Thermodynamics of Natural Systems*, 2nd Edition, Cambridge University Press, Cambridge, 2005: pp. 473–497.

14. J.W. Johnson, E.H. Oelkers, H.C. Helgeson, SUPCRT92: A software package for calculating the standard molal thermodynamic properties of minerals, gases, aqueous species, and reactions from 1 to 5000 bars and 0 to 1000°C, *Comput. Geosci.* 18 (1992) 899–947.

15. B.S. Krumgalz, Temperature dependence of mineral solubility in water. Part 3. Alkaline and alkaline earth sulfates, *J. Phys. Chem. Ref. Data* 2018. doi:10.1063/1.5031951.

16. E. Djamali, W.G. Chapman, K.R. Cox, A systematic investigation of the thermodynamic properties of aqueous barium sulfate up to high temperatures and high pressures, *J. Chem. Eng. Data* 61 (2016) 3585–3594. doi:10.1021/acs.jced.6b00506.

17. B. Johnson, L. Kanagy, J. Rodgers, J. Castle, Chemical, physical, and risk characterization of natural gas storage produced waters, *Water, Air Soil Pollut* 191 (2008) 33–54.

18. S.N. Lvov, N.N. Akinfiev, A.V. Bandura, F. Sigon, G. Perboni, Multisys: Computer code for calculating multicomponent equilibria in high-temperature subcritical and supercritical aqueous systems, in: P.R. Tremaine, P.G. Hill, D.E. Irish, P.V. Palakrishnan (Eds.), *Steam, Water, and Hydrothermal Systems: Physics and Chemistry Meeting the Needs of Industry*, NRC Press, Ottawa, 2000: pp. 866–873.

19. C. Monnin, A thermodynamic model for the solubility of barite and celestite in electrolyte solutions and seawater to 200°C and to 1 kbar, *Chem. Geol.* 153 (1999) 187–209. doi:10.1016/S0009-2541(98)00171-5.

20. A.R. Felmy, D. Rai, J.E. Amonette, The solubility of barite and celestite in sodium sulfate: Evaluation of thermodynamic data, *J. Solution Chem.* 19 (1990) 175–185. doi:10.1007/BF00646611.

21. S.N. Lvov, *Introduction to Electrochemical Science and Engineering*, 2nd edition, CRC Press Inc, Boca Raton, FL, 2022.

22. D.M. Hall, J.R. Beck, E. Brand, M. Ziomek-Moroz, S.N. Lvov, Copper-copper sulfate reference electrode for operating in high temperature and high pressure aqueous environments, *Electrochim. Acta* 221 (2016) 96–106. doi:10.1016/j.electacta.2016.10.143.

23. D.M. Hall, N.N. Akinfiev, E.G. LaRow, R.S. Schatz, S.N. Lvov, Thermodynamics and Efficiency of a CuCl(aq)/HCl(aq) Electrolyzer, *Electrochim. Acta* 143 (2014) 70–82. doi:10.1016/j.electacta.2014.08.018.

24. R. Feng, J. Beck, D.M. Hall, A. Buyuksagis, M. Ziomek-Moroz, S.N. Lvov, Effects of CO_2 and H_2S on corrosion of martensitic steels in brines at low temperature, *Corrosion* 74 (2018) 276–287. doi: 10.5006/2406.

25. V. Valyashko, *Hydrothermal Properties of Materials*, Wiley-VCH, Weinheim, 2008.

26. G.W. Morey, J.M. Hesselgesser, The solubility of some minerals in superheated steam at high pressures, *Econ. Geol.* 46 (1951) 821–835.

27. B. Strübel, Zur Kenntnis und genetischen Bedeutung des Systems $BaSO_4 - NaCl - H_2O$. (To the knowledge and genetic importance of the system $BaSO_4 - NaCl - H_2O$.), *Neues Jahr. Miner. Monatsh.* 7/8 (1967) 223–233.

28. H. Gundlach, D. Stoppel, G. Strubel, The hydrothermal solubility of barite, *Proc. 24th Intern. Geol. Congr.* 10 (1972) 291–229.

29. C. Blount, Barite solubilities and thermodynamic quantities up to 300°C and 1400 bars, *Amer. Mineral* 62 (1977) 942–957.

30. F.W. Dickson, C.W. Blount, G. Tunell, Use of hydrothermal solution equipment to determine the solubility of anhydrite in water from 100°C to 275°C and from 1 bar to 1,000 bars pressure, *Amer. J. Sci.* 261 (1963) 61–78.

31. S.N. Lvov, D.M. Hall, A.V. Bandura, I.K. Gamwo, A semi-empirical molecular statistical thermodynamic model for calculating standard molar Gibbs energies of aqueous species above and below the critical point of water, *J. Mol. Liq.* 270 (2018) 62–73. doi:10.1016/j.molliq.2018.01.074.

32. S.N. Lvov, D.M. Hall, I.K. Gamwo, Molecular Statistical Thermodynamics to Model Quartz Solubility in Ultra High-Enthalpy Geothermal Systems, in: *4th Annual International Conference on Geology & Earth Science (GEOS 2015), Global Science and Technology Forum (GSTF)*, 2015. doi:10.5176/2251-3353_GEOS15.44.

33. T. Phi, R. Elgaddafi, M. Al Ramadan, K. Fahd, R. Ahmed, C. Teodoriu, Well integrity issues: Extreme high-pressure high-temperature wells and geothermal wells a review, in: *Society of Petroleum Engineers - SPE Thermal Well Integrity and Design Symposium 2019, TWID 2019.* (2019). doi:10.2118/198687-ms.

34. A.V. Bandura, M.F. Holovko, S.N. Lvov, The chemical potential of a dipole in dipolar solvent at infinite dilution: Mean spherical approximation and Monte Carlo simulation, *J. Mol. Liq.* 270 (2018) 52–61. doi:10.1016/j.molliq.2018.01.015.

35. S.N. Lvov, V.A. Umniashkin, A. Sharygin, M.F. Holovko, The molecular statistical theory of infinitely dilute solutions based on the ion-dipole model with Lennard-Jones interaction, *Fluid Phase Equilib.* 58 (1990) 283–305. doi:10.1016/0378-3812(90)85137-Y.

36. S.N. Lvov, D.D. Macdonald, Thermodynamic computer simulation of hydrothermal synthesis of oxides in supercritical aqueous environments, in: K.E. Spear (Ed.), *High Temperature Materials Chemistry*, Electrochemical Society, Pennington, NJ, 1997: pp. 472–479.

37. G.A. Mansoori, N.F. Carnahan, K.E. Starling, T.W. Leland Jr., Equilibrium thermodynamic properties of the mixture of hard spheres, *J. Chem. Phys.* 54 (1971) 1523–1525. doi:10.1063/1.1675048.

38. L. Blum, F. Vericat, W.R. Fawcett, On the Mean Spherical approximation for hard ions and dipoles, *J. Chem. Phys.* 96 (1992) 3039.

39. M.S. Wertheim, Exact solution of the mean spherical model for fluids of hard spheres with permanent electric dipole moments, *J. Chem. Phys.* 55 (1971) 4291. doi:10.1063/1.1676751.

40. R. Richert, Solvation energy of ions and dipoles in a finite number of solvent shells, *J. Phys. Condens. Matter.* 8 (1996) 6185–6190. doi:10.1088/0953-8984/8/34/008.

41. K.V. Ragnarsdóttir, J.V. Walther, Experimental determination of corundum solubilities in pure water between 400–700°C and 1–3 kbar, *Geochim. Cosmochim. Acta.* 49 (1985) 2109–2115. doi:10.1016/0016-7037(85)90068-7.

6 Membrane Technologies and Applications for Produced Water Treatment

Xiaoyi Chen and Haiqing Lin
University at Buffalo

Fan Shi, Kevin Resnik, and Shouliang Yi
National Energy Technology Laboratory,
NETL Support Contractor

CONTENTS

6.1   INTRODUCTION

The last decade has witnessed tremendous growth in hydraulic fracturing of oil and gas wells, increasing energy production, and transforming the energy industry in the United States. However, such practices consume a large quantity of freshwater.[1,2] Many hydraulic fracturing wells are in drought-prone areas, such as Colorado, Texas, and California, where the growth of energy production itself has also been

DOI: 10.1201/9781003091011-6

constrained by the lack of water. While consuming freshwater, hydraulic fracturing also produces significant amounts of wastewater or produced water, which accounts for 10%–25% of the water injected into the wells.

The produced water generally contains inorganic ions (such as Na^+, Ca^{2+}, and Mg^{2+}) and organic additives used in hydraulic fracturing such as foaming agents, biocides, and corrosion inhibitors.[3–5, 40, 41] The composition of the produced water is specific to each reservoir and varies with production time.[44] For example, the produced water from Marcellus shale contains 4–7600 mg L^{-1} suspended solid, 680–345,000 mg L^{-1} total dissolved solid, 1–1530 mg L^{-1} total organic carbon, and 4.6–802 mg L^{-1} oil/grease.[3,6–8]

Currently, produced water is usually pretreated and trucked away for disposal or deep well reinjection, which is energy-intensive and expensive and presents a potential impact on freshwater quality.[9–11] Membranes have emerged as one of the leading technologies to treat the produced water for reuse due to high energy efficiency, small footprint, and low cost.[12,13] Membranes allow pure water to permeate through for reuse and reject contaminants such as dissolved solids and organic matter. The retentate with significantly reduced volume can then be disposed or reinjected into deep wells. Membranes with a range of pore sizes have been explored, including microfiltration (MF), ultrafiltration (UF), nanofiltration (NF), and reverse osmosis (RO), as shown in Table 6.1. Membrane processes for the treatment of produced water can be classified based on different driving forces, such as hydraulic pressure, osmotic pressure, and temperature. Most membranes for water treatment are made of polymers, and ceramic membranes are an attractive alternative in harsh conditions. Based on the requirement for the permeate quality and the constituents of the produced water, a membrane process can be used individually or combined with other processes.[43]

TABLE 6.1
Overview of Membrane Technologies for Produced Water Treatment, Including Separation Mechanism and Maturation Status

Membrane Technology	Separation Mechanism	Driving Force	Maturation
Nanofiltration (NF)/reverse osmosis (RO)	Solution-diffusion	Hydraulic pressure	Bench scale[14–20] Pilot scale[21,22] Commercial scale[23,24]
Microfiltration (MF)/ ultrafiltration (UF)	Size-sieving	Hydraulic pressure	Bench scale[15,16,20,25] Pilot scale[22]
Forward osmosis (FO)	Solution-diffusion	Osmotic pressure	Bench scale[26–28] Pilot scale[29–33]
Membrane distillation (MD)	Evaporation, diffusion	Temperature	Bench scale[25,34–36] Pilot scale[37]

6.2 PRESSURE-DRIVEN MEMBRANE SEPARATION

6.2.1 MF AND UF MEMBRANES

MF and UF membranes operate at low pressures, where the hydraulic pressure drives the water through the membrane while the pollutants larger than the membrane pore size are retained.[39] MF membranes have pore sizes of 0.1–3 μm and remove suspended solids, while UF membranes having pore sizes of 1–50 nm can remove large molecules such as proteins, starch, viruses, colloid silica, grease, oil, and other organic matters.

Polymeric and ceramic MF membranes have been widely used to reduce water turbidity and suspended particles. Polymeric membranes show a high removal efficacy and low energy requirement and are generally cost-effective. Some drawbacks include the ease of fouling by oil (which decreases the flux and filtration capacity) and sensitivity to both polar and chlorinated solvents (which limit their applications).[42] By contrast, ceramic MF membranes have great mechanical strength, excellent separation performance, and great thermal and chemical resistance.[39,46–52] For example, ceramic membranes composed of ZrO_2 were used to treat produced water with flocculation as pretreatment.[52] The pretreatment decreased the membrane fouling and increased water flux and permeate quality. The oil concentration reduced from 200 mg L^{-1} in the feed to 8.7 mg L^{-1} in the permeate, and the treated produced water met the National Discharge Standard. A pilot-scale system containing a range of multichannel ceramic MF membranes were conducted to treat the produced water from the Arabian Gulf.[49] Silicon carbide (SiC)-based membranes provided a higher permeability than the titanium dioxide (TiO_2)-based membranes. Chemical cleaning-in-place (CIP) was applied between each run, and the effect of the various cleaning protocols on the flux was studied.[53] Backflushing during the CIP was shown to be the most influencing factor, and it retained the flux, though the membrane became hydrophobic with increasing number of the filtration cycles. Ozonation and photoperoxidation were also used as pretreatment of the produced water to migrate the membrane fouling.[47] The 3-h exposure to UV radiation of the produced water improved the permeate flux from 84 to 182 L m^{-2}h (LMH).

UF membrane systems can be mobile for on-site treatment. Common UF membrane materials include polyacrylonitrile (PAN), polyethersulfone (PES), polysulfone (PSf), cellulose acetate (CA), poly(vinylidene fluoride) (PVDF), aromatic polyamides, and poly(vinyl pyrrolidone) (PVP). One of the great challenges for UF membranes is fouling, resulting in pore-blocking and cake layer formation and thus decreased water permeance.[54,55] The fouling behavior depends on the foulants in the feed, membrane surface properties (such as pore sizes and interaction with the foulants), and operating conditions. The fouled membranes need periodic cleaning and eventually replacement to maintain productivity, which increases the operation cost. Various strategies have been explored to alleviate fouling, including feed pretreatment, selection of membranes without affinity toward foulants, and membrane surface modification to repel foulants.

One common strategy to improve antifouling properties of UF membranes is to modify the surface with hydrophilic materials, such as poly(ethylene glycol) (PEG), zwitterionic materials, and hydrophilic graphene oxide (GO), as the hydrophilic surface forms a hydration layer, a barrier for foulants to deposit onto the surface.[22,56–59] Figure 6.1a shows the covalent grafting of PEG-NH$_2$ using bio-adhesive polydopamine (PD) onto PSf membranes. The grafting often decreases pure-water permeance due to the additional transport resistance. However, with careful optimization of the coating layer thickness and improved antifouling properties, the modification dramatically improved water flux during the filtration of oil emulsion as a model foulant (cf. Figure 6.1b).[60] This facile modification with aqueous solutions was adapted to post-modify commercial UF membrane modules, which were then used to treat the produced water from the Barnett shale gas basin.[22] At the beginning of the filtration, both modules showed similar water fluxes of ~43 LMH (cf. Figure 6.1c), and the modified module exhibited water permeance 20% higher than the unmodified module (cf. Figure 6.1d). On the other hand, after the 60-h operation, the modified

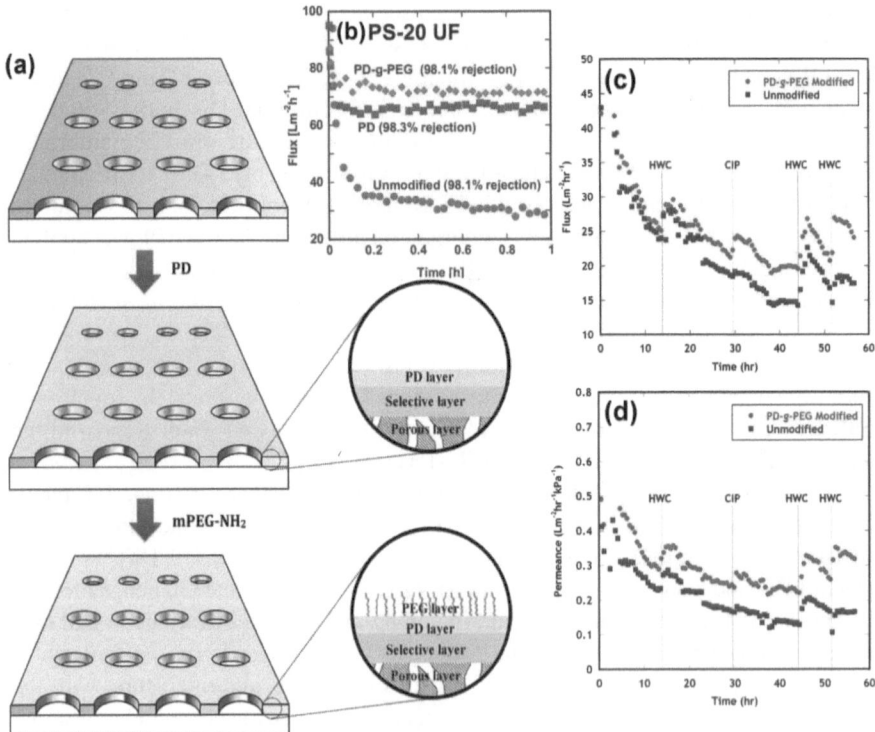

FIGURE 6.1 (a) Schematic illustration of membrane surface modification by PD-*g*-PEG. (b) Comparison of fouling behavior of the unmodified and modified PSf UF membranes in the oil emulsion filtration.[60] (c) Water flux and (d) permeance over the operating time for the unmodified and PD-*g*-PEG-modified UF modules in a pilot test. HWC and HCP denote chemically enhanced hot water cleaning and clean-in-place, respectively.[22] (Copyright 2012 and 2013 Elsevier.)

module showed water permeance 50% higher than the unmodified module due to the improved antifouling properties from the PD-*g*-PEG grafting.

PEG can also be incorporated into the polymer during membrane formation to increase hydrophilicity. For example, PVDF membranes were fabricated using a vapor-induced phase separation (VIPS) technique with PEG as an additive in the casting solutions.[61] Figure 6.2a shows that the introduction of PEG changed the membranes from asymmetric, microvoid-containing structure to symmetric, microvoid-free structure. Increasing the air exposure time increased the granular sizes and the mean pore size because of the increased PEG diffusion to the membrane/water interface. The introduction of PEG increased pure-water permeance (cf. Figure 6.2b), and PVDF/PEG-30 exhibited the highest water permeance. When challenged with 1 g L^{-1} oil/water emulsion prepared from crude oil, PVDF/PEG-30 exhibited the highest permeance in a multicycle filtration with water cleaning every 30 minutes due to the improved hydrophilicity and increased pore size (Figure 6.2c).

Zwitterionic materials have been widely used for membrane surface modification to mitigate fouling owing to their strong hydrophilicity.[56,58,62,63] These materials contain both cations and anions (such as sulfobetaine, carboxybetaine, and phosphorylcholine) and form hydration layers via electrostatic interaction with water molecules. For example, zwitterionic 3-(3–4-dihydroxyphenyl)-l-alanine (L-DOPA) was coated on PES UF membranes by dip coating.[64] Figure 6.3a shows that the modification improved surface hydrophilicity (as indicated by the decreased water contact angle),

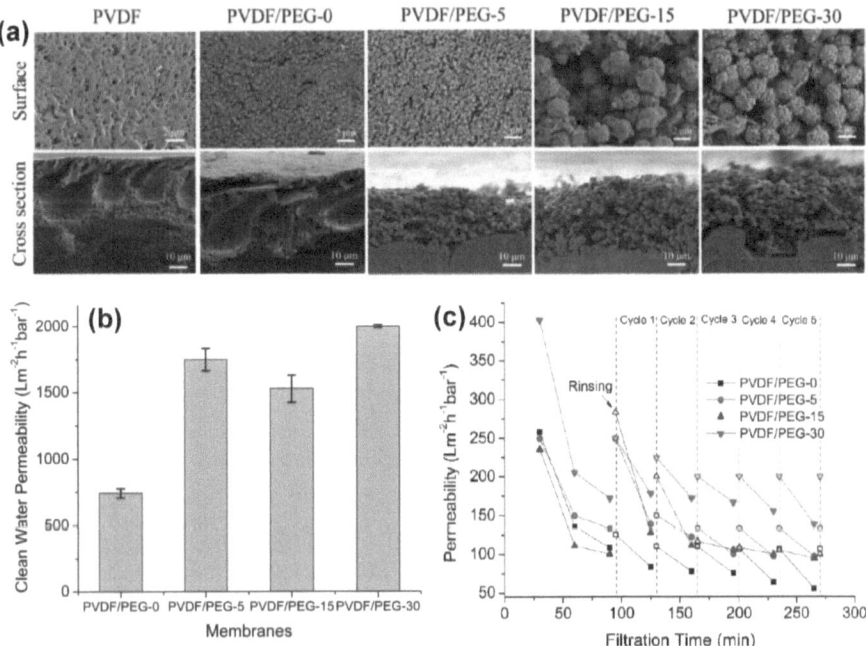

FIGURE 6.2 (a) SEM images of the surface and cross-section of the PVDF membranes fabricated with different PEG contents in the casting solutions: (b) pure-water permeance and (c) produced water permeance of the PVDF/PEG membranes.[61]

and increasing the coating time increased the hydrophilicity. Figure 3b displays that the modification increased negative charges on the surface. When challenged by produced water, the modified membranes showed higher normalized water permeance than the unmodified membrane (cf. Figure 6.3c), validating the improved antifouling properties. Figure 6.3d shows that surface modification also increased the flux recovery after 30-minutes of backflushing with deionized (DI) water.

Produced water may contain highly charged bituminous clays, causing severe fouling through the interaction with the charged membranes. To this end, the membrane surface can be modified with electrically neutral and hydrophilic chemicals. For example, tubular ceramic membranes were modified with hydrophilic poly(ethylene oxide)-based silane (PEOTMS, as shown in Figure 6.4a–c).[65] The surface modification increased the flux recovery ratio (FRR defined as the ratio of the first recorded flux after backflush to the last recorded flux before the next backflush) and water flux because of the reduced foulant adhesion to the membrane surface, as shown in Figure 6.4d and e.

6.2.2   NF AND RO MEMBRANES

NF and RO membranes are widely used to desalinate the produced water,[20,38] and state-of-the-art commercial membranes are thin-film composite (TFC) membranes with a polyamide selective layer, which rejects ions and dissolved organic carbon (DOC). NF membranes have an excellent rejection of divalent ions, such as over 90% for Ca^{2+}, Mg^{2+}, and SO_4^{2-}, and low rejection of monovalent ions, such as ~23% for Cl^- and Na^+.[21] By contrast, RO membranes can reject monovalent ions by 99% or higher, and thus, they can be used as the last step to obtain reusable water from the produced water.[2,38]

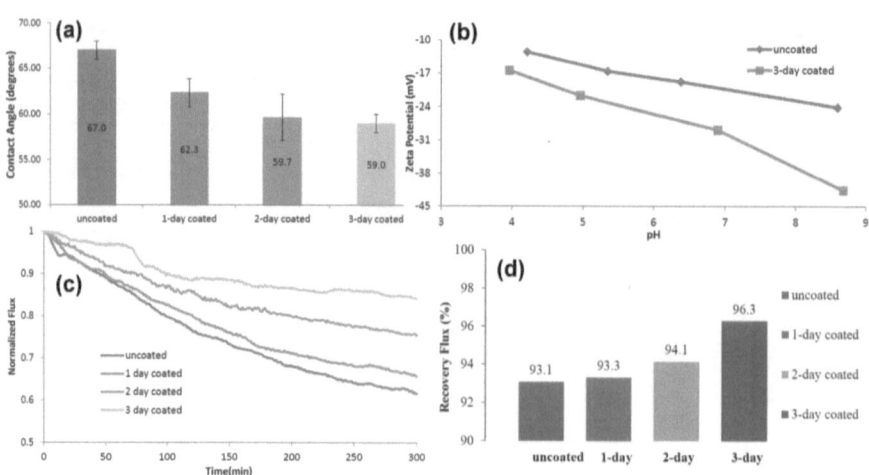

FIGURE 6.3 (a) Water contact angles of the uncoated and zwitterion-coated membranes. (b) Zeta potential of the uncoated and 3-day coated membranes at pH values from 4 to 9. Comparison of (c) the normalized flux and (d) recovery of water flux after cleaning for the uncoated and coated membranes when challenged by the produced water.[64]

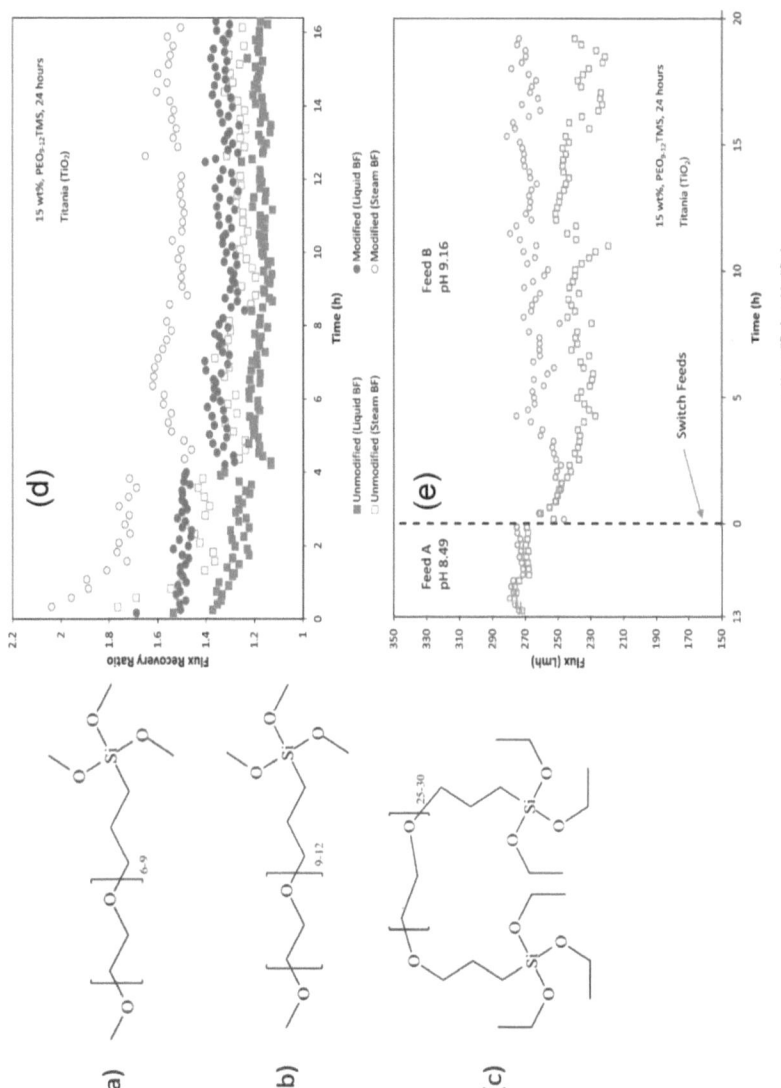

FIGURE 6.4 Chemical structure of (a) $PEO_{6-9}TMS$, (b) $PEO_{9-12}TMS$, and (c) $BIS-PEO_{25-30}TES$. (d) FRR and (e) permeate flux over the operating time for the modified membrane when challenged with produced water.[65] Copyright (2019) American Chemical Society.

Figure 6.5a shows an example UF-RO integrated process consisting of a submerged hollow fiber PVDF UF membrane and RO membrane to treat produced water.[66] The UF membrane was operated in a constant flux mode with periodic backwashing while the transmembrane pressure (TMP) was monitored. Figure 6.5b displays the change of TMP in the multicycle filtration of produced water at different fluxes. At low fluxes of 20 and 35 LMH, the TMP increased slowly and remained low, indicating that the fouling was mild and the backflushing was able to clean the membrane. At fluxes higher than 50 LMH, the TMP increased rapidly due to severe fouling.

Figure 6.5c shows the water flux of the RO membrane as a function of the pressure and water recovery rate. Water flux increased with increasing operating pressure (or driving force) and decreased with increasing water recovery rate due to the increased osmosis pressure of the retentate. For example, water flux at 2.5 MPa decreased to near zero at a recovery rate of 50%. Figure 6.5d shows that increasing the recovery rate decreased permeate quality due to the increased salinity in the retentate. Nevertheless, the produced water treated with the combined UF-RO process at over 2.5 MPa met the U.S. standards for surface water discharge.

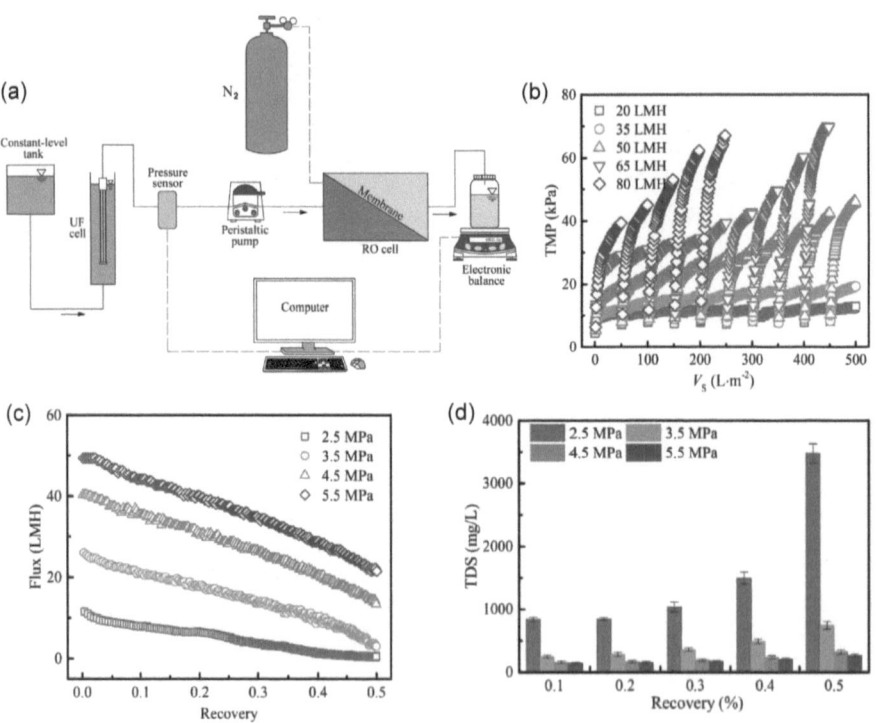

FIGURE 6.5 (a) Schematic diagram of the UF-RO system for treating Weiyuan shale gas flowback and produced water.[66] (b) TMP of the UF membrane as a function of the cumulative filtration volume normalized by membrane area at different fluxes. (c) Water flux and (d) TDS of the permeate of the RO membrane as a function of the feed pressure and water recovery rate. (Copyright 2018 RSC.)

Similar to UF membranes, NF/RO membranes are also subject to fouling, which can be mitigated by incorporation of hydrophilic materials to increase surface hydrophilicity.[67,68] Figure 6.6a illustrates the schematic of fabricating membranes from polybenzimidazole (PBI) and hydrophilic GO in three steps, including solution preparation, blade casting, and phase inversion.[59,67] The incorporation of GO in PBI improved surface hydrophilicity, surface roughness, and water permeance. Figure 6.6b–e compares the membrane surface of PBI-based membrane (M_{PBI}) and PBI-GO-based membrane $\left(M_{PBI}^{GO\,1.00}\right)$ after immersing in water for 45 days and

180 days. $M_{PBI}^{GO\,1.00}$ showed a virtually clean surface while M_{PBI} showed significant biofouling on the surface. In the multicycle filtration of an oil-in-water emulsion, $M_{PBI}^{GO\,1.00}$ exhibited 53% higher FRR than M_{PBI}. Though both membranes showed comparable water permeance at the beginning, $M_{PBI}^{GO\,1.00}$ exhibited water permeance

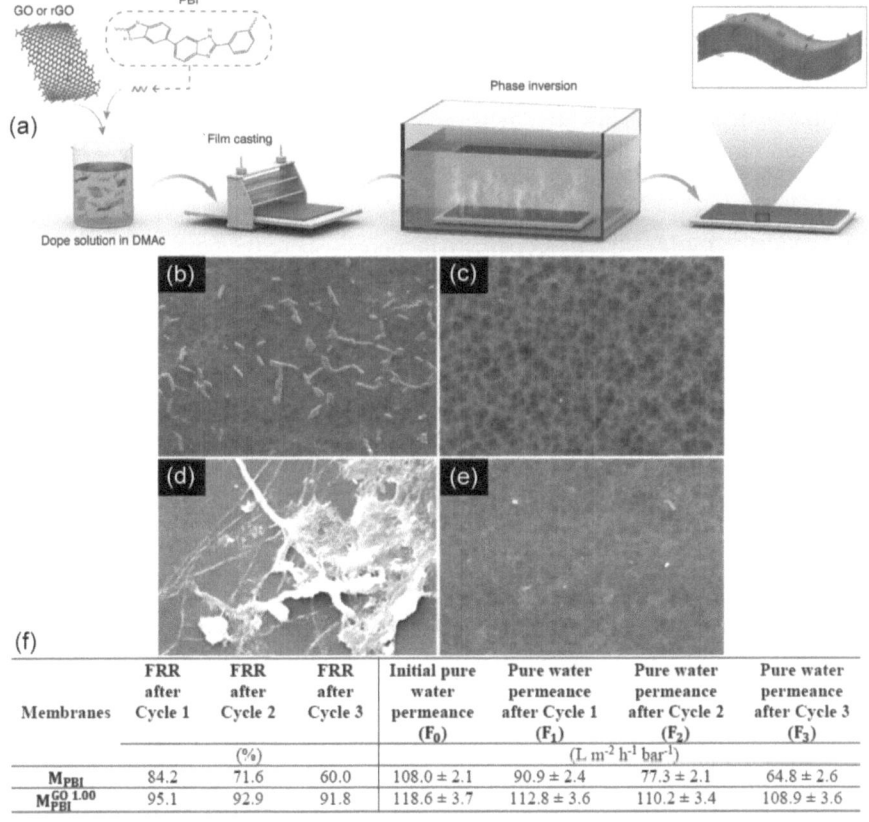

Membranes	FRR after Cycle 1	FRR after Cycle 2	FRR after Cycle 3	Initial pure water permeance (F_0)	Pure water permeance after Cycle 1 (F_1)	Pure water permeance after Cycle 2 (F_2)	Pure water permeance after Cycle 3 (F_3)
	(%)			(L m^{-2} h^{-1} bar^{-1})			
M_{PBI}	84.2	71.6	60.0	108.0 ± 2.1	90.9 ± 2.4	77.3 ± 2.1	64.8 ± 2.6
$M_{PBI}^{GO\,1.00}$	95.1	92.9	91.8	118.6 ± 3.7	112.8 ± 3.6	110.2 ± 3.4	108.9 ± 3.6

FIGURE 6.6 (a) Schematic of membrane fabrication, including solution mixing, casting, and phase inversion. SEM images at 1600× magnification of the fouled membrane surface for (b) M_{PBI} after immersing in water for 45 days, (c) $M_{PBI}^{GO\,1.00}$ after immersing in water for 45 days, (d) M_{PBI} after immersing in water for 180 days, and (e) $M_{PBI}^{GO\,1.00}$ after immersing in water for 180 days. (f) FRR and water permeance over the multicycle filtration.[67]

68% higher than M_{PBI} after the third cycle of filtration, demonstrating the improved antifouling properties by incorporating GO.

Another strategy to mitigate fouling is to design responsive membranes with properties tuned reversibly by external stimuli (such as pH, temperature, light, and ionic strength).[69–71] For example, magnetically responsive membranes were demonstrated to mitigate fouling when treating coalbed methane produced water.[72–74] Figure 6.7a shows the proposed mechanism to improve antifouling properties, where superparamagnetic nanoparticles (carboxyl shell Fe_3O_4) were covalently bonded to poly(2-hydroxyethyl methacrylate) (polyHEMA) grafted from NF-270 membrane surface. In an oscillating magnetic field, the orientation of the magnetic nanoparticles changed accordingly, creating small-scale mixing and alleviating the fouling. Figure 6.7b shows that the responsive membranes exhibited a slower decline of water permeance than the unmodified membrane. Figure 6.7c compares the permeate quality of the unmodified and responsive membranes. The permeate quality of the unmodified membrane dramatically decreased in the first 400 minutes, and the responsive membranes showed a much more stable rejection rate than the unmodified ones. The next step to move this technology for practical applications is to demonstrate its feasibility by long-term field tests of the responsive membranes with oscillating magnetic fields.

Novel membranes have also been developed as an alternative to conventional polyamide-based NF membranes, such as TFC membranes based on bicontinuous cubic (Q_I), lyotropic liquid crystals (LLC).[68] Figure 6.8a shows the schematic of the cross-linked Q_I LLC with hydrophilic nanopores, where the pore wall is generated by the cationic imidazolium head groups in the LLC monomer. Because of the high charge density and small pore size, the TFC Q_I membrane rejects ions while providing the passage for low-molecular-weight neutral organic solutes. The TFC Q_I membrane outperformed the NF90 membrane for treating the produced water. Figure 6.8b shows that the TFC Q_I membranes exhibited a much smaller decrease of water permeance than NF membranes. While the final water flux of the TFC Q_I membrane was lower than that of the NF90 at high pressures, TFC Q1 exhibited much higher water permeability than the polyamide layer of the NF90 (cf. Figure 6.8c). Figure 6.8d also compares the rejection of total dissolved solids (TDS) and DOC between the TFC Q_I and NF90 membranes. The salt-organic separation (SOS) efficiency is defined as the ratio of the TDS rejection to the DOC rejection. TFC Q1 membranes showed lower TDS rejection but higher SOS efficiency than NF90, which could be beneficial if the DOC is valuable to recover.

6.3   OSMOTICALLY DRIVEN MEMBRANE SEPARATION

While pressure-driven RO has been widely used in water purification treatment, osmotically driven forward osmosis (FO) has been recently applied in many industrial applications, including water/wastewater treatment and desalination.[75–79] In a typical FO process, a draw solution with high osmotic potential is used to draw water across a semipermeable membrane from a feed source. Advantages of FO over conventional RO membrane technology include lower hydrodynamic pressure and higher water recovery.

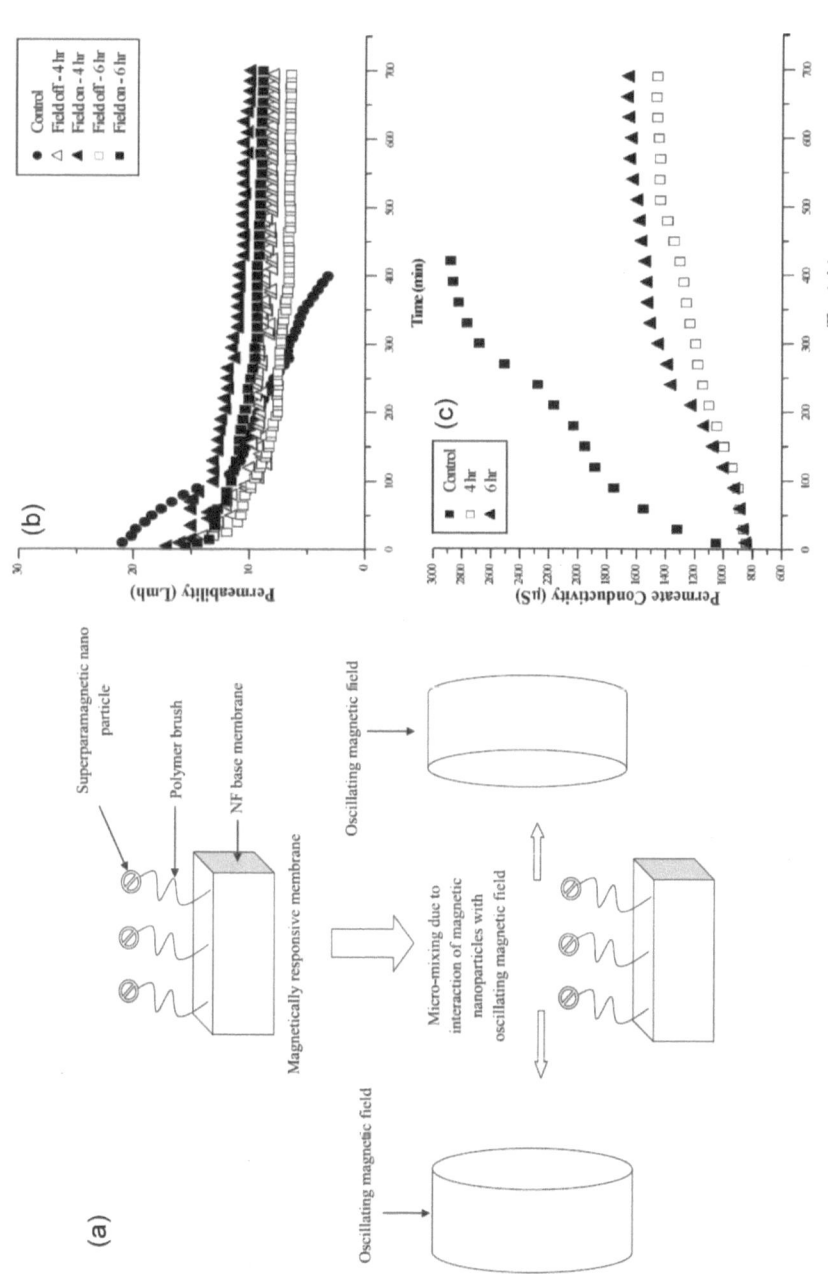

FIGURE 6.7 (a) Schematic of the antifouling effect of magnetically responsive NF membranes. (b) Water permeance and (c) permeate conductivity for the unmodified and responsive membranes when treating produced water. The magnetic field was intermittently applied for 10s in each 130s. (Copyright 2019 Elsevier.)

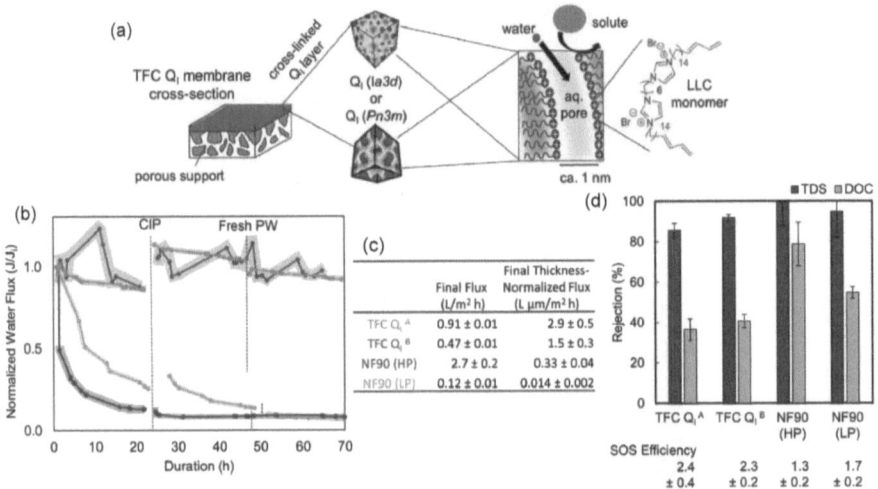

FIGURE 6.8 (a) Schematic of TFC Q_l membrane, including a porous support and a selective layer of cross-linked Q_l LLC. (b) Comparison of normalized water flux over time and (c) final flux after 66-h operation between TFC Q1 and NF90 membranes when treating the produced water. (d) Rejection of TDS and DOC by TFC Q_l membranes at 400 psi and NF90 membranes at 400 psi (HP) and 160 psi (LP). (Copyright Elsevier 2019.)

6.3.1 FUNDAMENTALS

Figure 6.9 shows the FO process, where water permeates across a semipermeable membrane between two aqueous solutions with different solute concentrations. The osmotic pressure difference, $\Delta\pi$, drives water from the solution with the lower osmotic potential to the solution with the higher osmotic potential in both FO and pressure-retarded osmosis (PRO), as opposed to the pressure-driven RO process.

For a typical FO system, a draw solution, such as brine, presents higher osmotic pressure than a feed with lower salinity. Then, the $\Delta\pi$ drives water flux without external hydraulic pressure ($\Delta p = p_{draw} - p_{feed} = 0$), as shown in Figure 6.10. The water flux first decreases with increasing Δp, under PRO ($\Delta p < \Delta\pi$), till $\Delta p = \Delta\pi$, where there is no water flux. As Δp keeps increasing ($\Delta p > \Delta\pi$), the reversed water flux increases (RO).[80] Water transport driving force in RO process need to overcome the osmotic pressure gradient that favors water transport from the permeate to the feed. Hence RO process requires more energy than FO.

The water flux in terms of osmotic and hydraulic pressures can be described by the following equation:

$$J_w = A\left(\Delta\pi - \Delta p\right)$$

where J_w is water flux and A is the hydraulic permeability of the membrane. In practical operations, J_w depends on complicated operating parameters, including the nature of the membrane,[81–83] types of feed,[45,78,84–86] and draw solutions[87–89] as well as the fluid dynamics within the process itself.[90]

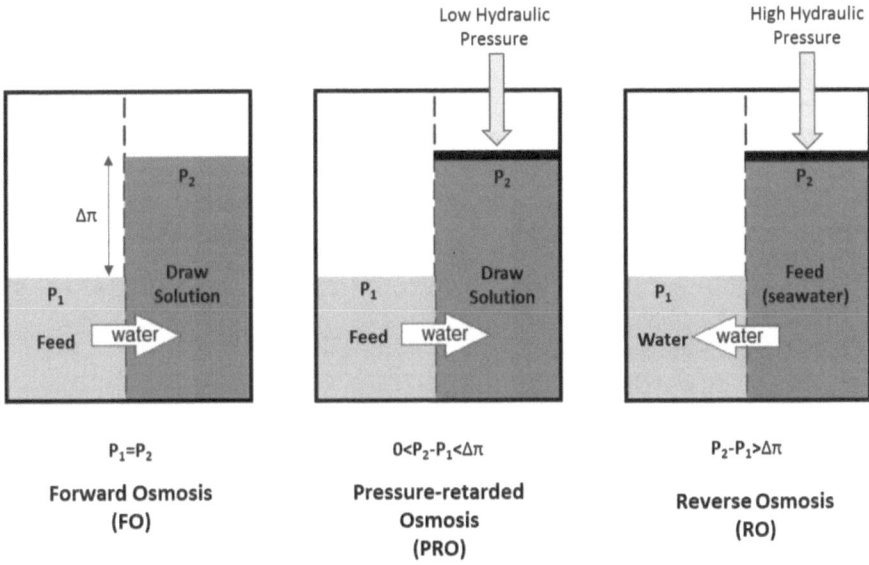

FIGURE 6.9   The principle of the osmotically driven membrane processes.

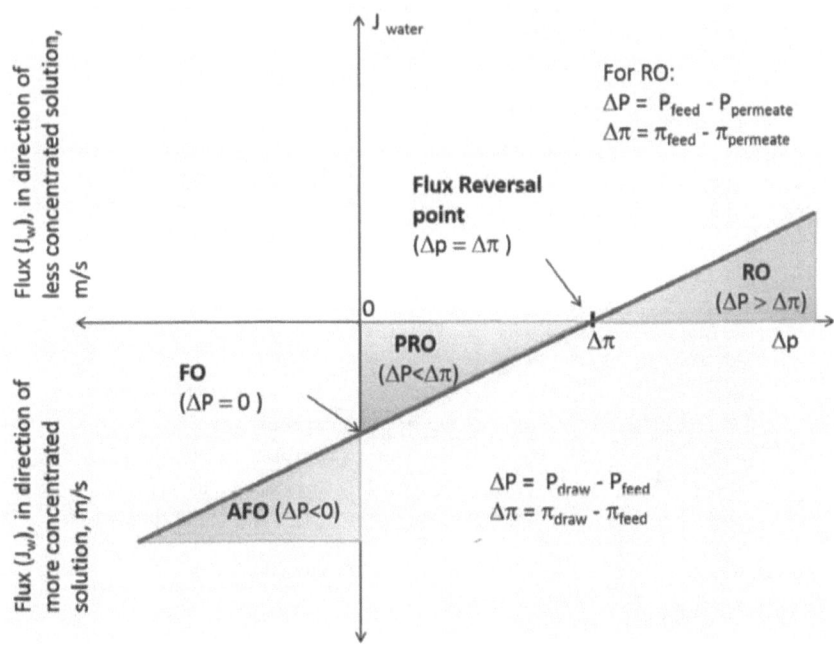

FIGURE 6.10   Correlation of water flux with external hydraulic pressure of osmotic processes.[80]

6.3.2 Draw Solutes Development for FO Process

The FO process requires a draw solution with high osmotic pressure. In many industrial applications, synthetic draw solutions are used and recycled. Generally, an FO unit always combines with a draw solution regeneration procedure. Therefore, the practical operation of the FO process heavily depends on the selection of an effective draw solution. The criteria of draw solution selection include, but not limited to, osmotic pressure, viscosity, internal concentrative polarization, availability, costs, regeneration, and toxicity. Potential synthetic draw solutions are:[87]

- gases and volatile compounds, e.g., NH_3-CO_2,
- inorganic draw solutions, e.g., salts,
- organic draw solutions, e.g., organic ionic liquids, organic ionic salts, polyelectrolytes, polymers, hydrogels, and
- functionalized nanoparticles (NPs).

Draw solutions can be categorized as nonresponsive and responsive types according to their regeneration behaviors.[88] During the regeneration, nonresponsive draw solutions (i.e., inorganic salts) do not have significant changes, while the responsive draw solutions (such as volatile liquids) can have phase changes. Nowadays, responsive draw solutions are widely used because of their easy regeneration. Figure 6.11 shows a hybrid FO system with draw solution regeneration and examples of different types of draw solutions.

In addition to draw solutions, many recent studies also optimized FO membranes to increase water flux, including thickness, porosity, and tortuosity. More details can be found in recent review papers.[81,83,91–93]

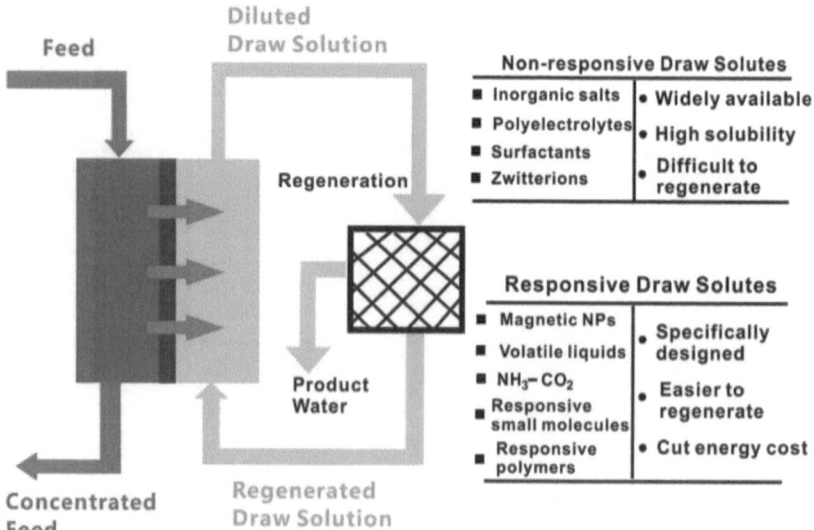

FIGURE 6.11 A hybrid FO and draw solution regeneration process and different types of draw solutions.[88] (Copyright 2016 Elsevier.)

6.3.3 FO APPLICATIONS TO PRODUCED WATER TREATMENT

FO technology has been applied from lab-scale setup to full-scale industrial implementation.[80,86,94] As shown in Figure 6.12,[94] in the early stage of FO, osmotic hydration bags were commercial products in 1970s. In the past decade, FO technologies were used in food and beverage industries, mainly to treat the products (milk, juice, and whey).[95] After 2015, many novel FO technologies have been developed for wastewater reclamation, brackish water/seawater desalination, and other industrial processes, including industrial scale-up of zero liquid discharge concept. Bell et al. systematically studied the membrane fouling and performance for an extended time of treating produced water using cellulose triacetate (CTA) and polyamide TFC membranes.[30] While both membranes had over 90% rejection of neutral hydrophobic compounds, the polyamide membrane exhibited a higher rejection of small organic molecules than CTA. FO is also subject to fouling, and the pretreatment of complex streams is necessary to protect the FO membranes from excessive fouling and degradation.

6.4 THERMALLY DRIVEN MEMBRANES

While the RO membrane exhibits high salt rejection, it requires high pressures to overcome the osmotic pressure difference, and thus, the RO process is not economical if the produced water has a TDS over $45,000\,\text{mg}\,\text{L}^{-1}$.[96–99] However, the produced water from most shale gas sources may contain TDS of $98,900–394,600\,\text{mg}\,\text{L}^{-1}$.[8,57,100–102] To this end, the Membrane distillation (MD) process has been extensively investigated, where water in the feed is heated and evaporates, diffuses through an air-filled porous hydrophobic membrane, and then condenses on the cool permeate side of the membrane. The membranes can have almost 100% rejection of non-volatile solutes such as ions. The MD process circumvents the osmotic pressure issue and is capable of treating the brine containing up to $350,000\,\text{mg}$ TDS. Furthermore, it can utilize low-grade or waste heat to drive the distillation, lowering the costs.

MD can operate in various modes, such as direct contact MD (DCMD), air gap MD (AGMD), and sweeping gas MD. The key to this technology is thermally stable, hydrophobic membranes resistant to pore wetting, fouling, and scaling.[103] Commercial MD membranes have symmetric and porous structure and are often made of hydrophobic polymers such as polytetrafluoroethylene (PTFE), PVDF, and polypropylene (PP).[13,104] The wetting causes the leakage of the high-salinity feed, reducing the salt rejection. The salts can precipitate from the highly saline feed at high water recovery, causing scaling formation and pore-plugging. Both fouling and scaling decrease water permeance. More importantly, they can reduce the hydrophobicity of the membrane surface, leading to wetting and thus reduced salt rejection.

Various strategies have been developed to mitigate wetting by improving hydrophobicity of the membrane surface, such as grafting nanoparticles and coating with low surface energy materials.[105,106] For example, Figure 6.13a shows omniphobic PVDF membranes prepared by a four-step process.[107] First, PVDF membranes were treated using an alkaline solution to generate hydroxyl groups on the surface. Second, positively charged (3-aminopropyl)triethoxysilane (APTES) was grafted onto the treated PVDF membrane. Third, the negatively charged silica nanoparticles

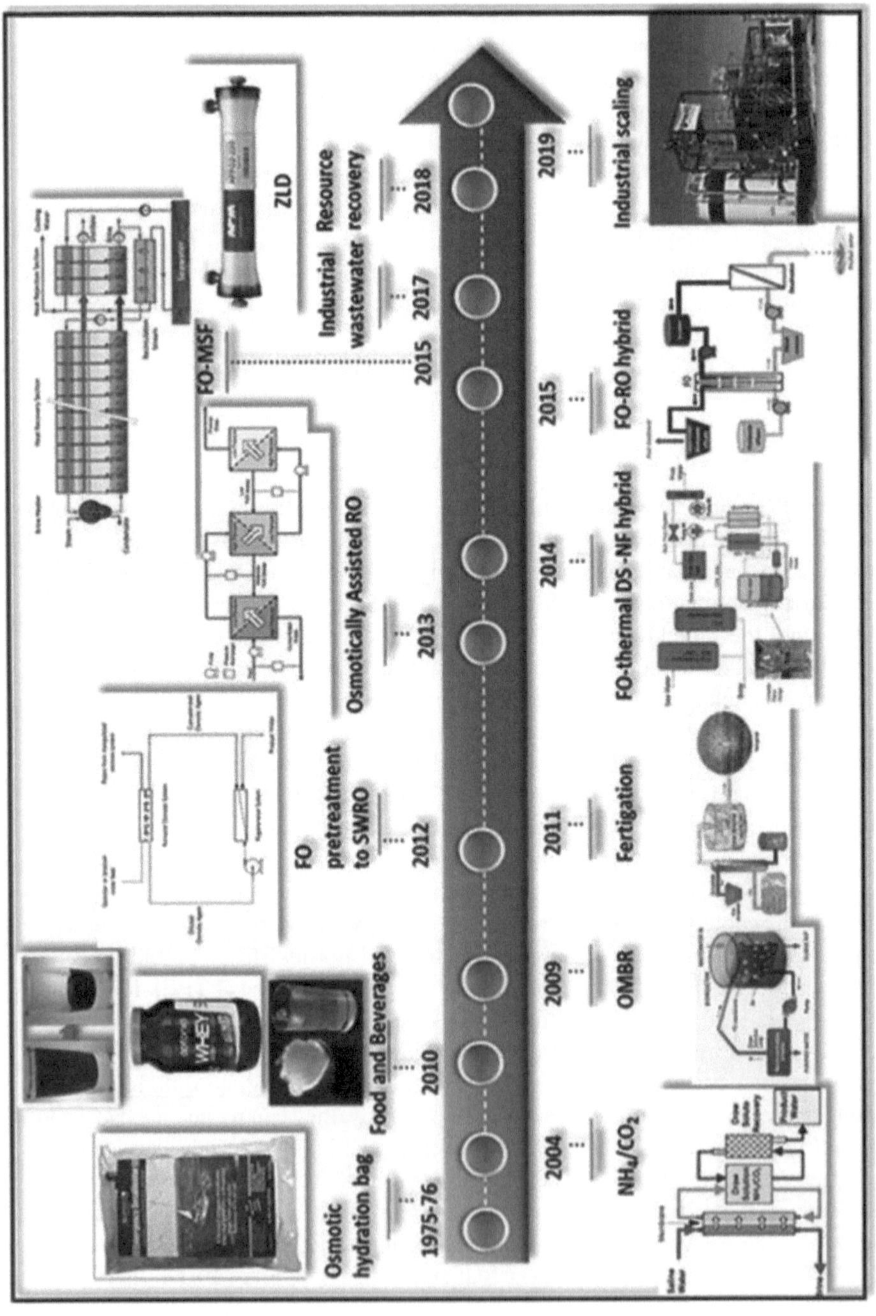

FIGURE 6.12   FO for industrial applications.[94] (Copyright 2020 Elsevier.)

(SiNPs) were coated. Finally, perfluoro-decyl trichlorosilane (FDTS) was coated on the SiNPs-modified surface. Figure 6.13b and c confirm that the SiNPs coating increased the membrane surface roughness.

Figure 6.13d compares the performance of the control and modified membrane in the DCMD using saline feed containing varying concentrations of oil. The omniphobic membrane showed stable water flux and ~100% salt rejection when the oil concentration increased up to 0.01% v/v. By contrast, the unmodified membrane suffered from a dramatic reduction of salt rejection due to the wetting when challenged with 0.005% oil in feed. Figure 6.13e directly compares the wettability of the control and

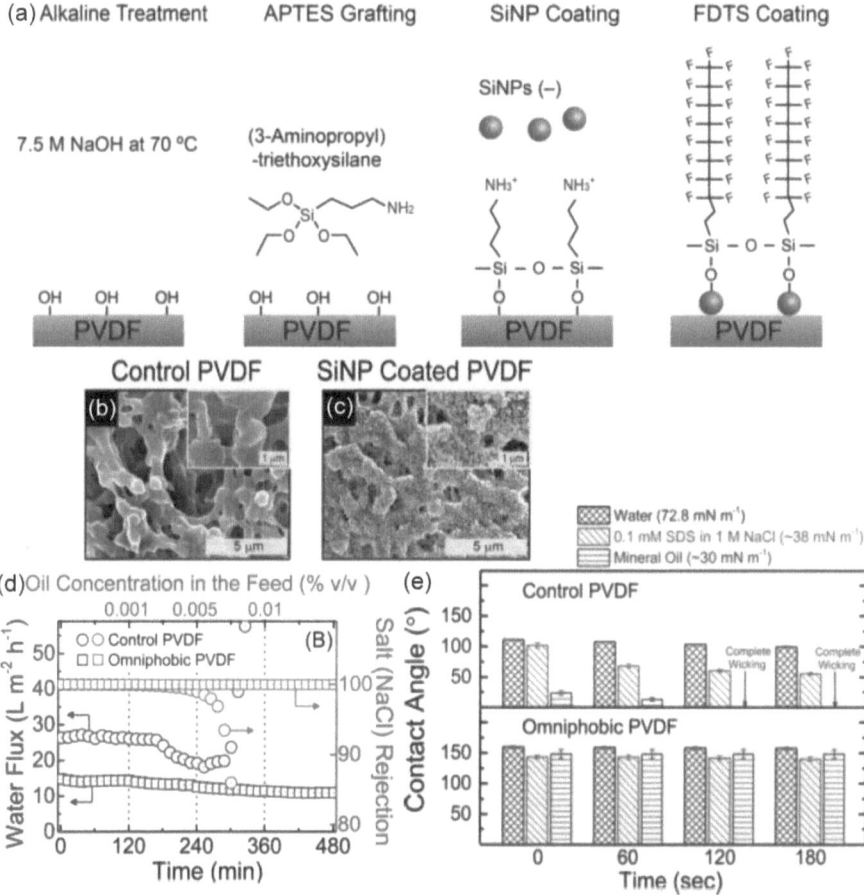

FIGURE 6.13 (a) Schematic of fabricating omniphobic PVDF membranes, including alkaline treatment, APTES grafting, SiNP coating, and FDTS coating. SEM image of the membrane surface for (b) control PVDF and (c) SiNP coated PVDF. (d) Water flux and NaCl rejection of the control and omniphobic PVDF in DCMD at 60°C. The feed contained 1M NaCl and mineral oil with increasing concentration over time. DI water at 20°C was used as a permeate solution. (e) Contact angles of the control and omniphobic membranes over time measured with water, SDS in NaCl solution, and mineral oil (Copyright 2016 ACS.)

modified membranes by measurement of contact angles for water, sodium dodecyl sulfate (SDS) solution, and mineral oil. The contact angles of PVDF decreased with time while the modified membrane showed stable contact angles.

Electrospun membranes consisting of ultrafine fibers with high porosity, controllable pore size, and good mechanical strength have attracted significant interest for MD applications. Furthermore, the electrospun membranes can be treated with CF_4 plasma to achieve omniphobic surface.[108] Figure 6.14a shows the replacement of the CH_2-CF_2 groups in PVDF electrospun nanofiber membranes (ENMs) by the CF_2-CF_2 and CF_3 groups and the decreased surface free energy, as evidenced by the spherical to semi-spherical droplets of liquids. In the 24-h AGMD test with produced water, the modified membranes exhibited higher water flux than the neat

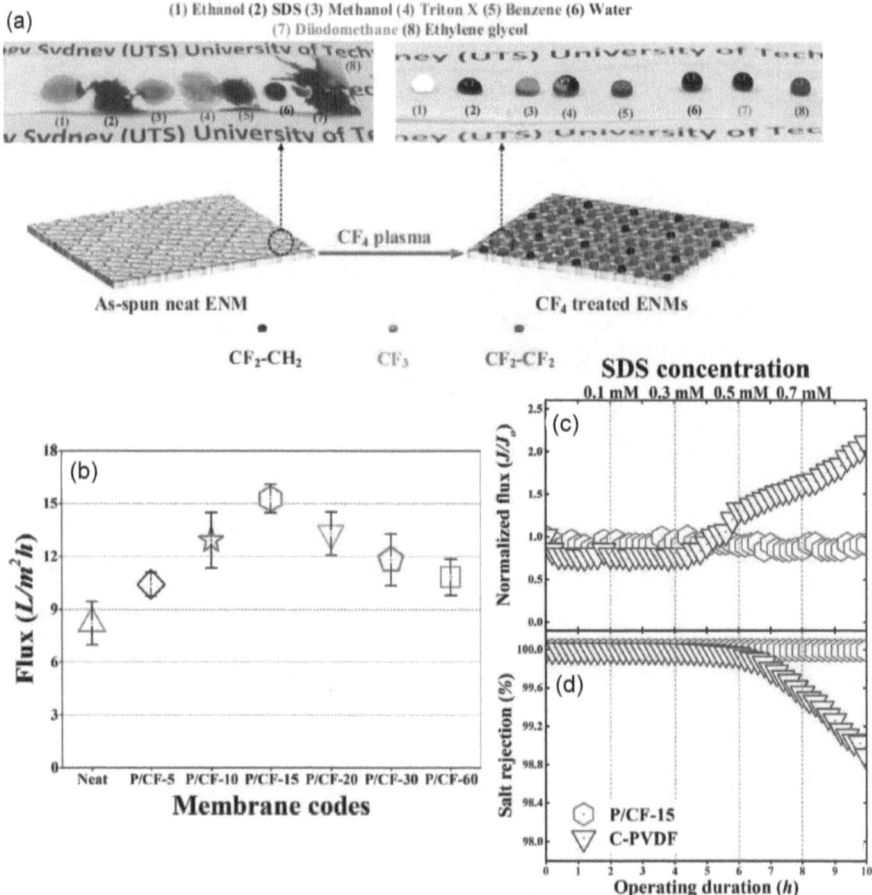

FIGURE 6.14 (a) Schematic of the neat and CF_4 plasma-treated ENMs and their surface hydrophobicity with various liquids.[108] (b) Comparison of water flux in the neat and CF_4 plasma-treated ENMs in 24-h tests of AGMD with produced water. (c) Normalized water flux and (d) salt rejection of the commercial PVDF membrane (C-PCDF) and CF_4 plasma-treated membrane (P/CF-15) challenged by produced water-containing SDS. (Copyright 2017 Elsevier.)

ENMs (cf. Figure 6.14b). The membrane with 15-min plasma treatment (P/CF-15) showed water flux 80% higher than the neat ENMs. Furthermore, when SDS solutions of 0.1–0.7 mM were added to the feed, the P/CF-15 exhibited stable water flux and nearly 100% salt rejection. By contrast, commercial PVDF membrane (C-PVDF) showed an increased water flux and decreased salt rejection when 0.3 mM SDS solution was added because of its lack of resistance to the SDA wetting.

MD process also requires the membrane with low heat conductivity to retain the temperature difference between the feed and permeate or driving force for water transport.[109,110] Figure 6.15 shows Janus composite hollow fiber membranes (J-HFMs) with both high water permeance and low heat loss.[109,111–116] As shown in Figure 6.15a and e, J-HFMs consisted of a thin hydrophobic outer layer of PVDF and superhydrophobic silica nanoparticles (PVDF/Si-R) and a thick hydrophilic inner

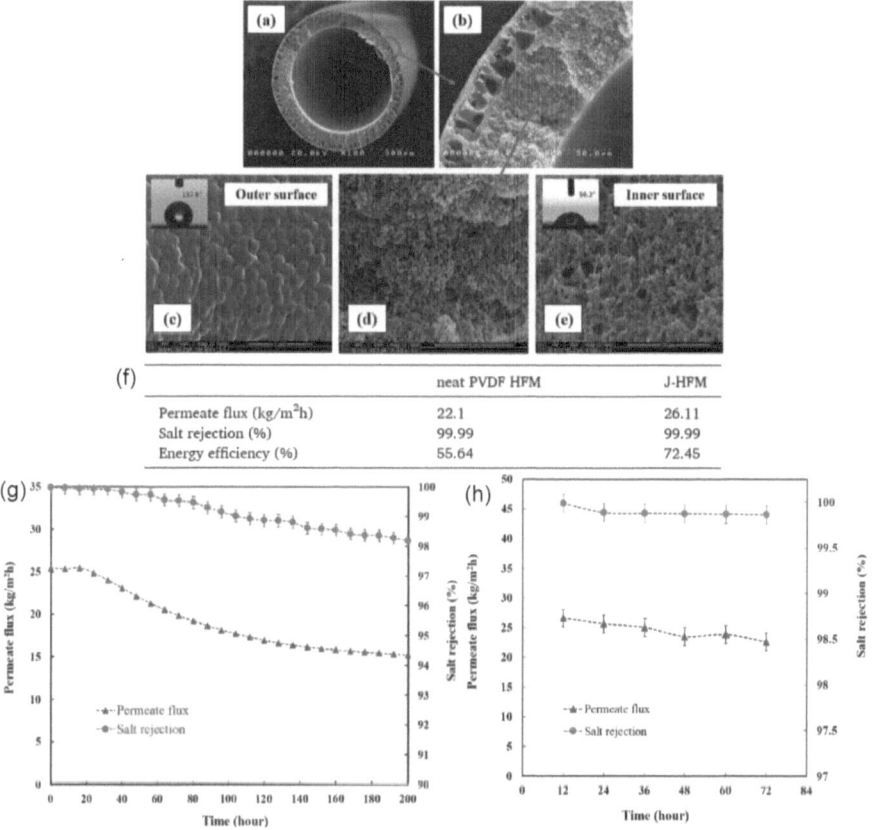

FIGURE 6.15 SEM images of the Janus-HFMs, including (a-b) and (d) cross-section, (c) hydrophobic PVDF/Si-R outer surface, and (e) hydrophilic PVDF/PEG inner surface.[109] (f) Performance comparison between the neat PVDF HFMs and J-HFMs. Permeate flux and salt rejection of J-HFMs (g) during a continuous 200-h DCMD and (h) during a 72-h DCMD with the membrane regeneration every 12 h. The produced water was sampled from Permian Basin with TDS of 154,220 mg L^{-1}. (Copyright 2020 Elsevier.)

layer of PVDF/PEG. The hydrophobic outer layer functioned as a barrier to protect the membrane from wetting while the porous hydrophilic layer facilitated the water vapor transfer through the membrane. When challenged with a high-salinity feed containing 35,000 mg L^{-1} NaCl, the J-HFMs outperformed a neat PVDF HFMs in terms of permeance and energy efficiency, as shown in Figure 6.15f. Additionally, after the 200-h DCMD with produced water, the Janus-HFMs exhibited the permeate flux of 15.2 kg/(m²hour) and salt rejection of 98.4% with good resistance to fouling and scaling (Figure 6.15g). Figure 6.15h shows that the performance of the J-HFMs during DCMD can be effectively regenerated by cleaning every 12 hours, as evidenced by the stable flux and high salt rejection (i.e., 99.8%) at the end of the 72-h operation.

6.5   CONCLUSION AND FUTURE OUTLOOK

This chapter provides an updated overview of advanced membranes for various processes for the treatment of produced water, including hydraulic pressure-driven, osmotic pressure-driven, and thermal-driven processes. For membrane processes to be competitive, the key is to develop membranes with high water permeance, superior rejection, and excellent long-term stability against the produced water, which is the focus of this short review. Different membrane systems can also be integrated together to combine their advantages to improve separation efficiency. However, it is beyond the scope of this review to discuss the processes and techno-economic analysis.

In the pressure-driven processes, MF and UF membranes remove large particles such as grease and suspended solids; NF membranes reject divalent ions and most organics, generating permeate reusable for the oil industry, irrigation, and livestock; and RO membranes provide potable water. Membranes can be combined to improve treatment efficiency. One of the great challenges of these membranes is fouling by the contaminants in the produced water, which can be mitigated by membrane surface modification to improve hydrophilicity (such as grafting with PEG, zwitterions, and GO) without significantly increasing transport resistance to water. Chemical modification to the surface includes two major categories of "grafting to" and "grafting from". In the former method, materials with superior antifouling properties are covalently bonded to the surface to achieve superior properties. The challenge would be to control the grafting density. The deposition of PD on the membrane surface also belongs to this category. In the "grafting from" approach, an initiator is often grafted on the surface, and then the monomers can be polymerized. The grafting density and the polymer chain length can be well controlled while the challenge is to functionalize the surface with the initiators with desired quantities.

In the osmotically driven processes, FO has emerged as an exciting technology for produced water treatment, though it is faced with challenges, such as salinity build-up, low membrane flux, membrane fouling, and system scale-up. The strategies to overcome these challenges rely heavily on the improvements of membrane separation properties (high water permeance, minimal concentration polarization, and great fouling resistance), draw solutions (high osmotic pressure, low reverse solute flux, and easy regeneration), and the integration of the FO with other processes such as RO, NF, and MD.

Thermally driven MD membranes provide a unique approach for treating high-salinity produced water, particularly if waste heat is available. The challenge is to develop membranes with resistance to wetting (leading to salt leakage), scaling (lowering water permeance and promoting wetting), and fouling. Hydrophobic membranes are used, and membranes are often modified to improve omniphobicity to mitigate wetting.

REFERENCES

1. Rahm, B. G.; Bates, J. T.; Bertoia, L. R.; Galford, A. E.; Yoxtheimer, D. A.; Riha, S. J., Wastewater management and Marcellus Shale gas development: Trends, drivers, and planning implications, *J. Environ. Manage.* **2013**, *120*, 105–113.
2. Shaffer, D. L.; Chavez, L. H. A.; Ben-Sasson, M.; Castrillon, S. R. V.; Yip, N. Y.; Elimelech, M., Desalination and reuse of high-salinity shale gas produced water: Drivers, technologies, and future directions, *Environ. Sci. Technol.* **2013**, *47* (17), 9569–9583.
3. Orem, W.; Tatu, C.; Varonka, M.; Lerch, H.; Bates, A.; Engle, M.; Crosby, L.; McIntosh, J., Organic substances in produced and formation water from unconventional natural gas extraction in coal and shale, *Int. J. Coal Geol.* **2014**, *126*, 20–31.
4. Liu, Y.; Zou, C.; Li, C.; Lin, L.; Chen, W., Evaluation of β-cyclodextrin–polyethylene glycol as green scale inhibitors for produced-water in shale gas well, *Desalination* **2016**, *377*, 28–33.
5. Kahrilas, G. A.; Blotevogel, J.; Stewart, P. S.; Borch, T., Biocides in hydraulic fracturing fluids: A critical review of their usage, mobility, degradation, and toxicity, *Environ. Sci. Technol.* **2015**, *49* (1), 16–32.
6. Barbot, E.; Vidic, N. S.; Gregory, K. B.; Vidic, R. D., Spatial and temporal correlation of water quality parameters of produced waters from devonian-age shale following hydraulic fracturing, *Environ. Sci. Technol.* **2013**, *47* (6), 2562–2569.
7. Blondes, M. S.; Gans, K. D.; Rowan, E. L.; Thordsen, J. J.; Reidy, M. E.; Engle, M.; Kharaka, Y.; Thomas, B., US geological survey national produced waters geochemical database v2. 2 (PROVISIONAL) documentation, *USGS Energy Resources Program: Produced Waters, USGS* **2016**, *16*. doi: 10.5066/F7J964W8.
8. Shih, J.-S.; Saiers, J. E.; Anisfeld, S. C.; Chu, Z.; Muehlenbachs, L. A.; Olmstead, S. M., Characterization and analysis of liquid waste from marcellus shale gas development, *Environ. Sci. Technol.* **2015**, *49* (16), 9557–9565.
9. Zhao, S.; Hu, S.; Zhang, X.; Song, L.; Wang, Y.; Tan, M.; Kong, L.; Zhang, Y., Integrated membrane system without adding chemicals for produced water desalination towards zero liquid discharge, *Desalination* **2020**, *496*, 114693.
10. Fakhru'l-Razi, A.; Pendashteh, A.; Abdullah, L. C.; Biak, D. R. A.; Madaeni, S. S.; Abidin, Z. Z., Review of technologies for oil and gas produced water treatment, *J. Hazard. Mater.* **2009**, *170* (2–3), 530–551.
11. Igunnu, E. T.; Chen, G. Z., Produced water treatment technologies, *Int. J. Low-Carbon Tec.* **2014**, *9* (3), 157–177.
12. Kundua, P.; Mishra, I. M., Treatment and reclamation of hydrocarbon-bearing oily wastewater as a hazardous pollutant by different processes and technologies: A state-of-the-art review, *Rev. Chem. Eng.* **2019**, *35* (1), 73–108.
13. Chang, H.; Li, T.; Liu, B.; Vidic, R. D.; Elimelech, M.; Crittenden, J. C., Potential and implemented membrane-based technologies for the treatment and reuse of flowback and produced water from shale gas and oil plays: A review, *Desalination* **2019**, *455*, 34–57.

14. Dischinger, S. M.; Rosenblum, J.; Noble, R. D.; Gin, D. L.; Linden, K. G., Application of a lyotropic liquid crystal nanofiltration membrane for hydraulic fracturing flowback water: Selectivity and implications for treatment, *J. Membr. Sci.* **2017**, *543*, 319–327.

15. Guo, C.; Chang, H.; Liu, B.; He, Q.; Xiong, B.; Kumar, M.; Zydney, A. L., A combined ultrafiltration–reverse osmosis process for external reuse of Weiyuan shale gas flowback and produced water, *Environ. Sci. Water Res. Technology* **2018**, *4* (7), 942–955.

16. Kong, F.; Sun, G.; Chen, J.; Han, J.; Guo, C.; Tong, Z.; Lin, X.; Xie, Y. F., Desalination and fouling of NF/low pressure RO membrane for shale gas fracturing flowback water treatment, *Sep. Purif. Techn.* **2018**, *195*, 216–223.

17. Regnery, J.; Coday, B. D.; Riley, S. M.; Cath, T. Y., Solid-phase extraction followed by gas chromatography-mass spectrometry for the quantitative analysis of semi-volatile hydrocarbons in hydraulic fracturing wastewaters, *Anal. Methods* **2016**, *8* (9), 2058–2068.

18. Riley, S. M.; Ahoor, D. C.; Oetjen, K.; Cath, T. Y., Closed circuit desalination of O&G produced water: An evaluation of NF/RO performance and integrity, *Desalination* **2018**, *442*, 51–61.

19. Riley, S. M.; Ahoor, D. C.; Regnery, J.; Cath, T. Y., Tracking oil and gas wastewater-derived organic matter in a hybrid biofilter membrane treatment system: A multi-analytical approach, *Sci. Total Environ.* **2018**, *613–614*, 208–217.

20. Riley, S. M.; Oliveira, J. M. S.; Regnery, J.; Cath, T. Y., Hybrid membrane bio-systems for sustainable treatment of oil and gas produced water and fracturing flowback water, *Sep. Purif. Techn.* **2016**, *171*, 297–311.

21. Eboagwu, U. Evaluation of Membrane Treatment Technology to Optimize and Reduce Hypersalinity Content of Produced Brine for Reuse in Unconventional Gas Wells. Texas A & M University, 2012.

22. Miller, D. J.; Huang, X.; Li, H.; Kasemset, S.; Lee, A.; Agnihotri, D.; Hayes, T.; Paul, D. R.; Freeman, B. D., Fouling-resistant membranes for the treatment of flowback water from hydraulic shale fracturing: A pilot study, *J. Membr. Sci.* **2013**, *437*, 265–275.

23. Boschee, P., Handling produced water from hydraulic fracturing, *Oil Gas Facilities* **2012**, *1* (01), 22–26.

24. Myers, J. E. Chevron San Ardo Facility Unit (SAFU) beneficial produced water reuse for irrigation, *SPE International Conference on Health, Safety, and Environment, Society of Petroleum Engineers*, **2014**.

25. Cho, H.; Choi, Y.; Lee, S., Effect of pretreatment and operating conditions on the performance of membrane distillation for the treatment of shale gas wastewater, *Desalination* **2018**, *437*, 195–209.

26. Coday, B. D.; Almaraz, N.; Cath, T. Y., Forward osmosis desalination of oil and gas wastewater: Impacts of membrane selection and operating conditions on process performance, *J. Membr. Sci.* **2015**, *488*, 40–55.

27. Sardari, K.; Fyfe, P.; Lincicome, D.; Wickramasinghe, S. R., Aluminum electrocoagulation followed by forward osmosis for treating hydraulic fracturing produced waters, *Desalination* **2018**, *428*, 172–181.

28. Coday, B. D.; Xu, P.; Beaudry, E. G.; Herron, J.; Lampi, K.; Hancock, N. T.; Cath, T. Y., The sweet spot of forward osmosis: Treatment of produced water, drilling wastewater, and other complex and difficult liquid streams, *Desalination* **2014**, *333* (1), 23–35.

29. Coday, B. D.; Cath, T. Y., Forward osmosis: Novel desalination of produced water and fracturing flowback, *J. Am. Water Works Assoc.* **2014**, *106* (2), E55–E66.

30. Bell, E. A.; Poynor, T. E.; Newhart, K. B.; Regnery, J.; Coday, B. D.; Cath, T. Y., Produced water treatment using forward osmosis membranes: Evaluation of extended-time performance and fouling, *J. Membr. Sci.* **2017**, *525*, 77–88.

31. Maltos, R. A.; Regnery, J.; Almaraz, N.; Fox, S.; Schutter, M.; Cath, T. J.; Veres, M.; Coday, B. D.; Cath, T. Y., Produced water impact on membrane integrity during extended pilot testing of forward osmosis – reverse osmosis treatment, *Desalination* **2018**, *440*, 99–110.

32. McGinnis, R. L.; Hancock, N. T.; Nowosielski-Slepowron, M. S.; McGurgan, G. D., Pilot demonstration of the NH_3/CO_2 forward osmosis desalination process on high salinity brines, *Desalination* **2013**, *312*, 67–74.

33. Hutchings, N. R.; Appleton, E. W.; McGinnis, R. A. In Making high quality frac water out of oilfield waste, *SPE Annual Technical Conference and Exhibition, Society of Petroleum Engineers*, **2010**.

34. Du, X.; Zhang, Z.; Carlson, K. H.; Lee, J.; Tong, T., Membrane fouling and reusability in membrane distillation of shale oil and gas produced water: Effects of membrane surface wettability, *J. Membr. Sci.* **2018**, *567*, 199–208.

35. Kim, J.; Kim, J.; Hong, S., Recovery of water and minerals from shale gas produced water by membrane distillation crystallization, *Water Res.* **2018**, *129*, 447–459.

36. Sardari, K.; Fyfe, P.; Lincicome, D.; Ranil Wickramasinghe, S., Combined electrocoagulation and membrane distillation for treating high salinity produced waters, *J. Membr. Sci.* **2018**, *564*, 82–96.

37. Duong, H. C.; Chivas, A. R.; Nelemans, B.; Duke, M.; Gray, S.; Cath, T. Y.; Nghiem, L. D., Treatment of RO brine from CSG produced water by spiral-wound air gap membrane distillation — A pilot study, *Desalination* **2015**, *366*, 121–129.

38. Estrada, J. M.; Bhamidimarri, R., A review of the issues and treatment options for wastewater from shale gas extraction by hydraulic fracturing, *Fuel* **2016**, *182*, 292–303.

39. Jimenez, S.; Mico, M. M.; Arnaldos, M.; Medina, F.; Contreras, S., State of the art of produced water treatment, *Chemosphere* **2018**, *192*, 186–208.

40. Al-Ghouti, M. A.; Al-Kaabi, M. A.; Ashfaq, M. Y.; Da'na, D. A., Produced water characteristics, treatment and reuse: A review, *J. Water Process. Eng.* **2019**, *28*, 222–239.

41. Wei, X. C.; Zhang, S. C.; Han, Y. X.; Wolfe, F. A., Treatment of petrochemical wastewater and produced water from oil and gas, *Water Environ. Res.* **2019**, *91* (10), 1025–1033.

42. Jepsen, K. L.; Bram, M. V.; Pedersen, S.; Yang, Z. Y., Membrane fouling for produced water treatment: A review study from a process control perspective, *Water* **2018**, *10* (7), 28.

43. Ali, A.; Quist-Jensen, C. A.; Drioli, E.; Macedonio, F., Evaluation of integrated microfiltration and membrane distillation/crystallization processes for produced water treatment, *Desalination* **2018**, *434*, 161–168.

44. Camarillo, M. K.; Domen, J. K.; Stringfellow, W. T., Physical-chemical evaluation of hydraulic fracturing chemicals in the context of produced water treatment, *J. Environ. Manage.* **2016**, *183*, 164–174.

45. Munirasu, S.; Haija, M. A.; Banat, F., Use of membrane technology for oil field and refinery produced water treatment - A review, *Process Saf. Environ. Protect.* **2016**, *100*, 183–202.

46. Weschenfelder, S. E.; Fonseca, M. J. C.; Borges, C. P., Treatment of produced water from polymer flooding in oil production by ceramic membranes, *J. Pet. Sci. Eng.* **2021**, *196*, 108021.

47. Ferreira, A. D. F.; Coelho, D. R. B.; dos Santos, R. V. G.; Nascimento, K. S.; Presciliano, F. A.; da Silva, F. P.; Campos, J. C.; da Fonseca, F. V.; Borges, C. P.; Weschenfelder, S. E., Fouling mitigation in produced water treatment by conjugation of advanced oxidation process and microfiltration, *Environ. Sci. Pollut. Res.* **2021**, *28*, 12803–12816.

48. Maguire-Boyle, S. J.; Huseman, J. E.; Ainscough, T. J.; Oatley-Radcliffe, D. L.; Alabdulkarem, A. A.; Al-Mojil, S. F.; Barron, A. R., Superhydrophilic functionalization of microfiltration ceramic membranes enables separation of hydrocarbons from frac and produced water, *Sci. Rep.* **2017**, *7* (1), 12267.

49. Zsirai, T.; Al-Jaml, A. K.; Qiblawey, H.; Al-Marri, M.; Ahmed, A.; Bach, S.; Watson, S.; Judd, S., Ceramic membrane filtration of produced water: Impact of membrane module, *Sep. Purif. Tech.* **2016**, *165*, 214–221.

50. Ebrahimi, M.; Kovacs, Z.; Schneider, M.; Mund, P.; Bolduan, P.; Czermak, P., Multistage filtration process for efficient treatment of oil-field produced water using ceramic membranes, *Desalin. Water Treat* **2012**, *42* (1–3), 17–23.

51. Ebrahimi, M.; Willershausen, D.; Ashaghi, K. S.; Engel, L.; Placido, L.; Mund, P.; Bolduan, P.; Czermak, P., Investigations on the use of different ceramic membranes for efficient oil-field produced water treatment, *Desalination* **2010**, *250* (3), 991–996.

52. Zhong, J.; Sun, X.; Wang, C., Treatment of oily wastewater produced from refinery processes using flocculation and ceramic membrane filtration, *Sep. Purif. Tech.* **2003**, *32* (1), 93–98.

53. Zsirai, T.; Qiblawey, H.; Buzatu, P.; Al-Marri, M.; Judd, S. J., Cleaning of ceramic membranes for produced water filtration, *J. Pet. Sci. Eng.* **2018**, *166*, 283–289.

54. Miller, D. J.; Dreyer, D. R.; Bielawski, C. W.; Paul, D. R.; Freeman, B. D., Surface modification of water purification membranes, *Angew. Chem. Int. Ed.* **2017**, *56* (17), 4662–4711.

55. Shahkaramipour, N.; Tran, T. N.; Ramanan, S.; Lin, H., Membranes with surface-enhanced antifouling properties for water purification, *Membranes* **2017**, *7* (1), 13.

56. Kirschner, A.; Chang, C.; Kasemset, S.; Emrick, T.; Freeman, B. D., Fouling-resistant ultrafiltration membranes prepared via co-deposition of dopamine/zwitterion composite coatings, *J. Membr. Sci.* **2017**, *541* (1), 300–311.

57. Dreyer, D. R.; Miller, D. J.; Freeman, B. D.; Paul, D. R.; Bielawski, C. W., Perspectives on poly(dopamine), *Chem. Sci.* **2013**, *4* (10), 3796–3802.

58. Shahkaramipour, N.; Jafari, A.; Tran, T.; Stafford, C. M.; Cheng, C.; Lin, H., Maximizing the grafting of zwitterions onto the surface of ultrafiltration membranes to improve antifouling properties, *J. Membr. Sci.* **2020**, *601*, 117909.

59. Chen, X.; Deng, E.; Park, D.; Pfeifer, B. A.; Dai, N.; Lin, H., Grafting activated graphene oxide nanosheets onto ultrafiltration membranes using polydopamine to enhance antifouling properties, *ACS Appl. Mater. Interfaces* **2020**, *12* (42), 48179–48187.

60. McCloskey, B. D.; Park, H. B.; Ju, H.; Rowe, B. W.; Miller, D. J.; Freeman, B. D., A bioinspired fouling-resistant surface modification for water purification membranes, *J. Membr. Sci.* **2012**, *413–414*, 82–90.

61. Nawi, N. I. M.; Chean, H. M.; Shamsuddin, N.; Bilad, M. R.; Narkkun, T.; Faungnawakij, K.; Khan, A. L., Development of hydrophilic PVDF membrane using vapour induced phase separation method for produced water treatment, *Membranes* **2020**, *10* (6), 17.

62. Shahkaramipour, N.; Lai, C. K.; Venna, S. R.; Sun, H.; Cheng, C.; Lin, H., Membrane surface modification using thiol-containing zwitterionic polymers via bioadhesive polydopamine, *Ind. Eng. Chem. Res.* **2018**, *57* (6), 2336–2345.

63. Li, P.; Ge, Q., Membrane surface engineering with bifunctional zwitterions for efficient oil–water separation, *ACS Appl. Mater. Interfaces* **2019**, *11* (34), 31328–31337.

64. Babayev, M.; Du, H. B.; Botlaguduru, V. S. V.; Kommalapati, R. R., Zwitterion-modified ultrafiltration membranes for permian basin produced water pretreatment, *Water* **2019**, *11* (8), 15.

65. Atallah, C.; Mortazavi, S.; Tremblay, A. Y.; Doiron, A., Surface-modified multi-lumen tubular membranes for SAGD-produced water treatment, *Energy Fuel.* **2019**, *33* (6), 5766–5776.

66. Guo, C.; Chang, H. Q.; Liu, B. C.; He, Q. P.; Xiong, B. Y.; Kumar, M.; Zydney, A. L., A combined ultrafiltration-reverse osmosis process for external reuse of Weiyuan shale gas flowback and produced water, *Environ. Sci.-Wat. Res. Technol.* **2018**, *4* (7), 942–955.

67. Alammar, A.; Park, S.-H.; Williams, C. J.; Derby, B.; Szekely, G., Oil-in-water separation with graphene-based nanocomposite membranes for produced water treatment, *J. Membr. Sci.* **2020**, *603*, 118007.

68. Dischinger, S. M.; Rosenblum, J.; Noble, R. D.; Gin, D. L., Evaluation of a nanoporous lyotropic liquid crystal polymer membrane for the treatment of hydraulic fracturing produced water via cross-flow filtration, *J. Membr. Sci.* **2019**, *592*, 117313.

69. Ramanan, S. N.; Shahkaramipour, N.; Tran, T.; Zhu, L.; Venna, S. R.; Lim, C.-K.; Singh, A.; Prasad, P. N.; Lin, H., Self-cleaning membranes for water purification by co-deposition of photo-mobile 4, 4′-azodianiline and bio-adhesive polydopamine, *J. Membr. Sci.* **2018**, *554*, 164–174.

70. Himstedt, H. H.; Qian, X.; Weaver, J. R.; Wickramasinghe, S. R., Responsive membranes for hydrophobic interaction chromatography, *J. Membr. Sci.* **2013**, *447*, 335–344.

71. Kota, A. K.; Kwon, G.; Choi, W.; Mabry, J. M.; Tuteja, A., Hygro-responsive membranes for effective oil–water separation, *Nat. Commun.* **2012**, *3* (1), 1–8.

72. Himstedt, H. H.; Sengupta, A.; Qian, X. H.; Wickramasinghe, S. R., Magnetically responsive nano filtration membranes for treatment of coal bed methane produced water, *J. Taiwan Inst. Chem. Eng.* **2019**, *94*, 97–108.

73. Lin, L.; Jiang, W.; Xu, X.; Xu, P., A critical review of the application of electromagnetic fields for scaling control in water systems: Mechanisms, characterization, and operation, *NPJ Clean Water* **2020**, *3* (1), 25.

74. Himstedt, H. H.; Yang, Q.; Dasi, L. P.; Qian, X.; Wickramasinghe, S. R.; Ulbricht, M., Magnetically activated micromixers for separation membranes, *Langmuir* **2011**, *27* (9), 5574–5581.

75. Gwak, G.; Kim, D. I.; Hong, S., New industrial application of forward osmosis (FO): Precious metal recovery from printed circuit board (PCB) plant wastewater, *J. Membr. Sci.* **2018**, *552*, 234–242.

76. Ansari, A. J.; Hai, F. I.; Price, W. E.; Drewes, J. E.; Nghiem, L. D., Forward osmosis as a platform for resource recovery from municipal wastewater - A critical assessment of the literature, *J. Membr. Sci.* **2017**, *529*, 195–206.

77. Roy, D.; Rahni, M.; Pierre, P.; Yargeau, V., Forward osmosis for the concentration and reuse of process saline wastewater, *Chem. Eng. J.* **2016**, *287*, 277–284.

78. Valladares Linares, R.; Li, Z.; Sarp, S.; Bucs, S.; Amy, G.; Vrouwenvelder, J. S., Forward osmosis niches in seawater desalination and wastewater reuse, *Water Res.* **2014**, *66*, 122–139.

79. Yang, Q.; Wang, K. Y.; Chung, T. S., Dual-layer hollow fibers with enhanced flux as novel forward osmosis membranes for water production, *Environ. Sci. Tech.* **2009**, *43* (8), 2800–2805.

80. Korenak, J.; Basu, S.; Balakrishnan, M.; Hélix-Nielsen, C.; Petrinic, I., Forward osmosis in wastewater treatment processes, *Acta Chim. Slov.* **2017**, *64* (1), 83–94.

81. Kim, B.; Gwak, G.; Hong, S., Review on methodology for determining forward osmosis (FO) membrane characteristics: Water permeability (A), solute permeability (B), and structural parameter (S), *Desalination* **2017**, *422*, 5–16.

82. Zhao, S.; Huang, K.; Lin, H., Impregnated membranes for water purification using forward osmosis, *Ind. Eng. Chem. Res.* **2015**, *54* (49), 12354–12366.

83. Tran, T.; Pan, S.; Chen, X.; Lin, X.; Blevins, A. K.; Ding, Y.; Lin, H., Zwitterionic hydrogel-impregnated membranes with polyamide skin achieving superior water/salt separation properties, *ACS Appl. Mater. Interfaces* **2020**, *12* (43), 49192 49199.

84. Rastogi, N. K., Opportunities and challenges in application of forward osmosis in food processing, *Crit. Rev. Food Sci. Nutr.* **2016**, *56* (2), 266–291.

85. Qasim, M.; Darwish, N. A.; Sarp, S.; Hilal, N., Water desalination by forward (direct) osmosis phenomenon: A comprehensive review, *Desalination* **2015**, *374*, 47–69.

86. Coday, B. D.; Cath, T. Y., Forward osmosis: Novel desalination of produced water and fracturing flowback, *J. Am. Water Works Assoc.* **2014**, *106* (2), E55–E66.

87. Johnson, D. J.; Suwaileh, W. A.; Mohammed, A. W.; Hilal, N., Osmotic's potential: An overview of draw solutes for forward osmosis, *Desalination* **2018**, *434*, 100–120.

88. Cai, Y.; Hu, X. M., A critical review on draw solutes development for forward osmosis, *Desalination* **2016**, *391*, 16–29.

89. Luo, H.; Wang, Q.; Zhang, T. C.; Tao, T.; Zhou, A.; Chen, L.; Bie, X., A review on the recovery methods of draw solutes in forward osmosis, *J. Water Process. Eng.* **2014**, *4* (C), 212–223.

90. Lee, K. L.; Baker, R. W.; Lonsdale, H. K., Membranes for power generation by pressure-retarded osmosis, *J. Membr. Sci.* **1981**, *8* (2), 141–171.

91. Xu, W.; Chen, Q.; Ge, Q., Recent advances in forward osmosis (FO) membrane: Chemical modifications on membranes for FO processes, *Desalination* **2017**, *419*, 101–116.

92. Alsvik, I. L.; Hägg, M. B., Pressure retarded osmosis and forward osmosis membranes: Materials and methods, *Polymers* **2013**, *5* (1), 303–327.

93. Yip, N. Y.; Tiraferri, A.; Phillip, W. A.; Schiffman, J. D.; Elimelech, M., High performance thin-film composite forward osmosis membrane, *Environ. Sci. Technol.* **2010**, *44* (10), 3812–3818.

94. Suwaileh, W.; Pathak, N.; Shon, H.; Hilal, N., Forward osmosis membranes and processes: A comprehensive review of research trends and future outlook, *Desalination* **2020**, *485*, 114455.

95. Haupt, A.; Lerch, A., Forward osmosis application in manufacturing industries: A short review, *Membranes* **2018**, *8* (3), 1–33.

96. Chang, Y.; Myerson, A., The diffusivity of potassium chloride and sodium chloride in concentrated, saturated, and supersaturated aqueous solutions, *AIChE J.* **1985**, *31* (6), 890–894.

97. Lester, Y.; Ferrer, I.; Thurman, E. M.; Sitterley, K. A.; Korak, J. A.; Aiken, G.; Linden, K. G., Characterization of hydraulic fracturing flowback water in Colorado: implications for water treatment, *Sci. Total Environ.* **2015**, *512*, 637–644.

98. Semiat, R.; Hasson, D., Water desalination, *Rev. Chem. Eng.* **2012**, *28* (1), 43–60.

99. Beh, J. J.; Ooi, B. S.; Lim, J. K.; Ng, E. P.; Mustapa, H., Development of high water permeability and chemically stable thin film nanocomposite (TFN) forward osmosis (FO) membrane with poly(sodium 4-styrenesulfonate) (PSS)-coated zeolitic imidazolate framework-8 (ZIF-8) for produced water treatment, *J. Water Process. Eng.* **2020**, *33*, 15.

100. Blondes, M. S.; Gans, K. D.; Engle, M.; Kharaka, Y.; Reidy, M. E.; Saraswathula, V.; Thordsen, J. J.; Rowan, E. L.; Morrissey, E. A. *U.S. Geological Survey National Produced Waters Geochemical Database v2.3: U.S. Geological Survey data release*, https://doi.org/10.5066/F7J964W8; 2018.

101. Rosenblum, J.; Thurman, E. M.; Ferrer, I.; Aiken, G.; Linden, K. G., Organic chemical characterization and mass balance of a hydraulically fractured well: From fracturing fluid to produced water over 405 days, *Environ. Sci. Technol.* **2017**, *51* (23), 14006–14015.

102. Rosenblum, J. S.; Sitterley, K. A.; Thurman, E. M.; Ferrer, I.; Linden, K. G., Hydraulic fracturing wastewater treatment by coagulation-adsorption for removal of organic compounds and turbidity, *J. Environ. Chem. Eng.* **2016**, *4* (2), 1978–1984.

103. Liu, F.; Hashim, N. A.; Liu, Y.; Abed, M. M.; Li, K., Progress in the production and modification of PVDF membranes, *J. Membr. Sci.* **2011**, *375* (1–2), 1–27.

104. Deshmukh, A.; Boo, C.; Karanikola, V.; Lin, S. H.; Straub, A. P.; Tong, T. Z.; Warsinger, D. M.; Elimelech, M., Membrane distillation at the water-energy nexus: Limits, opportunities, and challenges, *Energy Environ. Sci.* **2018**, *11* (5), 1177–1196.

105. Tran, T.; Chen, X.; Doshi, S.; Stafford, C. M.; Lin, H., Grafting polysiloxane onto ultrafiltration membranes to optimize surface energy and mitigate fouling, *Soft Matter* **2020**, *16*, 5044–5053.

106. Tran, T.; Tu, Y.-C.; Hall-Laureano, S.; Lin, C.; Kawy, M.; Lin, H., "Nonstick" membranes prepared by facile surface fluorination for water purification, *Ind. Eng. Chem. Res.* **2020**, *59* (12), 5307–5314.

107. Boo, C.; Lee, J.; Elimelech, M., Omniphobic polyvinylidene fluoride (PVDF) membrane for desalination of shale gas produced water by membrane distillation, *Environ. Sci. Technol.* **2016**, *50* (22), 12275–12282.

108. Woo, Y. C.; Chen, Y.; Tijing, L. D.; Phuntsho, S.; He, T.; Choi, J. S.; Kim, S. H.; Shon, H. K., CF_4 plasma-modified omniphobic electrospun nanofiber membrane for produced water brine treatment by membrane distillation, *J. Membr. Sci.* **2017**, *529*, 234–242.

109. Zou, L.; Gusnawan, P.; Zhang, G.; Yu, J., Novel Janus composite hollow fiber membrane-based direct contact membrane distillation (DCMD) process for produced water desalination, *J. Membr. Sci.* **2020**, *597*, 117756.

110. Ullah, R.; Khraisheh, M.; Esteves, R. J.; McLeskey, J. T.; AlGhouti, M.; Gad-el-Hak, M.; Vahedi Tafreshi, H., Energy efficiency of direct contact membrane distillation, *Desalination* **2018**, *433*, 56–67.

111. Yang, H.; Wei, F.; Hu, K.; Lyu, J., Effects of mud slurry on flow resistance of cohesionless coarse particles, *Powder Technol.* **2017**, *310*, 1–7.

112. Figoli, A.; Ursino, C.; Galiano, F.; Di Nicolò, E.; Campanelli, P.; Carnevale, M. C.; Criscuoli, A., Innovative hydrophobic coating of perfluoropolyether (PFPE) on commercial hydrophilic membranes for DCMD application, *J. Membr. Sci.* **2017**, *522*, 192–201.

113. Puranik, A. A.; Rodrigues, L. N.; Chau, J.; Li, L.; Sirkar, K. K., Porous hydrophobic-hydrophilic composite membranes for direct contact membrane distillation, *J. Membr. Sci.* **2019**, *591*, 117225.

114. Khayet, M.; Mengual, J. I.; Matsuura, T., Porous hydrophobic/hydrophilic composite membranes: Application in desalination using direct contact membrane distillation, *J. Membr. Sci.* **2005**, *252* (1), 101–113.

115. Feng, X.; Jiang, L. Y.; Matsuura, T.; Wu, P., Fabrication of hydrophobic/hydrophilic composite hollow fibers for DCMD: Influence of dope formulation and external coagulant, *Desalination* 2017, *401*, 53–63.

116. Bonyadi, S.; Chung, T. S., Flux enhancement in membrane distillation by fabrication of dual layer hydrophilic–hydrophobic hollow fiber membranes, *J. Membr. Sci.* 2007, *306* (1), 134–146.

7 Assessment of Oil Fouling by Oil– Membrane Interaction Energy Analysis

Henry J. Tanudjaja
Nanyang Technology University

Jia W. Chew
Nanyang Technology University and Nanyang Environmental and Water Research Institute

CONTENTS

7.1 INTRODUCTION

Membrane filtration technology offers a greener solution to treat wastewater for reuse or to meet discharge requirement and has gained prominence especially for oily wastewater [1,2], whose volume is significant from various industries. In particular, conventional separation technologies, such as gravity-based, hydrocyclone, centrifugation, or dissolved air flotation (DAF), are inefficient for treating oily wastewater with oil emulsion sized at lower than 20 μm, whereas membrane process is highly feasible [1]. Despite the promising application of

DOI: 10.1201/9781003091011-7

membrane filtration technology, the inevitable membrane fouling issue remains to be resolved. The deposition of foulant(s) onto the membrane surface is detrimental to the separation performance, and thus needs to be understood comprehensively for systematically tackling this problem.

The study on membrane fouling during oil–water separation has benefited extensively in recent years from the methods that offer in-situ, nonobstructive, and real-time observation of the development of fouling. The direct observation through the membrane (DOTM) technique was employed by Tummons et al. and Tanudjaja et al. to observe how oil emulsions deposited and behaved on the membrane surface [3,4]. Tummons et al. observed how the oil emulsions attached to the membrane surface, deformed, and coalesced [3]. Subsequently, they also found that the presence of salt decreased the interfacial tension of the emulsion, which led to coalescence and worsened fouling [5]. Tanudjaja et al. characterized the critical fluxes of oil emulsion using DOTM images for various magnitudes of cross-flow velocity (CFV), oil concentration, and salinity [4]. Higher oil concentration and salinity increased the fouling propensity while higher CFV reduced. The optical coherence tomography (OCT) technique was employed by Trinh et al. to study and visualize internal fouling caused by oil emulsion [6], while UTDR (Ultrasonic time-domain reflectometry) was used by Silalahi et al. and Xu et al. to understand oil fouling on flat sheet and hollow fiber microfiltration (MF) membranes, respectively [7,8].

To understand oil fouling more comprehensively, the observation of the fouling should not be done only experimentally but theoretically, such as the interfacial interaction energy analysis between membrane and foulant [9]. Molecular dynamics simulations have provided some insights congruent with DOTM observations [10], while the well-known DLVO and XDLVO models are much more popular [11,12]. The Derjaguin–Landau–Verwey–Overbeek (DLVO) model is well-used by many studies for various foulants but is inadequate with respect to the shorter-range separation distance between two surfaces [13]. For tackling this problem, the Lewis acid–base (AB) interaction, which is based on the electron–donor and electron–acceptor interaction at shorter distances (<10 nm), was incorporated into the DLVO model to give the extended DLVO (XDLVO) model [14]. The XDLVO model was found to predict fouling more accurately than DLVO for three colloid particles and three reverse osmosis (RO) membranes [13], which also agreed well with the atomic force microscopy (AFM) measurement of the interaction energy between two surfaces [15]. Ahmad et al. and Kuhnl et al. employed XDLVO analysis to explain the fouling mechanisms of microalgae and casein micelle [16, 17], respectively, which proved that these models are applicable to non-rigid and deformable particles as well.

Generally, XDLVO has been used more often than DLVO for understanding colloidal fouling on the membrane surface, but the know-how for apolar foulants like oil is not as well understood. In the following, the description of DLVO and XDLVO interaction energy analysis is given. For the subsequent section, we will give a summary of the fouling studies that have incorporated these models, with more emphasis on the oil fouling studies in recent years. Finally, we will discuss the challenges and improvements that can be done for DLVO/XDLVO analyses for membrane fouling by oil emulsion.

7.2 DLVO AND XDLVO INTERACTION ENERGY

The classical DLVO theory stipulates that the total interaction energy is the sum of the Lifshitz–van der Waals (LW) and electrostatic (EL) components while the XDLVO theory incorporates an additional Lewis AB component for short separation distances (<10 nm) based on the donor–acceptor and electron–acceptor (hydrophobic and hydrophilic) interaction. DLVO and XDLVO total interaction energies (U_{mlo}^{DLVO} and U_{mlo}^{XDLVO}) are shown in these following equations [14]:

$$U_{mlo}^{DLVO} = U_{mlo}^{LW} + U_{mlo}^{EL} \tag{7.1}$$

$$U_{mlo}^{XDLVO} = U_{mlo}^{LW} + U_{mlo}^{EL} + U_{mlo}^{AB} \tag{7.2}$$

where U_{mlo}^{LW}, U_{mlo}^{EL}, and U_{mlo}^{AB} are interaction energy of LW, electrostatic (EL), and Lewis AB components, respectively. The subscripts m, l, and o denote, respectively, the membrane, liquid environment in the bulk feed and oil emulsion. U as the interaction energy is a function of separation distance between two surfaces: zeta potential and surface tension. The DLVO and XDLVO interaction energies can be expressed in terms of U or free energy of adhesion per unit area (ΔG), the latter of which is not a function of separation distance and usually uses the minimum separation distance (0.158 nm) in the calculation [13], as shown in these equations:

$$\Delta G^{DLVO} = \Delta G_{y0}^{LW} + \Delta G_{y0}^{EL} \tag{7.3}$$

$$\Delta G^{XDLVO} = \Delta G_{y0}^{LW} + \Delta G_{y0}^{EL} + \Delta G_{y0}^{AB} \tag{7.4}$$

Details of each component of the interaction energy will be discussed in the next section. Specifically, U is a function of separation distance and can present the development in the foulant–membrane interactions as the oil approaches the membrane surface. As the separation distance of the two surfaces increases, the interaction energy decays. Negative interaction energy is associated with attractive force and vice versa. The original DLVO and XDLVO models assume the interaction of two infinite planar surfaces, but this approach is not relevant for membrane fouling. For tackling this issue, Brant and Childress [13] proposed the idea to convert the interaction between a sphere (in this case, oil emulsion) and a planar surface (i.e., membrane) interaction via the Derjaguin's approximation that will be discussed in detail in the next section.

7.2.1 LW INTERACTION

The nonpolar LW interaction between two surfaces, specifically in this case between the membrane and oil emulsion, can be represented as the free energy of adhesion per unit area (ΔG^{LW}) between two infinite planar surfaces [13]:

$$\Delta G_{y0}^{LW} = 2\left(\sqrt{\gamma_l^{LW}} - \sqrt{\gamma_m^{LW}}\right)\left(\sqrt{\gamma_o^{LW}} - \sqrt{\gamma_l^{LW}}\right) \tag{7.5}$$

where γ denotes the surface tension (i.e., surface free energy per unit area) of a medium (i.e., liquid or membrane or oil) and y_0 is the minimum separation distance of the two surfaces and is usually assigned the value of 0.158 nm. As mentioned earlier, for getting the relevant interaction energy between the membrane surface (assumed to be an infinite planar surface) and an oil droplet (assumed to be a sphere), the Derjaguin's technique can be applied to scale the LW free energy of adhesion as a function of separation distance (h) between the membrane and oil emulsion:

$$U_{mlo}^{LW}(h) = 2\pi\Delta G_{y0}^{LW}\frac{y_0^2 a_c}{h} \tag{7.6}$$

where a_c is the radius of the oil droplet.

7.2.2 EL INTERACTION

The EL free energy per unit area (ΔG^{EL}) between two infinite planar surfaces and the interaction energy as a function of the distance between the membrane surface and oil droplet are given by [13]:

$$\Delta G_{y0}^{EL} = \frac{\varepsilon_0\varepsilon_r\kappa}{2}\left(\zeta_m^2 + \zeta_o^2\right)\left[1 - \coth(\kappa h) + \frac{2\zeta_m\zeta_o}{\left(\zeta_m^2 + \zeta_o^2\right)}csch(\kappa h)\right] \tag{7.7}$$

$$U_{mlo}^{EL}(h) = \pi\varepsilon_0\varepsilon_r a_c\left[2\zeta_m\zeta_o \ln\left(\frac{1+e^{-\kappa h}}{1-e^{-\kappa h}}\right) + \left(\zeta_m^2 + \zeta_o^2\right)\ln\left(1-e^{-2\kappa h}\right)\right] \tag{7.8}$$

where ζ_m and ζ_o are the zeta potentials of the membrane and the oil, respectively, $\varepsilon_0\varepsilon_r$ is the dielectric permittivity of the liquid environment, and κ is the inverse Debye screening length, which is equal to 10^6 m for deionized (DI) water [18].

7.2.3 AB INTERACTION

The Lewis AB interaction is based on the polar interaction due to the electron–donor and electron–acceptor association, which is linked to the hydrophobicity and hydrophilicity of a surface. The free energy of adhesion per unit area between two infinite planar surfaces and the interaction energy (U_{AB}) from the Derjaguin correlation are expressed as a function of the polar components of the surface tension of the membrane (m), bulk feed solution (l), and oil droplet (o):

$$\Delta G_{yo}^{AB} = 2\sqrt{\gamma_l^+}\left(\sqrt{\gamma_m^-} + \sqrt{\gamma_o^-} - \sqrt{\gamma_l^-}\right) + 2\sqrt{\gamma_l^-}\left(\sqrt{\gamma_m^+} + \sqrt{\gamma_o^+} - \sqrt{\gamma_l^+}\right)$$

$$- 2\left(\sqrt{\gamma_m^+\gamma_o^-} + \sqrt{\gamma_m^-\gamma_o^+}\right) \tag{7.9}$$

$$U_{mlo}^{AB}(h) = 2\pi a_c \lambda \Delta G_{yo}^{AB} \exp\left[\frac{y_0 - h}{\lambda}\right] \tag{7.10}$$

where the superscripts $+$ and $-$ represent the electron–donor and electron–acceptor, respectively, while λ is the characteristic decay length of the AB interaction in water, with the commonly used value for aqueous medium of 0.6 nm [19]. Since oil emulsion is mostly nonpolar in nature, the value of γ_o^- and γ_o^+ could be assumed as zero [20], hence equation 7.9 is simplified to Eq. 7.11.

$$\Delta G_{yo}^{AB} = 2\sqrt{\gamma_l^+}\left(\sqrt{\gamma_m^-} - \sqrt{\gamma_l^-}\right) + 2\sqrt{\gamma_l^-}\left(\sqrt{\gamma_m^+} - \sqrt{\gamma_l^+}\right) \tag{7.11}$$

7.2.4 Surface Tension Component, Contact Angle, and Zeta Potential

The overall surface tension (γ) consists of the nonpolar (LW) and polar (AB) components (shown by Eq. 7.12), which is usually measured by a goniometer.

$$\gamma = \gamma^{LW} + \gamma^{AB} \tag{7.12}$$

where the nonpolar component of the surface tension is symbolized with γ^{LW} while the polar component with γ^{AB}, which is like the nonpolar (i.e., LW) and polar (i.e., AB) components of the DLVO and XDLVO models. The polar component can be broken down to two main parameters, namely the electron–donor (γ^+) and electron–acceptor (γ^-) components [21]:

$$\gamma^{AB} = 2\sqrt{\gamma^+\gamma^-} \tag{7.13}$$

In most studies, three parameters (namely γ^{LW}, γ^+, and γ^-), either of the membrane or the foulant (in this case, oil), are not readily available and cannot be measured directly. So, the extended Young equation is applied by measuring their contact angles (θ) with standard materials and calculating via Eq. 7.14:

$$(1 + \cos\theta)\gamma_l = 2\left[\sqrt{\gamma_s^{LW}\gamma_l^{LW}} + \sqrt{\gamma_l^+\gamma_l^-} + \sqrt{\gamma_s^-\gamma_l^+}\right] \tag{7.14}$$

where the subscripts s and l denote solid (membrane) and standard liquid, respectively. Since there are three unknown parameters (i.e., γ^{LW}, γ^+, and γ^-), three standard liquids are used, consisting of two nonpolar (e.g., diiodomethane and bromonaphthalene) and one polar (e.g., water, formamide, ethylene glycol, and glycerol) liquids [13,15,20,22]. Table 7.1 shows the summary of the surface tension components of these standard liquids.

TABLE 7.1

Surface Tension Components of Standard Liquids [13,15,20,22]

Liquid	γ	γ^{LW}	γ^{AB}	γ^+	γ^-
Diiodomethane	50.8	50.8	0	0	0
Bromonaphthalene	44.0	44.0	0	0	0
Water	72.8	21.8	51.0	25.5	25.5
Formamide	58.0	39.0	19.0	2.3	39.6
Ethylene glycol	48.0	29.0	19.0	1.9	47.0
Glycerol	64.0	34.0	30.0	3.9	57.4

Note: Units are in mJ m^{-2}

For the membrane surface tension component, the measurement can be done by captive bubble method with these three standard liquids while the measurement for oil surface tension component is different. Because oil is largely nonpolar, the polar component is assumed to be zero and only the nonpolar surface tension component (γ^{LW}) of the oil is measured by using the oil contact angle on a reference nonpolar solid material, such as parafilm, and calculating it with Eq. 7.14, thus giving Eq. 7.15 as follows [23]:

$$\gamma_o^{LW} = \frac{\left[(1+\cos\theta)\gamma\right]^2}{4\,\gamma_{parafilm}^{LW}} \tag{7.15}$$

where θ is the contact angle of the oil on the parafilm, γ is the overall surface tension, and $\gamma_{parafilm}^{LW}$ is the LW surface tension component of the parafilm (which is 25.5 mN m^{-1} [24]). He et al. used a flat sheet of polytetrafluoroethylene (PTFE) as the substitute for parafilm in their study for the measurement of crude oil surface tension components [20]. In both studies [20,23], it was shown that, for oil, the total surface tension and the nonpolar component of the surface tension had similar values, thus confirming the nonpolarity of the oils.

7.3　USAGE IN NONOIL MEMBRANE FOULING

DLVO and XDLVO models have been used in understanding membrane fouling by a variety of foulants and for different types of membranes. For filtration ranging from MF to RO, DLVO and XDLVO has been proven to give a more comprehensive understanding of membrane fouling and complements well with the observed fouling. Table 7.2 shows a summary of membrane fouling studies in the last two decades, and it can be observed that XDLVO is more popular in recent years due to the inclusion of shorter-ranged AB force that describes more comprehensively the membrane fouling mechanism. These models predict fouling well not only for rigid-type foulants such as polystyrene and silica particles but also the soft and deformable types of foulants like casein micelles, algae, sludge floc, and oil emulsion. Among the filtration modes, MF has been most popularly studied with DLVO and XDLVO models.

TABLE 7.2

Derjaguin–Landau–Verwey–Overbeek (DLVO) and Extended DLVO (XDLVO) for the Various Nonoil Membrane Fouling

	Interaction Model					
Year	DLVO	XDLVO	Foulant	Membrane material	Process	vRef.
1998	✓	✗		Al_2O_3	MF	[25]
2002	✓	✓	Silica	TFC	RO	[13]
			Al_2O_3	Cellulose triacetate/diacetate		
			Polystyrene	Cellulose triacetate/diacetate		
2002	✓	✓	Silica	Cellulose triacetate/diacetate	RO	[15]
			Al_2O_3	TFC		
			Polystyrene			
2007	✗	✓	NOM	PES	UF	[26]
2010	✓	✓	Casein	Al_2O_3	MF	[17]
2013	✗	✓	Polysaccharide	PVDF	MF	[27]
2013	✓	✓	Microalgae	Cellulose acetate	MF	[16]
2014	✗	✓	AOM	Mixed cellulose	MF	[28]
2014	✗	✓	AOM	Mixed cellulose	MF	[29]
2016	✗	✓	*E. coli*	Modified PVDF	MF	[30]
2016	✗	✓	Sludge floc and gelling foulant	PVDF	MF	[31]
2017	✗	✓	NOM	PVDF	MF	[32]
2017	✗	✓	Sludge floc and gelling foulant	PVDF	MF	[33]
2017	✗	✓	Sludge floc and gelling foulant	PVDF	MF	[34]
2018	✗	✓	APAM	PVDF and PTFE	MF	[35]
2018	✗	✓	[BMIM]Cl	PTFE and PAN	MD	[36]
2019	✗	✓	DOM	TFC	RO	[37]
2019	✗	✓	BSA	PANI	MF	[38]
2019	✗	✓	Sodium alginate	PVDF	MF	[39]
2019	✗	✓	BSA	TFC-Ca	FO	[40]
			Humic acid			
			Sodium alginate			
2020	✗	✓	SDS	PVDF and PTFE	MF	[41]

MF, microfiltration; UF, ultrafiltration; RO, reverse osmosis; MD, membrane distillation; FO, forward osmosis; PES, polyethersulfone; PAN, polyacrylonitrile; BSA, bovine serum albumin; APAM, anionic polyacrylamide; DOM, dissolved organic matter; NOM, natural organic matter; AOM, algogenic organic matter; TFC, thin-film composite; PVDF, polyvinylidene fluoride; PTFE, polytetrafluoroethylene; PANI, polyaniline; SDS, sodium dodecyl sulfate; [BMIM]Cl, 1-butyl-3-methylimidazolium chloride.

7.4   USAGE IN OIL MEMBRANE FOULING

The application of DLVO and XDVLO interaction energy analysis for membrane fouling by oil is relatively recent and not many studies are available to date. Such studies can be divided into two categories: the understanding of the oil fouling mechanism and development and assessment of antifouling characteristics of novel, modified membranes. The predictive capability of the DLVO and XDLVO models are shown by comparing the interaction force between the oil emulsion and membrane surface with fouling indicators, such as critical flux, transmembrane pressure (TMP), permeate flux stability, and oil contact angle [20,23,42–44].

7.4.1   Fouling Studies

Membrane filtration is an attractive technology to treat oily feeds, but membrane fouling hinders optimal performance. In efforts on understanding oil fouling on membrane surfaces, DLVO and XDLVO models are applied to provide insights on the fouling mechanism. He et al. studied the effect of ionic strength of the feed on the development of fouling by crude oil emulsion [20]. Three different salt concentrations were used, which varied the zeta potential and hence the EL force of the DLVO model, while the LW force remained the same. As the salinity increased, the EL force decreased and hence the oil–membrane repulsion was reduced, resulting in worsened fouling as confirmed by the TMP profiles (Figure 7.1a). The same trend of decreasing EL force was observed for oil emulsion–oil layer interaction energy, as for the case of oil emulsions interacting with a membrane already fouled with oil (Figure 7.1b), with the interaction being attractive at the highest NaCl concentration used (i.e., 0.1 M). The repulsion effect was greater for oil–membrane than oil–oil interaction, thus leading to slow TMP buildup at the beginning, then TMP jump at the later stage, as shown in Figure 7.1c.

Tanudjaja et al. used both DLVO and XDLVO models, along with the DOTM method which allowed the real-time and in-situ observation of fouling, to study the oil fouling mechanism by various type of oils [23]. The critical flux of hexadecane, soybean oil, fish oil, and crude oil emulsions were investigated via DOTM images in cross-flow MF. With respect to oil concentration and CFV, crude oil gave the lowest critical flux, which indicated greatest tendency to foul and was confirmed by the fastest TMP increase during the constant flux operation (Figure 7.2a). The DLVO energy profile for the oil–membrane and oil–oil layer interactions showed that the crude oil had the least repulsive force (Figure 7.2c and e, respectively), hence explaining why crude oil fouled the membrane easily. The interesting takeaway was the mismatch between the XDLVO analysis and fouling trends, because the inclusion of the AB interaction force masked the other two forces due to its significantly higher value as shown in Figure 7.2b and d for oil–membrane and oil–oil layer interactions. Due to the nonpolar nature of the oils, the AB force calculation only considered the membrane–water interaction and it was much greater than LW and EL forces, thus the total XDLVO interaction energies were similar for various oil emulsions and became inadequate for foretelling the different fouling tendencies. Therefore, the DLVO model, which discounted the AB component, was better correlated with the fouling.

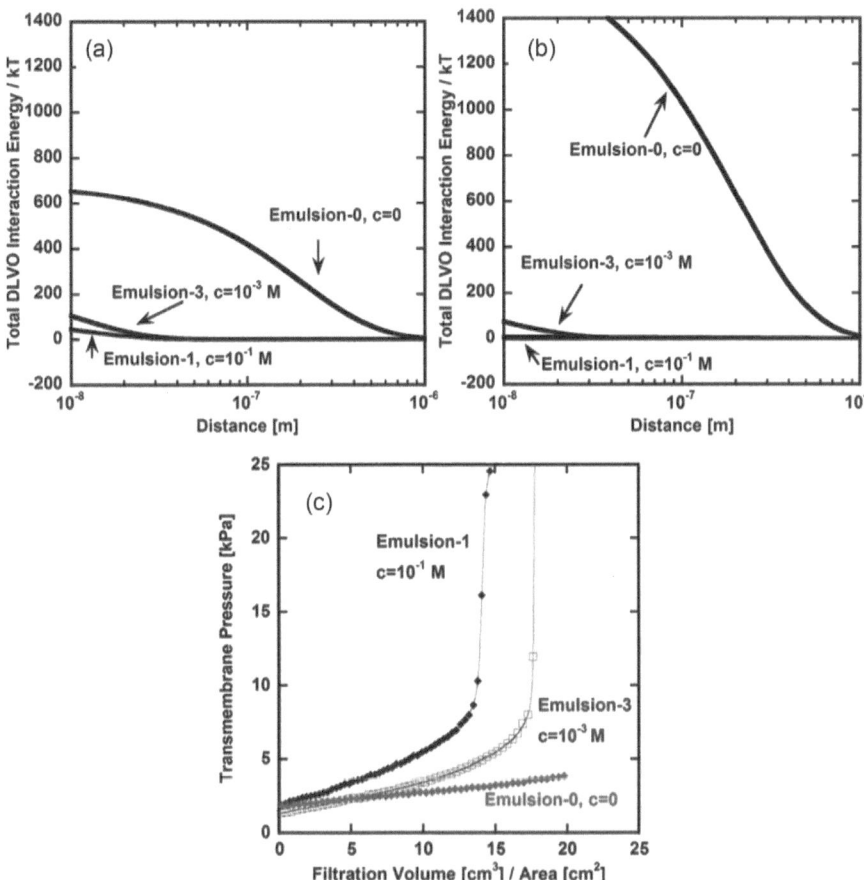

FIGURE 7.1 Total DLVO energy for oil–membrane interaction (a) and oil–oil layer interaction (b) with TMP profile during the filtration test with three salt concentration (c). (Reprinted with permission from [20].)

Furthermore, a more attractive XDLVO energy for oil emulsion–oil layer interaction than the oil emulsion–membrane interaction was tied to the moving cake layer phenomenon observed in this study and previous studies [4,45].

Tanudjaja et al. further used the DLVO model to correlate with oil fouling by a mixed feed consisting of casein micelle, oil emulsion, and lactose to mimic skim milk filtration with MF [42]. DOTM images were analyzed to obtain the critical fluxes, with the mixture of oil emulsion and casein micelle giving the lowest critical flux due to the increase of viscosity and smaller sizes of casein micelles. This agreed with the DLVO interaction energy profile that indicated that this mixture had the least repulsive force with the membrane, as shown by Figure 7.3. In another study, Tanudjaja et al. used the XDLVO interaction energy analysis to compare the fouling evolution of oil emulsions and polystyrene latex particles on hollow fiber membranes [43]. The TMP profile showed that oil fouling was relatively more severe

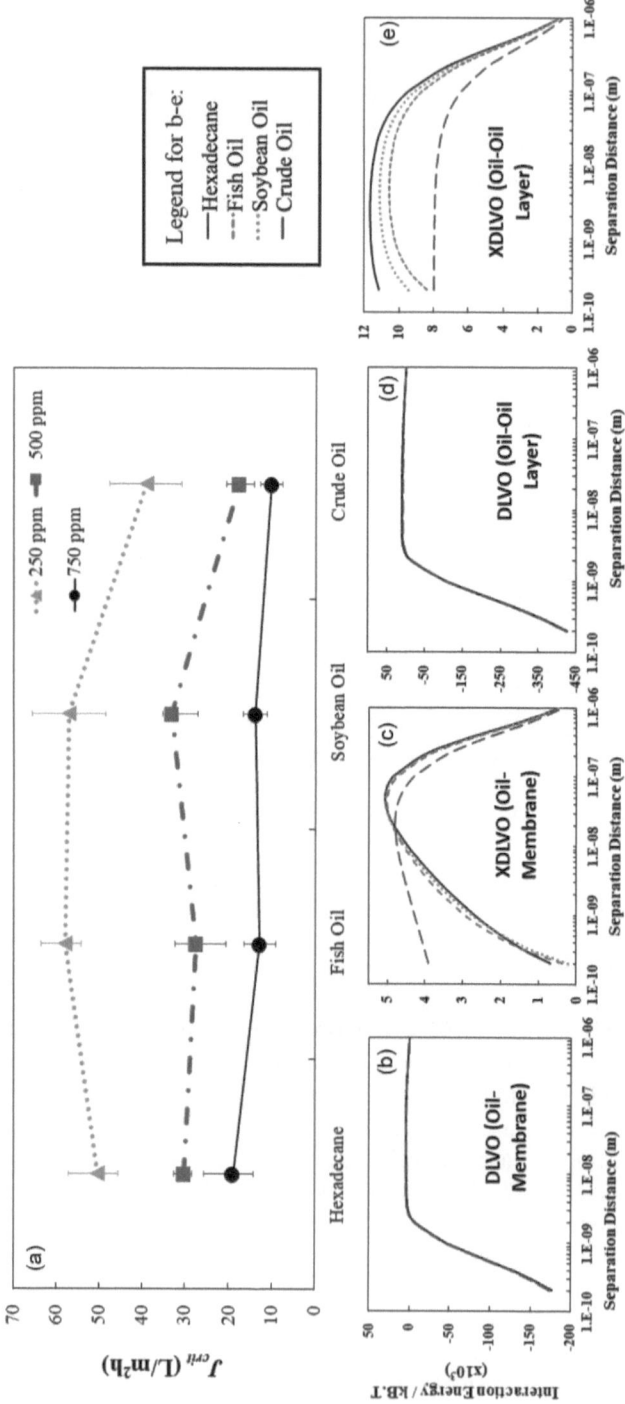

FIGURE 7.2 Critical flux of various oil as the effect of concentration (a), DLVO and XDLVO total interaction energy for oil–membrane interaction (b and c) and oil–oil layer interaction(d and e). (Reprinted with permission from [23].)

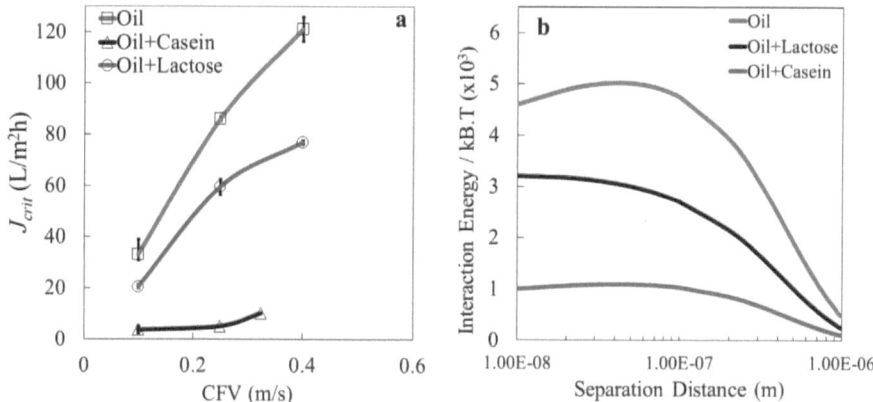

FIGURE 7.3 Critical flux of mixture feed of oil emulsion, oil–lactose, and oil–casein (a) with the total DLVO interaction energy of these mixture feeds (b). (Reprinted with permission from [42].)

while DOTM images showed that the oil deposition was relatively more uniform on the membrane surface. The XDLVO interaction energy for both the foulant–membrane and foulant–foulant interaction was more attractive for oil emulsions (as shown in Figure 7.4), which confirmed the more severe fouling phenomenon, and underlie why the deposition of oil emulsions was more uniform than latex particles. Due to the different polarity between oil and latex particle, the inclusion of the AB force gave different values of interaction energy and hence the XDLVO model correlated well with the fouling trends.

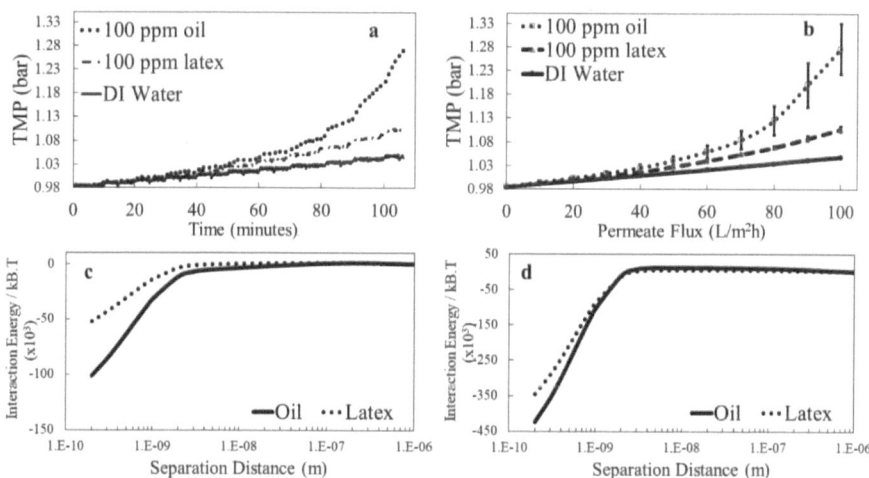

FIGURE 7.4 TMP profiles of DI water, 100 ppm oil emulsion, and latex with respect to time (a) and collected permeate flux (b). Total XDLVO interaction energy analysis of foulant–membrane (c) and foulant–foulant (d) interactions, where the foulants were oil emulsion and polystyrene latex particle. (Reprinted with permission from [43].)

Zhang et al. compared the interaction between oil emulsion and polymer (APAM) with a PTFE MF membrane via the XDLVO energy analysis during the treatment of alkali/surfactant/polymer (ASP) flooding oilfield wastewater [46]. The APAM or anion polyacrylamide had a less negative interaction energy with the membrane compared to oil–membrane interaction, which underlie the more significant fouling by the crude oil emulsion among the other foulants in ASP wastewater. As for foulant–foulant interaction energy, the oil–oil interaction was more attractive than that of other foulants, which agreed with the severe oil fouling at the later stage of the filtration. It was observed also that the presence of salinity and surfactant made the total interaction energy less attractive, which hence mitigated fouling. OCT was employed by Trinh et al. to study the effect of surfactant on oil fouling in conjunction with the DLVO-XDLVO energy analysis [22]. Anionic, cationic, and nonionic surfactants were used to give different charges on the oil emulsions, which were filtered with PVDF MF membranes in a dead-end setup. From the analysis of the OCT images, it was observed that the nonionic surfactant-stabilized oil emulsion (Tween 20) had the most extensive fouling in comparison to the other positively and negatively charged oil emulsion (cetrimonium bromide-CTAB and SDS, respectively) observed from the highest increase of the fouling voxel fraction along the filtration time, as shown in Figure 7.5a–c. All the DLVO of the oil–membrane interaction energies showed repulsive interaction, with the nonionic surfactant being the least repulsive, thus leading to more significant fouling. Although the XDLVO energies were attractive at separation distances below 5 nm, nonionic surfactant-stabilized oil emulsion had the least repulsion force, facilitating deposition, as shown in Figure 7.5d and e.

7.4.2 ANTIFOULING VALIDATION OF MODIFIED MEMBRANE

Like the membrane fouling studies, DLVO and XDLVO models are also used for validating the antifouling quality of novel membranes by focusing specifically on the oil–membrane interaction only rather than the oil–oil layer interaction. Membrane distillation (MD), a thermal-driven process, has emerged in recent years to treat oily wastewater, since it gives perfect rejection of oil by allowing only water vapor to pass through the membrane. This process utilizes low-grade heat in the plants and employs hydrophobic membranes to prevent the membrane pore-wetting phenomenon, but it is challenging for treatment of oily feeds especially due to the strong association of the hydrophobic membrane with hydrophobic oil emulsions. Zuo and Wang modified a PVDF membrane surface via plasma-induced grafting of polyethylene glycol and TiO_2 deposition to make it highly hydrophilic, thereby repelling the oil emulsions [47]. The virgin and modified membranes had mean pore sizes of 0.22 and 0.19 μm, respectively, and were tested with mineral oil emulsion with a mean pore diameter of 4.5 μm. Results showed that the modified membrane had a lower initial flux of 13% but flux remained stable at 6.3 kg m^{-2}h over 24 hours, whereas the flux of the virgin membrane dropped to 3.5 kg m^{-2}.h and salt breakthrough (permeate contained conductivity above 10 μS cm^{-1}) happened after 21 hours of operation. From the XDLVO interaction energy analysis, it was observed that the oil–membrane interaction for the virgin PVDF was attractive but repulsive for the modified membrane. The difference

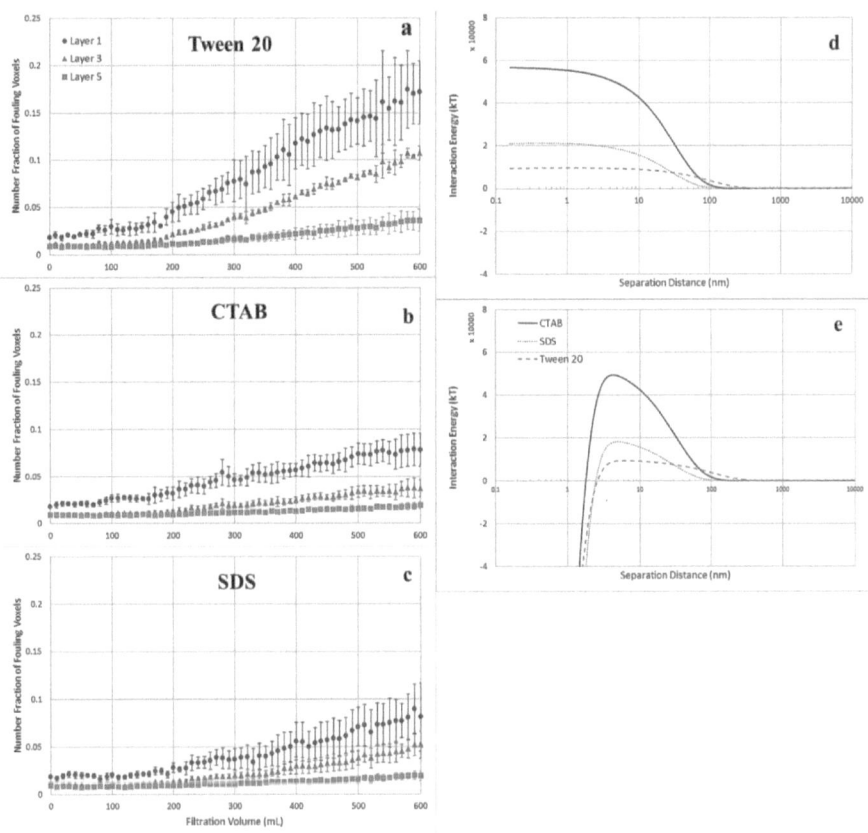

FIGURE 7.5 Profile of fouling voxels fraction at layer 1, 3, and 5 (approximately 1.5, 4.5, and 7.5 μm above feed–membrane interface) for the filtration oil emulsion stabilized with Tween 20 (a), CTAB (b), and SDS (c). Total DLVO (d) and XDLVO (e) interaction energy analysis of foulant–membrane of different charged oil emulsions. (Reprinted with permission from [22].)

stemmed from the AB values, with the modified membrane having a higher γ^- and lower γ^+, which in total became more polar than the virgin membrane.

Another study by Chen et al. took a different approach to tackle this issue by making the membrane superhydrophobic (water contact angle > 150°) via plasma surface treatment, and thus the membrane had a strongly negative charged surface [48]. Anionic and cationic surfactants, namely SDS and TDBAC, respectively, were used to stabilize the oil emulsion feed in a 10,000 ppm NaCl solution. No value of interaction energy was mentioned in this study, but the XDLVO approach was employed to explain the wetting mechanism by oil emulsion on these membranes. The virgin PVDF membrane was wetted by both oil emulsions, while the modified one was wetted only by cationic surfactant-stabilized oil emulsion. When the positively charged oil emulsion approached the membrane, the positive surfactants were attracted to the membrane, leading the hydrophobic oil to be also attracted to the hydrophobic membrane surface via electron–donor and electron–acceptor association. Meanwhile, the surfactant molecules from the negatively charged oil emulsion were repelled by the membrane surface via electrostatic repulsion and thus diffused away. From these scenarios, it can be observed that the negatively charged oil emulsion was more repelled due to electrostatics and repelled by the superhydrophobic membrane, which thus delayed or prevented the oil deposition. Another modification of the PVDF membrane was reported by Guo et al., in which graphene oxide (GO) nanosheets and alumina nanowires were coated on the membrane surface to make it more hydrophilic [49]. The modification enhanced the hydrophilicity and oleophobicity of the surface, which was confirmed by the XDLVO total energy of 0.63 and 0.9 kT for the virgin and modified membranes, respectively. The higher energy barrier led to higher repulsion, which thus prevented the oil fouling. The addition of alumina nanowires to the GO-modified membrane reduced γ^{LW} and enhanced γ^{AB}, which enhanced the polarity of the membrane surface and changed the total free adhesion energy between oil and membrane surface from negative to positive. The permeate flux increased from 942 to 29,142 LMH bar^{-1} with oil rejection of 99.4%. He et al. modified PAN membrane with polydopamine (PDA) micro- and nanospheres to make the surface more hydrophilic to prevent oil fouling [44]. These membranes were tested with toluene-in-water emulsion and the permeate flux drop was lower for the modified membrane. Specifically, while the flux of the modified membrane decreased from 32,556 to 11,666 LMH bar^{-1} in 2 hours with an oil rejection of 99.9%, that of the virgin PAN membrane dropped from 26,170 to 4260 LMH bar^{-1}. PDA has a strong association with water molecules, hence creates a stable hydration layer on the membrane surface to repel oil emulsion, as evident in the increase of underwater oil contact angle from 151° to 165°. The XDLVO interaction energy of the toluene-PAN membrane was more attractive, which agreed with to the higher oil fouling propensity and lower permeate flux. The AB force was more dominant than the LW force.

7.4.3 Consideration for Future Research

Studies have proven DLVO and XDLVO interaction energy analysis to be feasible in explaining oil–membrane interaction and thereby the oil fouling phenomenon. In oil fouling studies, the DLVO and XDLVO models are employed not only for the interaction between oil emulsion and membrane surface but also between oil emulsion and oil layer for the fouled membrane at the later filtration stage. The analysis from these interactions helps to understand the oil fouling from the early to later stages and explain why oil fouling behaves a certain way (e.g., moving cake layer) [4,23]. In explaining the fouling by various oils, the DLVO model, which is dominated by the EL force and discounts the AB force, is clearer in explaining the difference in fouling tendencies of various oils. Due to the nonpolar nature of oil emulsions, the AB force is calculated only from the surface tension values of membrane and water and hence has similar values that are orders of magnitude greater than that of EL and LW for all oil types. This does not mean that XDLVO gives false information regarding the oil fouling but that the DLVO model is more suitable for comparing oil fouling.

For comparison between fouling by oil and other colloids, the XDLVO model can give better clarity on the difference of foulant–membrane interactions that underlie fouling tendencies and antifouling qualities of modified membranes. The XDLVO model is used due to the significant difference in the polarity of the virgin and modified membranes that gives better clarity on the oil–membrane interaction. Since the focus is to differentiate the interaction energy value between oil with virgin and modified membranes to confirm the antifouling property, the XDLVO energy is not calculated as a function of separation distance but only using the minimum separation distance (0.158 nm). By comparing these values, predictions on how much better the modified membrane is in mitigating oil deposition can be obtained, which has been confirmed to relate well to stable permeate flux or increase in underwater oil contact angle.

He et al. mentioned that the oil fouling phenomenon is determined by a balance between the drag force from the permeate flux, membrane–oil emulsion surface interaction, and the shear force due to CFV [20]; hence, to understand oil fouling comprehensively, all these factors must be considered. DLVO and XDLVO interaction energy analysis are one of the main tools to explain the oil fouling mechanism, but it is a concept that cannot properly account for drag force and shear force effects. Another challenge for DLVO and XDLVO models is the need of having a universal standard measurement for unknown foulant surface tension components. He et al. and Tanudjaja et al. used PTFE membrane and parafilm to measure the nonpolar component of oil surface tension, respectively, due to the nonpolar nature of PTFE [20] and parafilm [23], while Brant and Childress deposited the silica, alumina, and polystyrene particles on the top of membrane and measured the contact angle of the three standard liquids atop the modified membranes [13]. A universal standard measurement of foulant surface tension component could reduce the error that is inherent from the various methods here and make DLVO-XDLVO models more reliable for different kinds of foulants.

7.5   CONCLUSION

Interaction energy analysis of Derjaguin–Landau–Verwey–Overbeek (DLVO) model and its extended version (XDLVO) has been successfully employed in explaining the oil fouling mechanisms in various types of membrane filtration processes. The results agreed with the fouling trends by oil observed via critical flux and TMP profiles, hence proving its feasibility to be applied in understanding oil–membrane interaction during the fouling. DLVO and XDLVO energy profiles are shown as a function of separation distance between two surfaces for giving a comprehensive view on the evolution of the interaction as the oil emulsion approaches the membrane surface. For the validation of antifouling membrane properties, the models are calculated using the minimum separation distance only. Although the XDLVO model is a more comprehensive model, in explaining the oil fouling, the AB force dominates the XDLVO total energy value, which drowns out LW and EL and makes unclear the fouling tendencies of different oil emulsions. Therefore, for comparing different oils, the DLVO model is superior. If either the polarity of oil or membrane is different or altered significantly, the XDLVO model could give clearer differences of the oil–membrane interactions, as found in the studies for validating antioil fouling property of modified membranes. As oil fouling on the membrane surface is a balance of permeate drag force, shear force, and oil–membrane interaction force, DLVO and XDLVO models are insufficient alone to explain the whole oil fouling story, thus understanding from other hydrodynamics factors shall be accounted for too.

REFERENCES

1. M. Cheryan, N. Rajagopalan, Membrane processing of oily streams. Wastewater treatment and waste reduction, *Journal of Membrane Science*, 151 (1998) 13–28.
2. J. Križan Milić, A. Murić, I. Petrinić, M. Simonič, Recent developments in membrane treatment of spent cutting-oils: A review, *Industrial & Engineering Chemistry Research*, 52 (2013) 7603–7616.
3. E.N. Tummons, V.V. Tarabara, J.W. Chew, A.G. Fane, Behavior of oil droplets at the membrane surface during crossflow microfiltration of oil–water emulsions, *Journal of Membrane Science*, 500 (2016) 211–224.
4. H.J. Tanudjaja, V.V. Tarabara, A.G. Fane, J.W. Chew, Effect of cross-flow velocity, oil concentration and salinity on the critical flux of an oil-in-water emulsion in microfiltration, *Journal of Membrane Science*, 530 (2017) 11–19.
5. E.N. Tummons, J.W. Chew, A.G. Fane, V.V. Tarabara, Ultrafiltration of saline oil-in-water emulsions stabilized by an anionic surfactant: Effect of surfactant concentration and divalent counterions, *Journal of Membrane Science*, 537 (2017) 384–395.
6. T.A. Trinh, W. Li, Q. Han, X. Liu, A.G. Fane, J.W. Chew, Analyzing external and internal membrane fouling by oil emulsions via 3D optical coherence tomography, *Journal of Membrane Science*, 548 (2018) 632–640.
7. S.H.D. Silalahi, T. Leiknes, J. Ali, R. Sanderson, Ultrasonic time domain reflectometry for investigation of particle size effect in oil emulsion separation with crossflow microfiltration, *Desalination*, 236 (2009) 143–151.
8. X. Xu, J. Li, N. Xu, Y. Hou, J. Lin, Visualization of fouling and diffusion behaviors during hollow fiber microfiltration of oily wastewater by ultrasonic reflectometry and wavelet analysis, *Journal of Membrane Science*, 341 (2009) 195–202.
9. J.W. Chew, J. Kilduff, G. Belfort, The behavior of suspensions and macromolecular solutions in crossflow microfiltration: An update, *Journal of Membrane Science*, 601 (2020) 117865.
10. M.B. Tanis-Kanbur, S. Velioğlu, H.J. Tanudjaja, X. Hu, J.W. Chew, Understanding membrane fouling by oil-in-water emulsion via experiments and molecular dynamics simulations, *Journal of Membrane Science*, 566 (2018) 140–150.
11. P. Janknecht, A.D. Lopes, A.M. Mendes, Removal of industrial cutting oil from oil emulsions by polymeric ultra- and microfiltration membranes, *Environmental Science & Technology*, 38 (2004) 4878–4883.
12. K.J. Howe, M.M. Clark, Fouling of microfiltration and ultrafiltration membranes by natural waters, *Environmental Science & Technology*, 36 (2002) 3571–3576.
13. J.A. Brant, A.E. Childress, Assessing short-range membrane–colloid interactions using surface energetics, *Journal of Membrane Science*, 203 (2002) 257–273.
14. C.J. Van Oss, *Interfacial Forces in Aqueous Media*, CRC Press, Boca Raton, FL, 2006.
15. J.A. Brant, A.E. Childress, Membrane–Colloid interactions: Comparison of extended DLVO predictions with AFM force measurements, *Environmental Engineering Science*, 19 (2002) 413–427.
16. A.L. Ahmad, N.H. Mat Yasin, C.J.C. Derek, J.K. Lim, Harvesting of microalgal biomass using MF membrane: Kinetic model, CDE model and extended DLVO theory, *Journal of Membrane Science*, 446 (2013) 341–349.

17. W. Kühnl, A. Piry, V. Kaufmann, T. Grein, S. Ripperger, U. Kulozik, Impact of colloidal interactions on the flux in cross-flow microfiltration of milk at different pH values: A surface energy approach, *Journal of Membrane Science*, 352 (2010) 107–115.

18. F. Zamani, A. Ullah, E. Akhondi, H.J. Tanudjaja, E.R. Cornelissen, A. Honciuc, A.G. Fane, J.W. Chew, Impact of the surface energy of particulate foulants on membrane fouling, *Journal of Membrane Science*, 510 (2016) 101–111.

19. S. Bhattacharjee, A. Sharma, P.K. Bhattacharya, Estimation and influence of long range solute. Membrane interactions in ultrafiltration, *Industrial & Engineering Chemistry Research*, 35 (1996) 3108–3121.

20. Z. He, S. Kasemset, A.Y. Kirschner, Y.-H. Cheng, D.R. Paul, B.D. Freeman, The effects of salt concentration and foulant surface charge on hydrocarbon fouling of a poly(vinylidene fluoride) microfiltration membrane, *Water Research*, 117 (2017) 230–241.

21. C.J. Van Oss, Lifshitz–van der Waals (LW) Interactions, in: *Interfacial Forces in Aqueous Media*, CRC Press, Boca Raton, FL, 2006.

22. T.A. Trinh, Q. Han, Y. Ma, J.W. Chew, Microfiltration of oil emulsions stabilized by different surfactants, *Journal of Membrane Science*, 579 (2019) 199–209.

23. H.J. Tanudjaja, J.W. Chew, Assessment of oil fouling by oil-membrane interaction energy analysis, *Journal of Membrane Science*, 560 (2018) 21–29.

24. M.-C. Michalski, S. Desobry, M.-N. Pons, J. Hardy, Adhesion of edible oils to food contact surfaces, *Journal of the American Oil Chemists' Society*, 75 (1998) 447.

25. D. Elzo, I. Huisman, E. Middelink, V. Gekas, Charge effects on inorganic membrane performance in a cross-flow microfiltration process, *Colloids and Surfaces A: Physicochemical and Engineering Aspects*, 138 (1998) 145–159.

26. S. Lee, S. Kim, J. Cho, E.M.V. Hoek, Natural organic matter fouling due to foulant–membrane physicochemical interactions, *Desalination*, 202 (2007) 377–384.

27. Y. Ding, Y. Tian, Z. Li, H. Wang, L. Chen, Interaction energy evaluation of the role of solution chemistry and organic foulant composition on polysaccharide fouling of microfiltration membrane bioreactors, *Chemical Engineering Science*, 104 (2013) 1028–1035.

28. W. Huang, H. Chu, B. Dong, J. Liu, Evaluation of different algogenic organic matters on the fouling of microfiltration membranes, *Desalination*, 344 (2014) 329–338.

29. W. Huang, H. Chu, B. Dong, Understanding the fouling of algogenic organic matter in microfiltration using membrane–foulant interaction energy analysis: Effects of organic hydrophobicity, *Colloids and Surfaces B: Biointerfaces*, 122 (2014) 447–456.

30. X. Zhang, Z. Wang, M. Chen, M. Liu, Z. Wu, Polyvinylidene fluoride membrane blended with quaternary ammonium compound for enhancing anti-biofouling properties: Effects of dosage, *Journal of Membrane Science*, 520 (2016) 66–75.

31. Q. Lei, F. Li, L. Shen, L. Yang, B.-Q. Liao, H. Lin, Tuning anti-adhesion ability of membrane for a membrane bioreactor by thermodynamic analysis, *Bioresource Technology*, 216 (2016) 691–698.

32. J. Liu, J. Tian, Z. Wang, D. Zhao, F. Jia, B. Dong, Mechanism analysis of powdered activated carbon controlling microfiltration membrane fouling in surface water treatment, *Colloids and Surfaces A: Physicochemical and Engineering Aspects*, 517 (2017) 45–51.

33. L. Zhao, F. Wang, X. Weng, R. Li, X. Zhou, H. Lin, H. Yu, B.-Q. Liao, Novel indicators for thermodynamic prediction of interfacial interactions related with adhesive fouling in a membrane bioreactor, *Journal of Colloid and Interface Science*, 487 (2017) 320–329.

34. L. Shen, X. Wang, R. Li, H. Yu, H. Hong, H. Lin, J. Chen, B.-Q. Liao, Physicochemical correlations between membrane surface hydrophilicity and adhesive fouling in membrane bioreactors, Journal of Colloid and Interface Science, 505 (2017) 900–909.

35. Y. Zhu, S. Yu, B. Zhang, J. Li, D. Zhao, Z. Gu, C. Gong, G. Liu, Antifouling performance of polytetrafluoroethylene and polyvinylidene fluoride ultrafiltration membranes during alkali/surfactant/polymer flooding wastewater treatment: Distinctions and mechanisms, *Science of the Total Environment*, 642 (2018) 988–998.

36. H. Wu, F. Shen, J. Wang, Y. Wan, Membrane fouling in vacuum membrane distillation for ionic liquid recycling: Interaction energy analysis with the XDLVO approach, *Journal of Membrane Science*, 550 (2018) 436–447.

37. W. Yin, X. Li, S.R. Suwarno, E.R. Cornelissen, T.H. Chong, Fouling behavior of isolated dissolved organic fractions from seawater in reverse osmosis (RO) desalination process, *Water Research*, 159 (2019) 385–396.

38. K. Wang, L. Xu, K. Li, L. Liu, Y. Zhang, J. Wang, Development of polyaniline conductive membrane for electrically enhanced membrane fouling mitigation, *Journal of Membrane Science*, 570–571 (2019) 371–379.

39. R. Li, Y. Lou, Y. Xu, G. Ma, B.-Q. Liao, L. Shen, H. Lin, Effects of surface morphology on alginate adhesion: Molecular insights into membrane fouling based on XDLVO and DFT analysis, *Chemosphere*, 233 (2019) 373–380.

40. X. Hao, S. Gao, J. Tian, S. Wang, H. Zhang, Y. Sun, W. Shi, F. Cui, New insights into the organic fouling mechanism of an in situ Ca^{2+} modified thin film composite forward osmosis membrane, *RSC Advances*, 9 (2019) 38227–38234.

41. D. Hou, Z. Yuan, M. Tang, K. Wang, J. Wang, Effect and mechanism of an anionic surfactant on membrane performance during direct contact membrane distillation, *Journal of Membrane Science*, 595 (2020) 117495.

42. H.J. Tanudjaja, J.W. Chew, Critical flux and fouling mechanism in cross flow microfiltration of oil emulsion: Effect of viscosity and bidispersity, *Separation and Purification Technology*, 212 (2019) 684–691.

43. H.J. Tanudjaja, J.W. Chew, In-situ characterization of cake layer fouling during crossflow microfiltration of oil-in-water emulsion, *Separation and Purification Technology*, 218 (2019) 51–58.

44. B. He, Y. Ding, J. Wang, Z. Yao, W. Qing, Y. Zhang, F. Liu, C.Y. Tang, Sustaining fouling resistant membranes: Membrane fabrication, characterization and mechanism understanding of demulsification and fouling-resistance, *Journal of Membrane Science*, 581 (2019) 105–113.

45. H.J. Tanudjaja, M.B. Tanis-Kanbur, V.V. Tarabara, A.G. Fane, J.W. Chew, Striping phenomenon during cross-flow microfiltration of oil-in-water emulsions, *Separation and Purification Technology*, 207 (2018) 514–522.

46. B. Zhang, R. Zhang, D. Huang, Y. Shen, X. Gao, W. Shi, Membrane fouling in microfiltration of alkali/surfactant/polymer flooding oilfield wastewater: Effect of interactions of key foulants, *Journal of Colloid and Interface Science*, 570 (2020) 20–30.

47. G. Zuo, R. Wang, Novel membrane surface modification to enhance anti-oil fouling property for membrane distillation application, *Journal of Membrane Science*, 447 (2013) 26–35.

48. Y. Chen, M. Tian, X. Li, Y. Wang, A.K. An, J. Fang, T. He, Anti-wetting behavior of negatively charged superhydrophobic PVDF membranes in direct contact membrane distillation of emulsified wastewaters, *Journal of Membrane Science*, 535 (2017) 230–238.

49. F. Guo, C. Zhang, Q. Wang, W. Hu, J. Cao, J. Yao, L. Jiang, Z. Wu, Modification of poly(vinylidene fluoride) membranes with aluminum oxide nanowires and graphene oxide nanosheets for oil–water separation, *Journal of Applied Polymer Science*, 136 (2019) 47493.

8 Enrichment of Rare Earth Element (REE) Minerals from Different Sources in the Coal Value Chain by Froth Flotation

Fan Shi, Tuo Ji, and Walter Christopher Wilfong
National Energy Technology Laboratory,
NETL Support Contractor

Yee Soong, Thomas J. Tarka, and McMahan Gray
National Energy Technology Laboratory

CONTENTS

DOI: 10.1201/9781003091011-8

8.1 INTRODUCTION

Rare earth elements (REEs) are a group of 17 chemical elements—cerium (Ce), dysprosium (Dy), erbium (Er), europium (Eu), gadolinium (Gd), holmium (Ho), lanthanum (La), lutetium (Lu), neodymium (Nd), praseodymium (Pr), promethium (Pm), samarium (Sm), scandium (Sc), terbium (Tb), thulium (Tm), ytterbium (Yb), and yttrium (Y). It was estimated that the United States imported \$170 million in rare earth compounds and metals in 2019, a 6% increase over 2018 REE imports [1]. REEs are used in a wide range of applications to manufacture magnets, catalysts, alloys, glasses, electronics, and other products [2–7]. Figure 8.1 shows the estimated consumption of rare earths by end use in the United States. REE use in catalysts, at 75%, was by far the largest application [1]. This was followed by metallurgical and alloys, ceramics and glass, polishing, and other uses. Products made from REEs are summarized in Table 8.1 [8].

However, supplies of REEs are limited by a small number of sources [9,10]. The challenge of meeting REE demand has thus generated global interests in research and technological development for obtaining REEs from new resources. As the pursuit of modern and emerging technologies in smart devices, energy, and medical fields increases, so does the need for a reliable and affordable domestic supply of REEs [11–13]. The development of an economically competitive supply of REEs is critical to maintain our nation's economic growth and national security.

Domestic coal is considered a tremendous resource. More than 750 million short tons of coal were mined in 2018 in the United States. Due to the low price of coal (~\$40/ton for the electric power sector in 2018 [14]), production from cheap domestic coals could potentially not only offer sustainable and affordable supplies of critical materials, including REEs, but also create new jobs in places where coal plays a significant economic role [15–17]. Numerous scientific investigations have identified coal as a potential alternative source of REEs, with estimated REE reserves in the

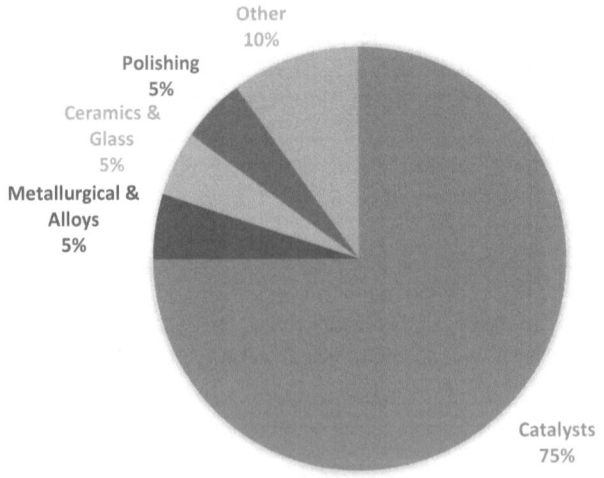

FIGURE 8.1 U.S. consumption of REEs, 2019 [1].

TABLE 8.1
Products Made from Rare Earth Elements [8]

Elements	Industrial Uses						Phosphors (Luminescent Materials)	Cell Phones (and Mobile Devices)	Example Products and Uses
	Catalysts	Ceramics	Defense	Glass and Polishing	Metal Alloys	Magnetics			
Sc									Aerospace aluminum alloys
Y		X	X		X		X	X	LCD displays, LED lights
La	X	X	X	X	X			X	Batteries, catalysts
Ce	X	X		X	X		X		Catalysts, glass polishers, steel
Pr	X	X	X	X	X	X	X	X	Strong magnets, aircraft engines
Nd	X	X	X	X	X	X	X	X	Strong magnets, lasers, speakers
Sm			X						Strong magnets, cancer treatments
Eu		X	X	X			X	X	LCD displays
Gd		X		X			X		MRIs, shielding in nuclear reactors
Tb			X			X	X	X	LCD displays, metal alloys
Dy		X	X			X		X	Computer hard drives, transducers
Ho				X					Strong magnets, cubic zirconia
Er							X		Optical fibers, lasers, glass coloring
Tm								X	Portable x-ray machine
Yb						X		X	Nuclear medicine, stainless steel
Lu	X	X							Catalysts, petroleum refining
Th	X					X			Arc welding, radiometric age dating

range of 50 million metric tons (MTs) [15–17]. The National Energy Technology Laboratory (NETL) has assessed the amounts of REEs in U.S. coal deposits and mineral matter, particularly clays. It was estimated that 6 million MTs of REEs were potentially recoverable from coal reserves in select Western state coal basins, while 4.9 million MTs were potentially available from coal deposits found in the Appalachian Basin coal region [18]. These estimates are based on projections of recoverable coal reserves in these regions with REE contents of more than 500 parts per million (ppm). Moreover, processing coal refuse and ash residues has received attention for the economic value of the REEs they contain [19–23]. Researchers found that Appalachian coal bed refuse represents one of the largest REE resources (591 ppm) in the United States [19].

As a result, opportunities to recover REEs from the whole coal value chain (from raw mineral to refuse material) appear possible. Scheme 8.1 illustrates the coal value chain as it relates to opportunities to recover REEs. Although a great opportunity is presented, the concentrations and dispersity of REEs are totally different depending on the source from specific areas of the coal value chain. Previous research has addressed the origin of REEs and their complex composition in coal and coal byproducts [24–27]. It has been reported that REE-mineral matter exists in coal and is associated with both organic and inorganic materials [28–30]. Based on the atomic radius, other than Sc and Y, REEs can be divided into a light REEs (LREEs) group of the following elements—La, Ce, Pr, Nd, Sm, Eu, and Gd—and a heavy REEs (HREEs) group consisting of Tb, Dy, Ho, Er, Tm, Yb, and Lu. Five REEs (Nd, Eu, Tb, Dy, and Y) were identified by the U.S. Department of Energy (DOE) as critical raw materials or critical REEs (CREEs), which are considered essential for U.S. economic security. Hu et al. investigated REE distribution in 50 coal samples and found that all REEs (except Y) have a strong correlation with ash yield. In addition, LREEs

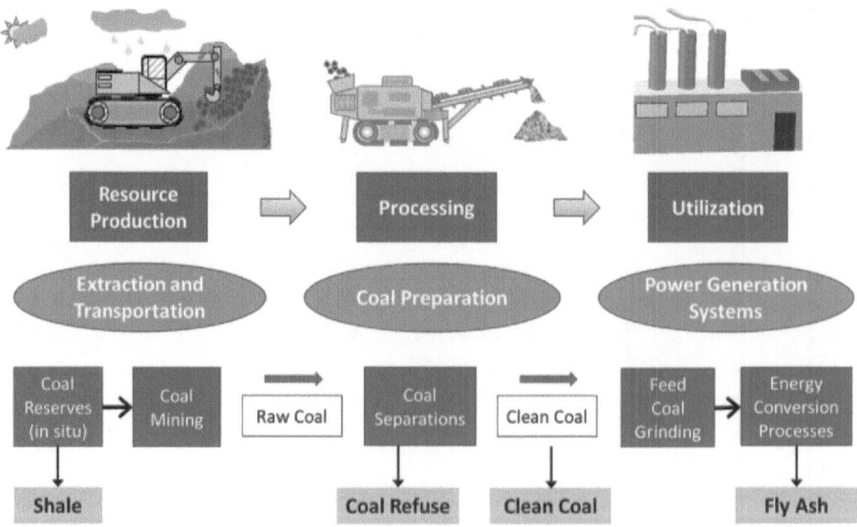

SCHEME 8.1   The scheme of locations of selected coal and coal byproducts in the coal value chain. (Modified from [36].)

had a stronger correlation with ash yield than HREEs in those samples [31]. Birk et al. demonstrated that the REEs in the bituminous coal of the Sydney Basin display a strong correlation with the ash content (ratio from 0.58% to 0.87%) [32]. Karayigit et al. reported that LREEs in 13 coal samples from Turkish coal, at a 95% confidence level, correlate positively with the ash yield while the HREEs show no correlation [33]. Besides, extensive studies revealed the existence of REEs (particularly HREEs) in organic compounds in coal [34]. For example, Gluskoter et al. calculated the organic affinity for REEs. The organic affinity for La to Ce ranges from 0.08 to 0.75, while for Yb and Lu it ranges from 0.29 to 0.75 [35]. Although the abundance and occurrence of REEs in coal have been investigated recently, further studies are needed to evaluate recovery of REE products from different parts in the coal value chain. Moreover, additional information and technology development are required to enrich REEs from large volumes of fly and bottom ash as well as other spent and low-concentration sources.

Flotation has been recognized as a promising physical separation technique due to its high separation efficiency, small footprint, low capital investment, and ease of operation [37–45]. Based on physicochemical characteristics of particles, hydrophobic particles captured by air bubbles ascend to the top and collect as the froth product, whereas hydrophilic particles remain in the pulp and are marked as tailings.

8.1.1 Factors Influencing Flotation

Flotation is a process to separate particles from liquid phase, usually aqueous solutions, with the help of air bubbles. First applied in mineral processing, fine particle flotation separations began in the early 20th century. The separation process involves feeding a solid material into a liquid-filled vessel with a gas—generally air—being bubbled up from the bottom of the vessel. The collisions between bubbles and particles results in certain particles floating based on their physical properties, the liquid medium in use, and various operating parameters. The fundamental principles of flotation are discussed in the following sections. These include floatability, the mechanics of droplet capture, how hydrodynamic forces impact flotation, and an overview of flotation equipment.

8.1.2 Floatability

For a given particle, floatability means its ability to attach to an air bubble and float to the surface of a liquid. It can be measured indirectly in terms of the surface properties of the particle, including hydrophobicity and zeta potential driven by surface charge, which can be expressed by following expression:

$$\text{Floatability} = f\left(\text{hydrophobicity}\right) = f\left(\text{contact angle, zeta potential, surface forces, etc.}\right)$$

Particles are said to be hydrophobic when they are easily attachable to gas bubbles, which is reflected by contact angle. Otherwise, they are hydrophilic. Therefore, the floatability of a particle is determined by the hydrophobicity of particles. Hydrophobic particles can be easily captured and carried upward by gas bubbles to the surface of the slurry, while hydrophilic particles cannot. Contact angle is the most commonly used parameter to indicate the hydrophobicity of particles. The measurement of contact angle in pure water is used to serve as a measurement of hydrophobicity of solid particles. Hydrophobic particles are associated with large contact angles, while hydrophilic particles are coupled with small contact angles. Usually, only particles with contact angles larger than 30 degrees can be separated efficiently by the flotation method [46].

Due to small size of suspended particles, surface force (electrostatic force) often plays a more important role in controlling particle behavior than gravity. Zeta potential is a measure of surface charge of solid particles, which is directly related to the ability of particles to form aggregates with themselves or with gas bubbles, namely, the stability of solid suspension. Generally, large zeta potential promotes stable particle suspensions.

8.1.3   MECHANICS OF DROPLET CAPTURE

Besides the floatability of the particles, other factors affect flotation efficiency. For a given type of particle and a set of operation conditions, the chance of flotation can be expressed in terms of three probabilities, as shown in following correlation [47]:

Chance of flotation = probability of particle/bubble collision

× probability of attachment × probability of retention of attachment

The probability of collision between particles and bubbles is controlled by hydrodynamic conditions of the flotation process. The three ways bubbles and particles make contact are collision, entrapment, and precipitation. Collision is induced by turbulent mixing or caused by direct contact between a rising bubble and a sedimenting particle. Entrapment and precipitation refer to bubbles that are entrapped in or when particles precipitate on a growing floc structure. For oil flotation, collision and entrapment play significant roles.

The probability of attachment is affected by interfacial properties, such as interfacial tensions. The retention of attachment is affected by turbulent condition and adhesive force of attached particles on bubbles. In general, the probability of attachment and the retention of attachment are considered as one factor called the efficiency of attachment. Figure 8.2 shows the aggregation of fine solid particles and air bubbles in a flotation process [48].

8.1.4   HYDRODYNAMICS IN FLOTATION

The effectiveness of flotation mostly depends on the interfacial properties of components in the mixtures to be separated, and these are driven by hydrodynamic conditions and can be modified through the use of surfactants, as indicated in Table 8.2. As alluded to above, interfacial properties, and therefore the hydrodynamic properties, can be modified by adding surface active reagents called surfactants. In a flotation process, typical surfactants are frother, collector, and modifier, as described in Table 8.2. Adding frother can change surface tension and therefore stabilize bubble

FIGURE 8.2 Attachment of fine particles on bubbles [48].

TABLE 8.2
Hydrodynamic Variables and Conditions

Variables		Description	Examples
Liquid		A continuous phase where solid particles are suspended and collided with gas bubbles.	Most systems are aqueous solutions, and some are non-polar solvents.
Minerals		The mineral particles most effectively removed are in the size range from 10 to 200 μm.	Ores include copper, lead, zinc, graphite, sulfur, talc, molybdenite, coals, etc.
Gas feed		Gas to produce bubbles with space velocity of 0.5–3 cm s^{-1}	Mostly air and N_2, and only a few of CO_2
Surfactant	Frother	Frothers are used to reduce surface tension in order to promote small bubble size and a stable froth	Methyl isobutyl carbinol (MIBC), 2-ethyl hexanol, aliphatic alcohols, polyglycols, etc.
	Collector	Collectors are polar chemicals adsorbed on particle's surface to make it hydrophobic and easy to attach to air bubbles	Long chain hydrocarbons: kerosene, fuel oil, etc.
	Modifier	Modifiers adjust the pH of the suspension and assist attachment of bubbles to specific particle species in the suspension	NaCl, Na_2SO_4, H_2SO_4, $Ca(OH)_2$, etc.
	Conditioner	Conditioners used include coagulants and polymer flocculants	Polymers, starch, dextrin, carboxy methyl cellulose, etc.

formation. Collectors alter particles hydrophobicity and are typically used in mineral processes. The functions of modifier and conditioner are to modify the physical properties of the solution to enhance the separation.

Bubble generation is an essential step in flotation because rising bubbles provide the lift force for components (in the form of fine particles or droplets) to separate. Understanding bubble generation parameters and the properties of those bubbles is critical to understanding the flotation mechanism. Therefore, the two major hydrodynamic parameters explored in this study are gas bubble size and gas holdup. Gas holdup is the ratio of gas phase volume to the total volume of liquid/gas phase in a bubble device. Gas holdup is the ratio of gas phase volume to the total volume of liquid/gas phase in a bubble device. Typically, small bubbles will achieve an excellent particle separation because small bubbles have long residence time and large number density in the suspension, which result in an increase in the probability of collision. High gas holdup will also increase the opportunities of contact and attachment, as described earlier.

Gas bubbles in a flotation column are relatively small and spherical with a typical bubble Sauter mean diameter as small as 1 mm [49], when frother concentration is 15 ppm and superficial gas velocity is greater than 2.0 cm s^{-1}. As discussed earlier, small bubbles can improve the chance of flotation. Small bubbles not only generate high gas holdup, they also increase opportunities for collisions between particles and bubbles. Therefore, flotation columns show a big advantage over conventional flotation devices which generate an average bubble size of 2~4 mm under similar operation conditions. The increase of gas holdup will increase the opportunities of contact and attachment between particles and gas bubbles. Gas holdup increases with the dosage of frother and superficial gas velocity. It is desirable in the flotation column to have high gas holdup because it provides a large interfacial area, which favors formation of bubble-particle aggregates. Typically, flotation columns have a high gas holdup up to 30% [50].

Exploration of surfactant and liquid medium combinations also remains a rich area of research. While researchers have tested a large number of surfactants to improve the flotation separation efficiency, new combinations remain intriguing [51,52].

8.1.5 Flotation Equipment

Although many different types of flotation devices are available, the mechanically agitated flotation cell and the flotation column are the two most widely used configurations in industry. A schematic of a laboratory mechanically agitated flotation cell, the Denver cell, is shown in Figure 8.3a [46]. After the slurry is fed into the cell, an impeller is installed. Air is flowed into the cell through a hollow shaft of an agitator, and then the air stream is broken by the agitating impeller so small bubbles are emitted from the end of the impeller blades. The rising bubbles, together with attached particles/droplets, form a foam layer on top of the dispersion phase. The foam layer is skimmed off mechanically from the top. Non-floated components and clean water are withdrawn from the bottom of the cell. The rotating impeller of the agitator not only introduces air bubbles into the flotation cell, but it also mechanically breaks them into small sizes. In addition, agitation induces turbulent mixing to promote

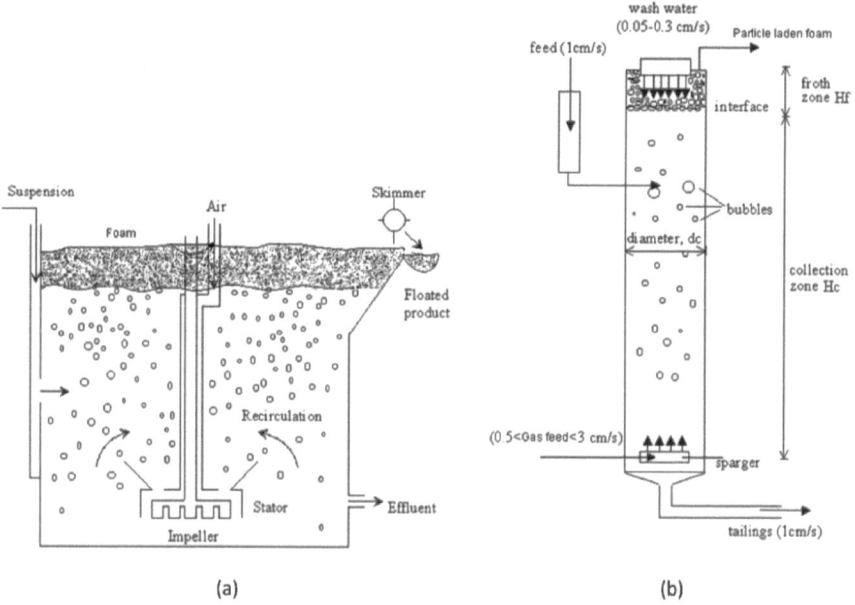

FIGURE 8.3   (a) Schematic diagram of a conventional flotation cell and (b) a flotation column [42].

particle bubble collisions. It can be used to duplicate all the operations in large-scale commercial flotation cells, such as the conditioning of the pulp with reagents, aeration of the pulp, and production of high-grade mineral froth, which can be removed at definite time intervals. The results obtained in testing can be compared to actual operating practice.

The flotation column has been recognized as a promising mineral/ore pretreatment technique because of its high separation efficiency, small footprint, low capital investment, and ease of operation. As shown in Figure 8.3b [42], processing mineral particles in a flotation column involves selective attachment of fine hydrophobic particles in liquid suspension to gas bubbles, which provides levitation for the particles. The resulting bubble-particle aggregates rise to the surface of the dispersion and form a particle-laden foam layer (froth zone), which is then skimmed off as froth products and separated from the dispersion phase.

8.1.6   FLOTATION IN RARE EARTH ELEMENTS PRODUCTION

The flotation method has been matured for mineral processing, especially in the treatment of fine rare earth industrial minerals [39,53,54]. Ren et al. studied separating bastnaesite from monazite using selective flotation while using benzoic acid and potassium alum as the collector and depressant, respectively [55]. Yu et al. found that 1,4-benzoyl hydroxamic acid (PDHA) can be used as both a collector and flocculant, which can significantly improve the recovery of ultrafine rare earth minerals [56].

Although flotation has been applied to separate REEs-mineral ore, there is limited information on selectively enriching REEs from coal-based feeds.

This work was part of the NETL's REEs R&D Program, an initiative aiming to address the current global REE separations market and process economics and to demonstrate the techno-economic feasibility of U.S. domestic REE separation technologies from the coal-based value chain [57–59]. In this work, the main objective is to develop a simple flotation process to enrich REE from coal-based products. To better understand REE variation and opportunities for REE recovery spanning the whole coal value chain, we investigated shale ore, clean coal, coal refuse, and fly ash as sources by using flotation, as indicated in Scheme 8.1. Effects of flotation operating parameters on REE enrichment, yield, and distribution, including LREE/HREE and organic phase/inorganic phase, were further studied. Effects of surfactants, i.e., frother, collector, and depressant, were also investigated.

8.2 EXPERIMENTAL

8.2.1 Materials, Sample Preparation, and Flotation Chemicals

Hundreds of coal-based product field samples have been collected and analyzed at NETL. Detailed information on these samples is available online [59]. Four samples were selected from the different parts in the coal value chain, as shown in Figure 8.4a–d. They are (1) a dry coal ash (ID 357, HF1) containing mostly bottom ash with a lesser amount of fly ash, which was collected from a pulverized coal power plant in Ohio, USA (coal seams unidentified); (2) a clean coal (ID 240, HF2); (3) a coal refuse (ID 145, HF3) collected from a coal preparation plant in Kentucky, USA (coal originally from the central Appalachian region); and (4) a shale (ID S2, HF4) sample collected from the Redstone floor in a coal mining site in Monongalia County, West Virginia, USA.

All samples were milled before the flotation experiments. Figure 8.4e showed the particle size distribution of four ground samples. The four samples have similar particle size and size distribution: d_{90} is in a range of 6–9 μm (HF1: 7.6 μm; HF2: 7.8 μm; HF3:7.7 μm; and HF4: 5.9 μm, respectively). The major elemental compositions of four samples are listed in Table 8.3.

Methyl isobutylcarbinol (MIBC, Fisher Scientific Inc.) was used as the frother in the flotation experiment. Sodium oleate (Fisher Scientific) was used as the collector for fly ash and shale samples while diesel was used for coal and coal refuse. The dosages of frother (MIBC) and collector were 150 ppm and 30 ppm, respectively, for all experiments, unless otherwise defined. Sodium metasilicate (Fisher Scientific) was used as dispersant [60]. Sodium carbonate (Fisher Scientific) was used to control the pH of the slurry.

8.2.2 Flotation Experiments

The flotation-release experiments were carried out using a bench top Metso Denver flotation cell (1–4 L), as shown in Figure 8.5. For each flotation experiment, 30 g of dry sample was added into the cell and mixed with 980 mL deionized (DI) water and

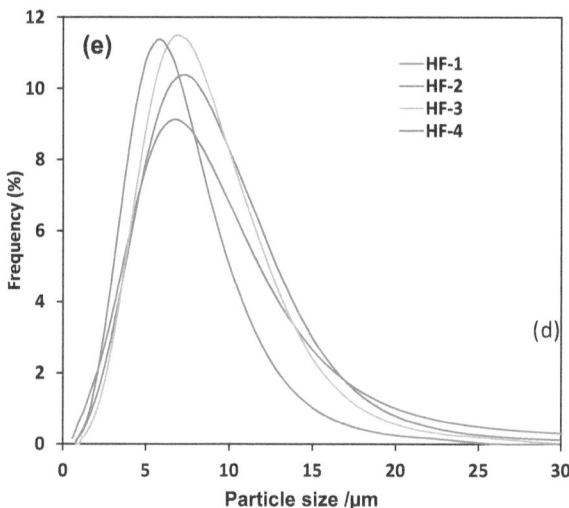

FIGURE 8.4   (a) Coal ash — HF1, (b) clean coal — HF2, (c) coal refuses — HF3 (d) shale – HF4, and (e) particle size distribution of four samples after grinding.

TABLE 8.3
Characteristics of Coal, Coal Ash, Coal Refuse, and Shale Samples

	Sample ID	Component	Dry Ash (%)	Carbon (%)	Fe (%)[a]	Al (%)[a]	Ti (%)[a]
HF1	357	Coal ash	97.8	2.2	18.5	11.2	0.5
HF2	240	Coal	7.72	77.5	3.3	15.4	1.1
HF3	145	Coal refuse	8.2	74.1	3.1	13.6	1.1
HF4	S2	Shale	93.5	0.96	1.9	11.7	0.5

[a]   Ash-based results

(a)   (b)

FIGURE 8.5   (a) A bench top Metso Denver flotation cell for (b) REEs enrichment tests with a mechanical skimmer.

heated to 65 degrees Celsius. After 5 minutes of conditioning, the desired amount of diesel or sodium oleate was added and mixed for 5 minutes. Sodium metasilicate and MIBC were added sequentially with 5 minutes of conditioning. The tailings and froth were collected and then washed with DI water several times. Concentrates and tailing were filtered and dried at 110 degrees Celsius overnight for further processing and analysis.

8.2.3   ANALYTICAL METHODS

For the dry and ash analysis, about 50 mg of sample from each product was placed in a LECO thermogravimetric analyzer (TGA) to obtain ash contents on a dry basis. The solids were dried at 100°C under nitrogen (N_2) for 2 hours and then ashed at 550°C for 4 hours under air. Concentrations of REEs in the dried coal samples were determined by the inductively coupled plasma-mass spectrometry (ICP-MS) method. Approximately 50 mg of ashed sample was mixed in a platinum crucible with 400 mg of lithium metaborate ($LiBO_2$) and fused at 1,100°C for 5 minutes in a microwave fusion oven. The fused glass was then added with stirring into a 5 wt% nitric acid (HNO_3) solution. The platinum crucibles were rinsed with a HNO_3 solution to ensure a complete digestion of the fused glass and finally diluted to 100 mL. For quality control purposes, certified reference solids, including coal standard reference material NIST 1632a and rare earth ore from Natural Resources Canada, were processed along with each batch of samples. Based on duplicate sample measurements, the precision of REEs content for each test can be generalized as < 5%. Based on the elemental recovery of standard reference materials, the measurements reported in this study are believed to be accurate within ± 20%.

8.3 RESULTS AND DISCUSSION

8.3.1 ENRICHMENT OF REEs

The REE concentrations in four feed materials on a whole coal basis were measured to be 483, 141, 112, and 810 ppm, respectively. Among them, REE content of clean coal sample, i.e., HF2, was 14% lower relative to North American Shale Composite (approximately 165 ppm, whole dry basis). Despite the slightly lower content, it was also recognized as a promising source for REE recovery. Based on their ash contents, as listed in Table 8.3, REE concentrations on a dry ash basis were calculated and listed in Table 8.4. After flotation treatment, REE concentrations in froth products and tailings were shown in Figure 8.6 and Table 8.4. It is clear that REE components have been concentrated in froth products, especially for two coal-based samples, i.e., HF2 and HF3. To quantify the enrichment effect of flotation, mass balance, enrichment percentage (EP), and yield have been calculated based on following Eqs. 8.1–8.3:

$$\text{Mass Balance}(\%) = \frac{\text{Mass of REEs in froth and tailing}}{\text{Mass of REEs in feed}} \times 100\% \qquad (8.1)$$

$$\text{Enrichment percentage}(EP) = \frac{C_{\text{Froth}} - C_{\text{Feed}}}{C_{\text{Feed}}} \times 100\% \qquad (8.2)$$

$$\text{Yield}(\%) = \frac{\text{Mass of REEs in froth}}{\text{Mass of REEs in feed}} \times 100\% \qquad (8.3)$$

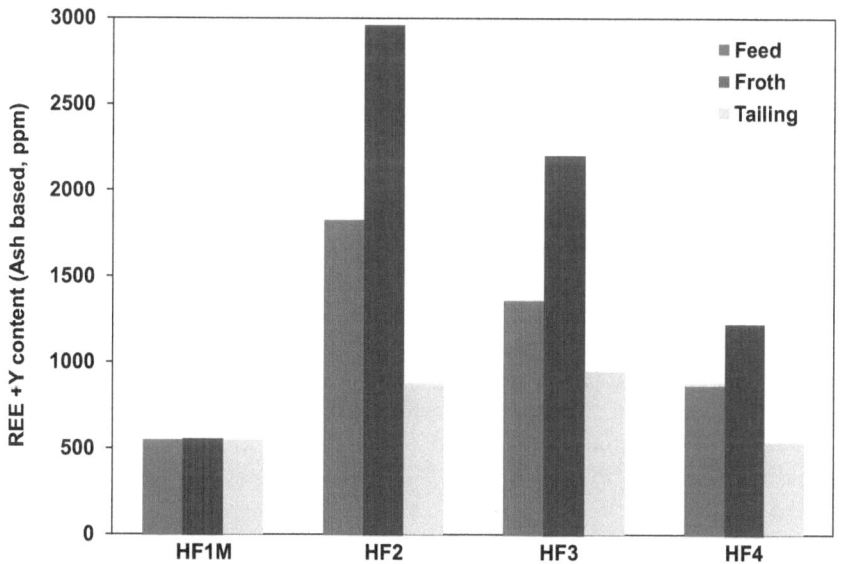

FIGURE 8.6 REEs concentrations of four samples after flotation enrichment.

TABLE 8.4
Calculated Rare Earth Elements (REEs) Mass Balance, Enrichment, and Yield Results

	REEs in Feed (Ash Basis, ppm)	REEs in Froth (Ash Basis, ppm)	REEs in Tailing (Ash Basis, ppm)	REEs Mass Balance (%)	Enrichment Percentage (%)	REEs Yield (%)
*HF1-M	549	554	547	94	<1	23
HF2	1,825	2,961	879	87	63.7	73
HF3	1,360	2,203	947	83	62.2	60
HF4	866	1,222	539	97	40.0	63

* HF1-M was magnetic separator treated HF1.

Results are summarized in Table 8.4. It is shown that the EP for HF2 and HF3 were 64% and 62%, respectively, while REEs yields were as high as 73% and 60%, respectively. It indicates that flotation can effectively concentrate REEs into the froth, probably due to the presence of natural hydrophobic carbon species, shown in Table 8.3, as REEs host sites in coal-based feeds, including coal and coal refuse [32]. These particles can be easily caught by rising bubbles and concentrated in froth. Relatively low EP values of HF1-M and HF4 indicated weak enrichment of REEs on high ash-content samples.

8.3.2 SELECTIVITY OF REEs GROUPS

In most coals, REEs could be resident in numerous types of minerals including monazite and xenotime. Concentrations of individual REE varied significantly, even in the same sample, as shown in Figure 8.7. Ce, Y, Nd, and La, mainly from monazite and xenotime, are by far the most common of the rare earths in these coal and coal byproducts [36]. Due to the same origination of HF2 and HF3, they have similar REE distributions. It was noted that flotation treatment did not change REE distributions in froth products compared to feed. The study also determined that more LREEs were concentrated in products than HREEs.

The individual REE contents in feed and froth sample were compared in Figure 8.7 to further reveal the effect of flotation on the selectivity of REEs. The study noted that the EP were different for each REE. For example, Nd and Tb, as major components of LREE and HREE, respectively, enriched 60% and 76%, respectively, in the HF2 froth product, while only 10% and 11%, respectively, in the HF1-M froth. To study the selective enrichment of different REEs in froth, especially LREE, HREE, and CREE, the selectivity was determined by using Eq. 8.4,

$$S_i = \frac{EP_i}{EP_{avg}}$$

(8.4)

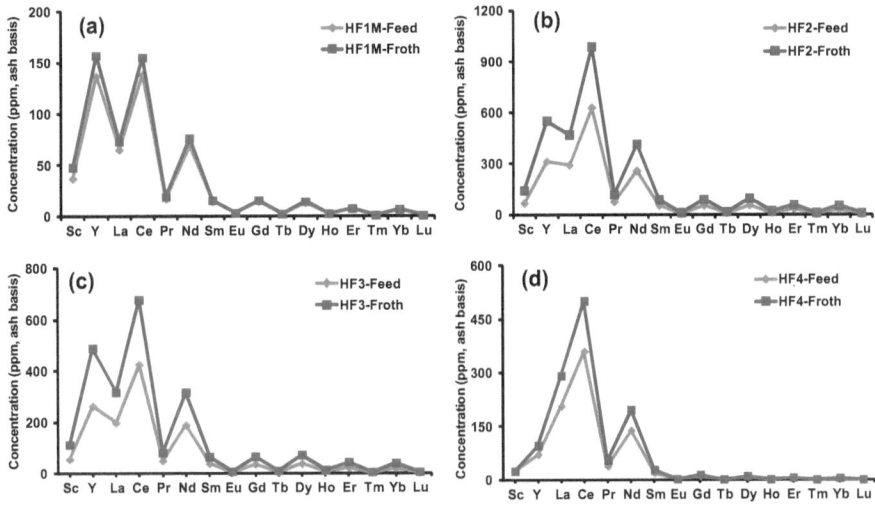

FIGURE 8.7   REEs concentrations (ash basis) and distribution in feed and froth for (a) HF1-M, (b) HF2, (c) HF3, and (d) HF4.

TABLE 8.5
Selectivity for Different Groups of REEs in Froth by Using Flotation Process

Froth Product	S_Y	S_{LREE}	S_{HREE}	S_{CREE}
HF1-M	1.28	0.94	0.85	1.01
HF2	1.19	0.97	1.26	1.10
HF3	1.21	0.97	1.36	1.16
HF4	0.86	1.10	1.19	1.06

where, S_i is the selectivity of i, $i =$ REE, or REE groups, i.e., LREE, HREE, and CREE; EP_i is the enrichment percentage of i; and EP_{avg} is the enrichment percentage of total REEs. Calculated results are summarized in Table 8.5.

As discussed earlier, flotation had negligible REE enrichment for HF1-M. Therefore, we only focused on the other three materials. It is interesting to note that the values of both S_{HREE} and S_{CREE} were greater than 1.0, especially for HF2 and HF3. This indicated that flotation process was favorable for HREEs and CREEs. However, the selectivity for LREEs for HF2 and HF3 were less than 1.0, indicating that flotation process was less favorable for LREEs in carbon-rich feeds, i.e., coal and coal refuse, even though the mass of LREEs product was a magnitude higher than that of the HREEs. For HF4, a shale feed, flotation was favorable to HREEs, LREEs, and CREEs, but not Y.

To analyze the selectivity of REEs in flotation fractions, the ratio of the concentration of LREEs over that of HREEs (LREE/HREE) was plotted against ash content, as shown in Figure 8.8. Results showed that LREE/HREE increased along with

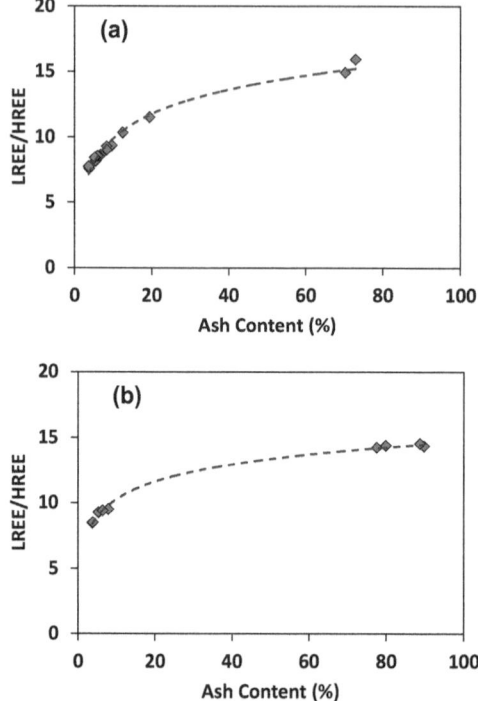

FIGURE 8.8 LREE/HREE ratio in (a) HF2 and (b) HF3 froth samples on ash basis as a function of the ash yield. Dash lines indicate the trend lines obtained by second-order polynomial regression.

ash content for both HF2 and HF3. The high ash-content fractions (or high-density component), i.e., tailing, were relatively enriched in LREE, while low ash-content fractions (or low-density component), i.e., froth, contained a higher ratio of HREE. This trend was consistent with previous discussion and literature that reports LREEs are mostly distributed in the ash component while HREEs existed in the low-density component [61]. Some studies attributed the distribution to coal formation processes. For instance, HREE may be more easily desorbed and concentrated into solution, which increases the likelihood to chemically bind with low-density fractions [62]. In any case, the flotation process selectively enriched HREE over LREEs, even at a low mass production rate.

8.3.3   Distribution of REEs in Organic/Inorganic Phases

It was reported that REEs could be associated with either organic or inorganic phases in coals. [32] A series of flotation tests have been performed for both HF2 and HF3 to study possible REE distributions in floated coal products. Figure 8.9 showed REE contents in froth products increased with the increase of ash content, which was similar to those of coal products treated by other physical separation methods [34].

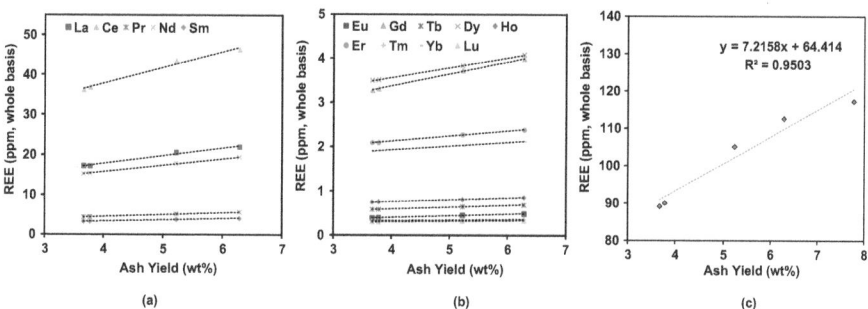

FIGURE 8.9 Comparison of experimental (solid symbols) and predicted (dash line) REEs concentrations (whole basis) in HF2 as a function of the ash yield. (a) La, Ce, Pr, Nd, and Sm elements; (b) Eu, Gd, Tb, Dy, Ho, Er, Tm, Yb, and Lu elements; and (c) total REE.

To quantitatively characterize the organic and inorganic association of REE in coal based on the relationship between REE content and ash content, we estimated the ratio in coal by following Eq. 8.5,

$$\text{REE}_{\text{whole}} = \text{REE}_a \frac{Y_a}{100} = \text{REE}_o + \frac{(\text{REE}_i - \text{REE}_o)}{100} Y_a \qquad (8.5)$$

where, $\text{REE}_{\text{whole}}$, REE_o, and REE_i are the concentrations (in ppm) in whole coal, in the organic phase and the inorganic phase, respectively; REE_a is the concentration in ash; and Y_a is the ash yield (wt%) on the dry basis. All variables were determined experimentally. As shown in Figures 8.9 and 8.10, REEs content, i.e., $\text{REE}_{\text{whole}}$ or ($\text{REE}_a\,Y_a$), increased linearly with the increase of ash yield, Y_a. This linear plot of ($\text{REE}_a\,Y_a$) versus Y_a gives the slope of the line, i.e., ($\text{REE}_i - \text{REE}_o$), and the y-intercept of the line, i.e., REE_o. Thus, values of both REE_o and REE_i can be calculated. Table 8.6 summarized the estimated concentrations of each element in the

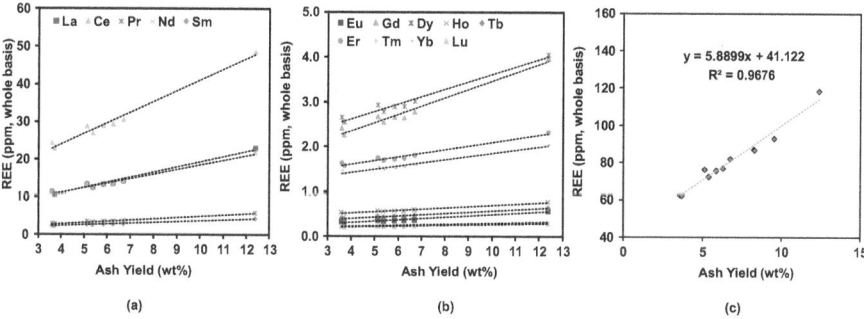

FIGURE 8.10 Comparison of experimental (solid symbols) and predicted (dash line) REEs concentrations (whole basis) in HF3 as a function of the ash yield. (a) La, Ce, Pr, Nd, and Sm elements; (b) Eu, Gd, Tb, Dy, Ho, Er, Tm, Yb, and Lu elements; and (c) total REE.

TABLE 8.6

Estimated Rare Earth Element (REE) Concentrations in the Organic and Inorganic Parts of HF2 and HF3

	Parameters	La	Ce	Pr	Nd	Sm	Eu	Gd	Tb	Dy	Ho	Er	Tm	Yb	Lu	Total
HF2	Slope	1.92	3.98	0.48	1.6277	0.316	0.037	0.275	0.041	0.225	0.041	0.115	0.016	0.084	0.012	
	Intercept	10.14	21.9	2.57	9.2071	2.048	0.258	2.273	0.429	2.668	0.599	1.671	0.26	1.596	0.259	
	REE_i (ppm)	199.1	405.36	55.33	169.67	38.52	10.57	34.43	11.04	29.36	11	18.4	8.51	15.34	8.09	1,014.74
	REE_o (ppm)	10.14	21.89	2.57	9.21	2.05	0.26	2.27	0.42	2.66	0.59	1.67	0.26	1.6	0.26	55.88
HF3	Slope	1.3631	2.8486	0.331	1.2234	0.221	0.029	0.186	0.028	0.165	0.028	0.082	0.011	0.072	0.009	
	Intercept	5.6728	12.668	1.623	6.2706	1.487	0.194	1.608	0.293	1.959	0.409	1.276	0.182	1.134	0.178	
	REE_i (ppm)	141.98	297.53	34.67	128.61	23.61	3.11	20.20	3.04	18.50	3.23	9.50	1.32	8.31	1.09	694.71
	REE_o (ppm)	5.67	12.67	1.62	6.27	1.49	0.19	1.61	0.29	1.96	0.41	1.28	0.18	1.13	0.18	34.96

organic and inorganic parts of HF2 and HF3 samples, respectively. The sum of REE concentrations of HF2 in the organic and inorganic parts are 55.88 and 1,014.74 ppm while the sum of REE concentrations of HF3 in the organic and inorganic parts are 34.96 ppm and 694.71 ppm, respectively. We have also estimated the distributions of organic and inorganic associated REEs using the concentrations of total REEs. As shown in Figures 8.9c and 8.10c, REE_o of HF2 and HF3 are 64.4 and 41.1 ppm, respectively, while REE_i of HF2 and HF3 are 786.0 and 630.0 ppm, respectively. These values are similar to the values of the sum of REE concentrations listed in Table 8.6. These findings verified that REEs were majorly associated with the inorganic phase in coals, which indicated that flotation could effectively concentrate REEs in coals.

Accordingly, the model of Eq. 8.5 is also able to quantify the amount of organically associated REEs in coals. Figure 8.11 showed a curve of REE_o/REE_{whole} as a function of Y_a. Obviously, overlapping curves of HF2 and HF3 samples indicated the same REE organic/inorganic association, which is attributed to HF2 and HF3 that share the same origin from the central Appalachian coal basin. Moreover, experimental data of feed, froth, and tailing matched fitted curve very well. The percentage of organically associated REEs in coal decreased from 0.62 to below 0.10 as the ash content increased from 3 up to 90 wt%, as listed in Table 8.7. The ratio indicated REEs were generally associated with mineral ores.

However, our studies found such calculation could introduce large error in predicting high ash samples (e.g., HF1 and HF4). This is possibly due to untraceable organic phase in ash and minerals samples.

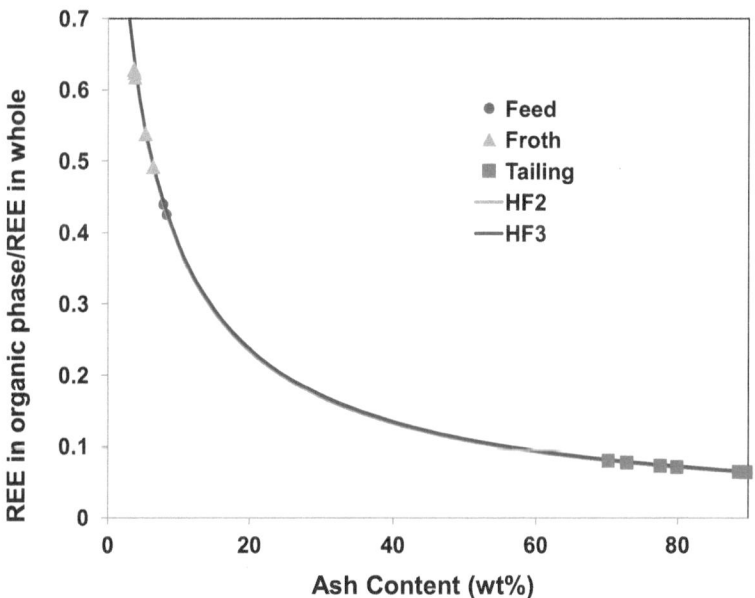

FIGURE 8.11 Predicted percentage of organically associated REEs in HF2 and HF3 as a function of ash yield.

TABLE 8.7

Distribution of Rare Earth Elements (REEs) in Organic Phase for Different Coal Samples (HF2 and HF3)

	REEs in Organic Phase (wt %)	Ash Content (wt %)
Feed	~45	~ 10
Froth	46~62	3~7
Tailing	<10	70~90

8.3.4 Effect of pH in Flotation

Level of pH is one of the most important operating parameters in flotation. The function of pH in flotation is to adjust the surface potential of the particles and affect the adsorption of collector on particle surface, thus further influencing the adhesion between mineral particles and bubbles. The effect of pH value on the enrichment of REEs for HF2 was studied by adjusting pH using Na_2CO_3. Results are shown in Figure 8.12. REE EP gradually decreased from 65% at pH around 6 down to 15% with increased pH value to greater than 10. Oats et al. studied the effect of pH on the zeta potential of coal and air bubbles [63]. They found the surface charge of coal changed from positive to negative when pH value increased from 4 to 10. Due to the negative charge on air bubbles, electrostatic force became unfavorable for the

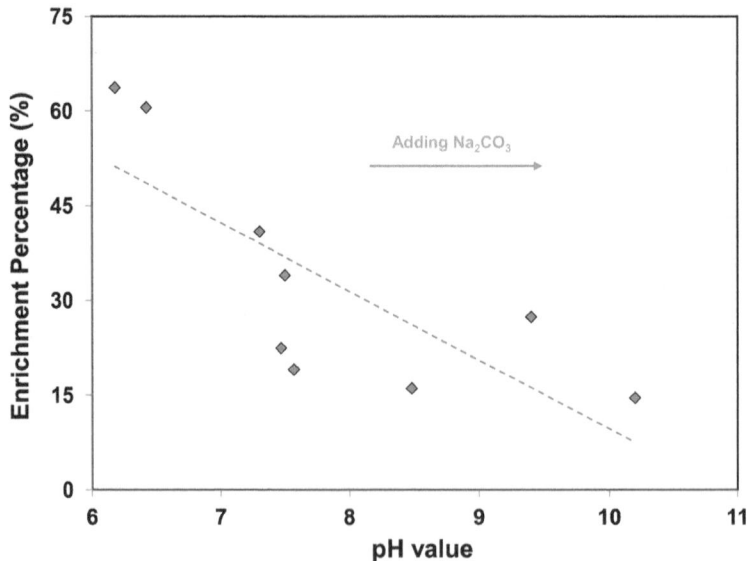

FIGURE 8.12   REE enrichment percentage of HF2 froth products as a function of pH level.

accumulation of air bubbles on the coal particles. This undesirable operating condition made flotation ineffective. As a result, ash content increased in froth product, which consequently reduced REE concentration, as discussed earlier. Therefore, the pH value needs to be optimized for enriching REEs from coal and coal byproducts using the flotation method. Typically, the pH for flotation operation should be neutral or of slightly acidity.

8.4 CONCLUSION

In summary, a simple flotation process was developed to enrich REEs from four samples—shale, clean coal, coal refuse, and coal ash—from different sections of the whole coal value chain. Results showed that REE components could be effectively concentrated in the froth products, especially for coal feeds. Among them, coal (HF2) and coal refuse (HF3) had noteworthy high REE yields at 73% and 60%, respectively. We also verified that REEs were primarily associated with the inorganic phase in coal samples. More importantly, the flotation process was found to be favorable for HREEs over LREEs, even at a low mass fraction. Neutral or slightly acidic pH level is recommended for flotation operation for enriching REEs from coals.

ACKNOWLEDGMENT

We thank R. Lin and E. Roth for technical discussions.

A portion of this technical effort was performed in support of the U.S. DOE's Fossil Energy Crosscutting Technology Research Program. The research was executed through the NETL Research and Innovation Center's Rare Earth Elements FWP. Research performed by Leidos Research Support Team staff was conducted under the RSS contract 89243318CFE000003. This research was supported in part by an appointment to the NETL Research Participation Program, sponsored by the U.S. DOE and administered by the Oak Ridge Institute for Science and Education.

DISCLAIMER

REFERENCES

1. U. S. Geological Survey, 2020. Mineral commodity summaries 2020.
2. C. Tunsu, M. Petranikova, M. Gergorić, C. Ekberg, T. Retegan, 2015. Reclaiming rare earth elements from end-of-life products: A review of the perspectives for urban mining using hydrometallurgical unit operations. *Hydrometallurgy*, 156: 239–258.
3. F. Xie, T.A. Zhang, D. Dreisinger, F. Doyle, 2014. A critical review on solvent extraction of rare earths from aqueous solutions. *Miner. Eng.*, 56: 10–28.
4. N. Rusman, M. Dahari, 2016. A review on the current progress of metal hydrides material for solid-state hydrogen storage applications. *Int. J. Hydrogen Energy*, 41: 12108–12126.
5. M. Escudero-Escribano, P. Malacrida, M.H. Hansen, U.G. Vej-Hansen, A. Velázquez-Palenzuela, V. Tripkovic, J. Schiøtz, J. Rossmeisl, I.E. Stephens, I. Chorkendorff, 2016. Tuning the activity of Pt alloy electrocatalysts by means of the lanthanide contraction. *Science*, 352: 73–76.
6. K. Binnemans, P.T. Jones, B. Blanpain, T. Van Gerven, Y. Pontikes, 2015. Towards zero-waste valorisation of rare-earth-containing industrial process residues: A critical review. *J. Cleaner Prod.*, 99: 17–38.
7. T. Ji, C. Liu, X. Lu, J. Zhu, 2018. Coupled chemical and thermal drivers in microwaves toward ultrafast HMF oxidation to FDCA. *ACS Sustain. Chem. Eng.*, 6: 11493–11501.
8. Kentucky geological survey, 2019. Rare Earth Eelements from Coal. https://www.uky.edu/KGS/coal/coal-for-rare.php. (accessed Jun, 2020).
9. E. Granite, E. Roth, M. Alvin, 2016. Recovery of rare earths from coal and by-products: A paradigm shift for coal research. *Bridges*, 46: 56–57.
10. Q. Wang, W.C. Wilfong, B.W. Kail, Y. Yu, M.L. Gray, 2017. Novel polyethylenimine–acrylamide/SiO$_2$ hybrid hydrogel sorbent for rare-earth-element recycling from aqueous sources. *ACS Sustain. Chem. Eng.*, 5: 10947–10958.
11. R. Blunt, S. Senate, 2014. 2006-National Rare Earth Cooperative Act of 2014.
12. K. Binnemans, P.T. Jones, B. Blanpain, T. Van Gerven, Y. Yang, A. Walton, M. Buchert, 2013. Recycling of rare earths: A critical review. *J. Cleaner Prod.*, 51: 1–22.
13. J.C. Hower, E.J. Granite, D.B. Mayfield, A.S. Lewis, R.B. Finkelman, 2016. Notes on contributions to the science of rare earth element enrichment in coal and coal combustion byproducts. *Minerals*, 6: 32.
14. U.S. Energy Information Administration, 2018. Coal prices and outlook. https://www.eia.gov/energyexplained/coal/prices-and-outlook.php. (accessed Jun, 2020).
15. R.K. Taggart, J.C. Hower, G.S. Dwyer, H. Hsu-Kim, 2016. Trends in the rare earth element content of US-based coal combustion fly ashes. *Environ. Sci. Technol.*, 50: 5919–5926.
16. W. Zhang, M. Rezaee, A. Bhagavatula, Y. Li, J. Groppo, R. Honaker, 2015. A review of the occurrence and promising recovery methods of rare earth elements from coal and coal by-products. *Int. J. Coal Prep. Util.*, 35: 295–330.
17. T.L. Bank, E.A. Roth, P. Tinker, E. Granite, National Energy Technology Lab.(NETL), Pittsburgh, PA,(United States), 2016. Analysis of Rare Earth Elements in Geologic Samples using Inductively Coupled Plasma Mass Spectrometry; US DOE Topical Report-DOE/NETL–2016/1794.
18. United States Department of Energy, 2017. Report on rare earth elements from coal and coal byproducts, report to congress. http://www.parc.xerox.com/istl/groups/iea/www/webmarkets.html (accessed Jun 2020).
19. R.K. Taggart, J.C. Hower, G.S. Dwyer, H. Hsu-Kim, 2016. Trends in the rare earth element content of U.S.-based coal combustion fly ashes. *Environmental Science & Technology*, 50: 5919–5926.

20. J.C. Hower, L.F. Ruppert, C.F. Eble, 1999. Lanthanide, yttrium, and zirconium anomalies in the fire clay coal bed, Eastern Kentucky. *Int. J. Coal Geol.*, 39: 141–153.

21. D.B. Mayfield, A.S. Lewis, 2013, Environmental review of coal ash as a resource for rare earth and strategic elements, *Proceedings of the 2013 World of Coal Ash (WOCA) Conference*, Lexington, KY, USA, 2013: 22–25.

22. V. Seredin, 2010. A new method for primary evaluation of the outlook for rare earth element ores. *Geol. Ore Depos.*, 52: 428–433.

23. V.V. Seredin, S. Dai, 2012. Coal deposits as potential alternative sources for lanthanides and yttrium. *Int. J. Coal Geol.*, 94: 67–93.

24. M. Pires, X. Querol, 2004. Characterization of Candiota (South Brazil) coal and combustion by-product. *Int. J. Coal Geol.*, 60: 57–72.

25. Z. Huang, M. Fan, H. Tiand, 2018. Coal and coal byproducts: A large and developable unconventional resource for critical materials–Rare earth elements. *J. Rare Earths*, 36: 337–338.

26. C. Zhao, B. Liu, L. Xiao, Y. Li, S. Liu, Z. Li, B. Zhao, J. Ma, G. Chu, P. Gao, 2017. Significant enrichment of Ga, Rb, Cs, REEs and Y in the Jurassic No. 6 coal in the IQE Coalfield, northern Qaidam Basin, China—a hidden gem. *Ore Geol. Rev.*, 83: 1–13.

27. W. Zhang, R.Q. Honaker, 2018. Rare earth elements recovery using staged precipitation from a leachate generated from coarse coal refuse. *Int. J. Coal Geol.*, 195: 189–199.

28. S. Dai, P. Xie, C.R. Ward, X. Yan, W. Guo, D. French, I.T. Graham, 2017. Anomalies of rare metals in Lopingian super-high-organic-sulfur coals from the Yishan Coalfield, Guangxi, China. *Ore Geol. Rev.*, 88: 235–250.

29. S. Dai, I.T. Graham, C.R. Ward, 2016. A review of anomalous rare earth elements and yttrium in coal. *Int. J. Coal Geol.*, 159: 82–95.

30. M. Stuckman, C. Lopano, E. Granite, 2018. Distribution and speciation of rare earth elements in coal combustion by-products via synchrotron microscopy and spectroscopy. *Int. J. Coal Geol.*, 195: 125–138.

31. J. Hu, B. Zheng, R.B. Finkelman, B. Wang, M. Wang, S. Li, D. Wu, 2006. Concentration and distribution of sixty-one elements in coals from DPR Korea. *Fuel*, 85: 679–688.

32. D. Birk, J.C. White, 1991. Rare earth elements in bituminous coals and underclays of the Sydney Basin, Nova Scotia: Element sites, distribution, mineralogy. *Int. J. Coal Geol.*, 19: 219–251.

33. A. Karayigit, R. Gayer, X. Querol, T. Onacak, 2000. Contents of major and trace elements in feed coals from Turkish coal-fired power plants. *Int. J. Coal Geol.*, 44: 169–184.

34. R. Lin, T.L. Bank, E.A. Roth, E.J. Granite, Y. Soong, 2017. Organic and inorganic associations of rare earth elements in central Appalachian coal. *Int. J. Coal Geol.*, 179: 295–301.

35. H.J. Gluskoter, 1977. Trace elements in coal: Occurrence and distribution. *Circular no. 499*.

36. The U.S. Department of Energy National Energy Technology Laboratory, NETL Rare Earth Elements Program. https://netl.doe.gov/coal/rare-earth-elements/program-overview/background. (accessed Jun, 2020).

37. G. Wang, A.V. Nguyen, S. Mitra, J. Joshi, G.J. Jameson, G.M. Evans, 2016. A review of the mechanisms and models of bubble-particle detachment in froth flotation. *Sep. Purif. Technol.*, 170: 155–172.

38. L. Wang, Y. Peng, K. Runge, D. Bradshaw, 2015. A review of entrainment: Mechanisms, contributing factors and modelling in flotation. *Miner. Eng.*, 70: 77–91.

39. W. Zhang, R. Honaker, J. Groppo, 2017. Concentration of rare earth minerals from coal by froth flotation. *Miner. Metall. Proc.*, 34: 132–137.

40. F. Shi, S. Chiang, 2003. Simultaneous removal of mixed pollutants from wastewater using multi-stage flotation column. *Fluid/Particle Sep. J.*, 15: 133–142.
41. F. Shi, X. Gu, C. Sh, 2002. A study of hydrodynamic behaviors in a multi-stage loop-flow flotation column, the fluid. *Particle Sep. J.*, 14: 185–198.
42. F. Shi, 2006. Removal of mixed contaminants from wastewater by multistage flotation process. University of Pittsburgh.
43. S. Chiang, F. Shi, X. Gu, 2003. A new development in flotation process. *J. Chin. Inst. Chem. Eng.*, 34: 7–15.
44. F. Shi, S. Chiang, R.W. Lai, 2003. Simultaneous removal of mixed pollutants from wastewater using multi-stage flotation column. *Adv. Filtr. Sep. Technol.*, 16: 750–766.
45. L. Chong, Y. Lai, M. Gray, Y. Soong, F. Shi, Y. Duan, 2017. Effects of frothers and oil at saltwater–air interfaces for oil separation: molecular dynamics simulations and experimental measurements. *J. Phys. Chem. B*, 121: 6699–6707.
46. L. Svarovsky, 1985. *Solid-Liquid Separation Processes and Technology*, Elsevier, Amsterdam.
47. J. Kitchener, 1984. The froth flotation process: Past, present and future-in brief. In Ives K.J. (eds.) *The Scientific Basis of Flotation*, Springer, Dordrecht, pp. 3–51.
48. M.C. Fuerstenau, J.D. Miller, M.C. Kuhn, 1985. *Chemistry of Flotation*, Society for Mining Metallurgy, Englewood, CO.
49. X. Gu, 1999. A novel multi-stage flotation column for oily wastewater treatment. University of Pittsburgh.
50. R.-H. Yoon, G.T. Adel, G.H. Luttrell, 1992. Apparatus and process for the separation of hydrophobic and hydrophilic particles using microbubble column flotation together with a process and apparatus for generation of microbubbles. US Patent, US5167798A.
51. L. Zhang, P. Somasundaran, V. Ososkov, C. Chou, 2001. Flotation of hydrophobic contaminants from soil. *Colloids Surf. A: Physicochem. Eng. Asp.*, 177: 235–246.
52. F. Puget, M. Melo, G. Massarani, 2000. Wastewater treatment by flotation. *Braz. J. Chem. Eng.*, 17: 407–414.
53. S. Al-Thyabat, P. Zhang, 2016. *Upgrading Phosphate Flotation Tailings for REE Extraction*. Society for Mining, Metallurgy & Exploration, Englewood, CO.
54. F. Zhou, L. Wang, Z. Xu, Q. Liu, M. Deng, R. Chi, 2014. Application of reactive oily bubbles to bastnaesite flotation. *Miner. Eng.*, 64: 139–145.
55. J. Ren, S. Song, A. Lopez-Valdivieso, S. Lu, 2000. Selective flotation of bastnaesite from monazite in rare earth concentrates using potassium alum as depressant. *Int. J. Miner. Process.*, 59: 237–245.
56. B. Yu, X. Che, Q. Zheng, 2014. Flotation of ultra-fine rare-earth minerals with selective flocculant PDHA. *Miner. Eng.*, 60: 23–25.
57. R. Lin, M. Stuckman, B.H. Howard, T.L. Bank, E.A. Roth, M.K. Macala, C. Lopano, Y. Soong, E.J. Granite, 2018. Application of sequential extraction and hydrothermal treatment for characterization and enrichment of rare earth elements from coal fly ash. *Fuel*, 232: 124–133.
58. R. Lin, Y. Soong, E.J. Granite, 2018. Evaluation of trace elements in US coals using the USGS COALQUAL database version 3.0. Part I: Rare earth elements and yttrium (REY). *Int. J. Coal Geol.*, 192: 1–13.
59. The U.S. Department of Energy National Energy Technology Laboratory, Integrated sample databases-collected samples spreadsheet v051515. https://edx.netl.doe.gov/ree/?page_id=1587. (accessed Jun, 2020).
60. W. Zhang, R.Q. Honaker, 2018. Flotation of monazite in the presence of calcite part II: Enhanced separation performance using sodium silicate and EDTA. *Miner. Eng.*, 127: 318–328.

61. R. Honaker, J. Hower, C. Eble, G. Weisenfluh, J. Groppo, M. Rezaee, A. Bhagavatula, G. Luttrell, R. Bratton, M. Kiser, 2014. Laboratory and bench-scale testing for rare earth elements. *Cell*, 724: 554–3652.
62. G. Eskenazy, 1999. Aspects of the geochemistry of rare earth elements in coal: An experimental approach. *Int. J. Coal Geol.*, 38: 285–295.
63. W.J. Oats, O. Ozdemir, A.V. Nguyen, 2010. Effect of mechanical and chemical clay removals by hydrocyclone and dispersants on coal flotation. *Miner. Eng.*, 23: 413–419.

9 Recent Advances for Solid–Liquid Separation by Crystallization

Alison E. Lewis
University of Cape Town

Torsten Stelzer
University of Puerto Rico

CONTENTS

9.1 INTRODUCTION

Before discussing the application of crystallization for the treatment of produced waters (byproducts from oil and gas extraction), it is necessary to briefly discuss the characteristics of produced waters and how these affect the choice of treatment method.

Produced water characteristics and physical properties can vary significantly as a function of their geographic location, the geological formations with which the produced water has been in contact and the type of hydrocarbon product being

DOI: 10.1201/9781003091011-9

extracted. In addition, produced water properties and volume from a single location can also vary over time. Although oil and grease have received the majority of attention, produced water contains other organic and inorganic compounds that are of interest. Some of these occur naturally as a result of the geology, while others are introduced via chemicals that are used in the extraction process. The components in produced water include (1) dispersed oil suspended in the aqueous phase; (2) dissolved or soluble organic components including naturally occurring hydrocarbons such as organic acids, polycyclic aromatic hydrocarbons (PAHs), phenols, and volatiles; (3) treatment chemicals such as biocides, reverse emulsion breakers, and corrosion inhibitors; (4) solids including precipitated solids, sand and silt, carbonates, clays, proppant, corrosion products, and other suspended solids derived from the producing formation and from wellbore operations; (5) precipitated scales (e.g., calcium carbonate, calcium sulfate, barium sulfate, strontium sulfate, and iron sulfate), which form when produced water, already supersaturated, is subjected to reduced pressures and temperatures during production; and (6) metals (e.g., zinc, lead, manganese, iron, and barium).[1]

This chapter will focus on treatment technologies that use the concept of crystallization. Crystallization has been used since the dawn of civilization to crystallize out the salt from seawater in large open ponds or from underground saline for easier handling, storing, and trading of the salt.[2] Therefore, it can surely be considered as one of the oldest separation technologies in chemical engineering. It represents the workhorse for separation tasks because it is often the most robust and cost-effective unit operation. Crystallization may be defined as a phase change in which a solid (crystalline) phase is generated from a liquid. This liquid feed can be either a solution composed of at least two or more solutes (species) in a solvent forming a homogeneous mixture or a melt that most correctly refers to a pure solid, molten above normal conditions. However, melts may be also homogeneous mixtures of more than one compound.

The purpose of this chapter is to provide the basic concepts that govern crystallization processes and discuss the practical aspects of conventional processes (Section 9.5) and latest technical advances (Section 9.6) related to the application of crystallization as a separation technology for produced waters. Particularly, the technologies described in Section 9.6 are aimed at near zero waste discharge by simultaneous recovery of water and the dissolved compounds as valuable products. Therefore, a brief understanding of the crystallization concepts is needed and will be discussed first. For the interested reader, detailed information can be found in specialist literature on industrial crystallization,[2–4] process monitoring and control[5,6] as well as in the various literature cited within this chapter.

9.2   BASIC CRYSTALLIZATION CONCEPTS

9.2.1   Solid–Liquid Equilibrium and Solubility

The knowledge about the equilibrium between a solid and a liquid phase is the thermodynamic foundation needed to develop, design, and control crystallization processes whether conducted from solution or melt. A phase diagram graphically

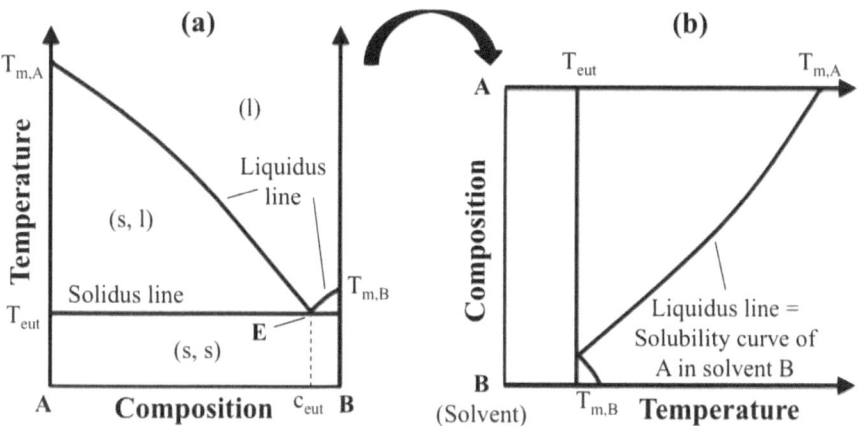

FIGURE 9.1   (a) Phase diagram for a hypothetical binary system composed of compound A and B and (b) graphical relationship with the solubility curve of compound A in solvent B. (Reproduced with permission from John Wiley and Sons, Inc.)[2]

displays this equilibrium between all possible thermodynamic states of a system (defined by temperature T, pressure p, composition c). In most systems, the effect of pressure on solid–liquid equilibria is negligible. Therefore, phase diagrams are typically plotted as composition *versus* temperature at constant pressure (Figure 9.1).

The area above the two liquidus lines in Figure 9.1a represents the thermodynamic state of the one-phase region (liquid) where the binary system is a homogeneous liquid. The two liquidus lines are formed by the melting temperature (T_m) of the compounds A and B with point E, respectively. Thus, cooling a liquid into the two-phase region (solid, liquid) below the liquidus line but above the solidus line leads to the formation of either A or B crystals in equilibrium with the liquid. Which crystals are formed depend on the location relative to point E, known as the eutectic composition (c_{eut}). Point E indicates where all three phases (solid A, solid B, and liquid AB) coexist at the eutectic temperature (T_{eut}). Thus, cooling a liquid below the solidus line the binary system consists of a heterogeneous mixture of pure crystals of A and B in this two-phase region (solid, solid).

For crystallization from solution, the equilibrium is commonly illustrated by a solubility curve, which depicts a small portion of the entire phase diagram. This relationship between solution and melt equilibria is immediately clear after exchanging the axes of temperature and composition (Figure 9.1b) and assuming that compound B is a solvent (e.g., water with its solid phase ice).

For clarification, the frequent discussions on solution and melt in the literature do not reflect different physical phenomena but the historical development of crystallization for different systems. Therefore, it was suggested that the expression *solution crystallization* is used whenever the mass-transfer effects dominates a process. However, it should be called *melt crystallization* whenever a process is dominated by heat transfer effects.[7,8] Due to the fact that the sustainable separation processes approaches for produced water also consider the recovery of dissolved compounds (Section 9.6), the following discussion of this section will focus on the subject of solution crystallization. The interested reader is also referred to specialist literature about the fundamentals and applications of solution and melt crystallization.[2–4,7,9,10]

The solubility of a compound in a solvent refers to the maximum amount that can be dissolved in a given amount of solvent at equilibrium. For most compounds, the solubility increases with increasing temperature. However, for sparingly soluble compounds, the solubility can also possess a negative temperature dependency (e.g., calcium hydroxide in water). A solution is defined as saturated when the maximum amount of solid that can be dissolved at a particular set of conditions is reached. The most convenient expression for solubility is mass of solute per mass of solvent because the units have no temperature dependency. Thus, solutions can be prepared by just weighing each species.[4]

Solubility curve and phase diagram are essential for crystallization process development because they provide much useful information including the calculation of the maximum theoretical yield of the crystals when moving from one set of conditions (liquid feed) to another in which crystals are formed (crystallization process).

9.2.2 SUPERSATURATION AND SUPERCOOLING

Crystallization is a rate process, which depends on time, and a thermodynamic driving force. This force is called *supersaturation* and *supercooling* when solution and melt crystallization are considered, respectively. Supersaturation/supercooling are strictly defined as the deviation from the thermodynamic equilibrium. The three most practical forms to express supersaturation are (1) concentration difference between the solution concentration and the saturation concentration, (2) concentration ratio, and (3) relative supersaturation.[2–4] In melt crystallization, the driving force is called *supercooling* and expressed by the temperature difference between the equilibrium temperature and the actual temperature.[2–4] Typically, three main methods are applied to generate supersaturation/supercooling for produced waters:

1. *Cooling* is the most common method applied if the compound solubility possesses a significant temperature dependency and if the residue solubility at the lower temperature is small. Cooling is also applied to generate supercooling in melt crystallization processes. For instance, applied in eutectic freeze crystallization (Section 9.6.2).
2. *Evaporation* removes the solvent to increase the solute concentration. Evaporation is typically applied for compounds with low temperature dependency of the solubility curve (e.g. sodium chloride in water), for instance, applied in evaporation ponds (Section 9.5.1).
3. *Precipitation*, also known as *reactive crystallization*, refers to the rapid generation of supersaturation by mixing two highly soluble reactants in a solution to form a new product with low solubility in the solvent. Supersaturation is generated as soon as the new product exceeds its solubility and crystals are formed. Generally, reactive crystallization is applied for sparingly soluble compounds as discussed in Section 9.6.1.

For more details, the interested reader is referred to specialist literature on the theory of crystallization including nucleation and crystal growth.[2–4]

9.3   CRYSTALLIZERS

Crystallizers, as applied in the unconventional crystallization technologies discussed in Section 9.6, enable the generation of an environment suitable for the nucleation and crystal growth of a compound with the desired properties (e.g., purity, crystal size distribution, and solid state). Therefore, crystallizers need the ability to (1) create driving force, (2) offer sufficient residence time for both the solution and crystals for nucleation and growth, and (3) provide some level of mixing via agitators or fluid pumping that permits ideally a uniform environment throughout the whole crystallizer volume with respect to temperature, concentration, and slurry density.

The simplest type of "crystallizers" are ponds, which have been historically developed to crystallize out the salt from seawater.[2] Ponds are till today employed as a conventional method for the industrial treatment of produced water as discussed in Section 9.5. For more sophisticated crystallizer designs, various equipment concepts exist.[2–4] Even more approaches on specialized crystallizers are documented.

9.4   BASIC CRYSTALLIZATION PROCESS DESIGN

In particular, the unconventional crystallization technologies discussed in Section 9.6 are strongly influenced by the individual characteristics of each processed compound, which requires a process design that needs to be evaluated on an individual basis. Their process aim is to recover the dissolved solutes as valuable products with defined properties besides purified water. Common starting points for the process design are product and process requirements (e.g., purity, crystal size, and production rate) as well as properties of the liquid feed (e.g., temperature and concentration).[11] Once the product requirements and the feed specifications have been defined, the most appropriate type of crystallization process can be chosen. Moreover, these two sets of information can be brought together in a process model to determine the optimum crystallization conditions or crystallization kinetic parameter.[11] For the latter, details are discussed in Chapter 7 on the example of calcium carbonate ($CaCO_3$). The interested reader is referred to specialist literature for further details[2–4,12,13] including the modeling of crystallization processes.[14]

9.5   CONVENTIONAL CRYSTALLIZATION METHODS

9.5.1   Evaporation Pond

Evaporation ponds are not strictly a treatment method but are included here as they are by far the most common disposal method that is based on evaporative crystallization. Some examples are documented in Ahmed et al.[15] and Lyman et al.[16] Generally, such ponds are considered to be an economical solution for the disposal of produced water.[17] However, in practice, they can be quite expensive because an evaporation pond is a large artificially generated pond, which requires a significant amount of land (space) and is designed to evaporate the water using solar energy.[18] Typically, evaporation ponds are designed either to prevent subsurface infiltration of water and/or the migration of water downward depending on the quality of the produced water. They can be very efficient in warm and dry climates because of the potential high rates of evaporation.[17] However, the costs can be significant and include the cost of large tracts of land, as well as the costs of lining to prevent ground water contamination. This disposal method is also increasingly being seen as undesirable and not consistent with sustainable process design, since all water is lost to the environment and there is no recovery of any other components within the produced water stream.

9.5.2 FREEZE-THAW EVAPORATION

Freeze-thaw evaporation (FTE®) is an extension of the evaporation ponds that addresses the drawbacks of their application in the warm weather season only by coupling evaporative with freeze (melt) crystallization.[19] Developed in 1992 by the University of North Dakota Energy & Environmental Research Centre and B.C. Technologies, Ltd., FTE® has been demonstrated as an effective and economic method to treat produced water.[17,19,20] It employs naturally occurring temperature swings all year round, which lead to alternating freezing, thawing, and evaporation of the produced water.[17] Consequently, the concentration of the dissolved solids increases and fresh water can be produced, suitable for various beneficial applications.[20] Specifically, due to the dissolved constituents, the freezing point of the produced water will be somewhat lower than 0°C.[17,19,20] This freezing point depression can be exploited to recover ice when the produced water is cooled below 0°C but not below its depressed freezing point. Relatively pure ice crystals are formed, while the remaining solution remains unfrozen with an increased concentration of the dissolved constituents compared to the original water. Due to the larger density of this brine compared to that of the ice, the purified ice and brine can easily be separated by draining the unfrozen solution.

The advantages of FTE® are reported as: (1) up to 55% of the produced water fed into the FTE® can be recovered as water from molten ice during the winter season, (2) can remove over 90% of heavy metals, volatile and semi-volatile organics, total dissolved solids, total suspended solids, and total recoverable petroleum hydrocarbons in the recovered water, and (3) does not require chemicals, infrastructure, or supplies other than a pond that is easy to operate and monitor and has a life expectancy of about 20 years.[17]

However, considering the entire year, about 30% of the water is lost by evaporation and/or sublimation and only 15% remains as concentrated brine.[17,20] Thus, during the winter off-season, with temperatures above 0°C, FTE® operates as a conventional evaporation pond and no water is recovered, since all the evaporated water is lost to the atmosphere.[17]

Another major drawback of FTE® is that heavy metals, volatile and semi-volatile organics, total dissolved solids, total suspended solids, and total recoverable petroleum hydrocarbons all remain behind when the water is either recovered through freezing or lost through evaporation. FTE® can also only be beneficially employed in a climate that has substantial number of days with temperatures below freezing and usually requires a significant amount of land. Moreover, waste disposal is an essential part when using the FTE® method because it generates a significant amount of concentrated brine and oil. It also cannot be used to treat produced water with a high methanol concentration.

Later, in Section 9.6.2, the unconventional technology of eutectic freeze crystallization is discussed, which exploits the drawbacks of FTE® to enable the recovery of both water and dissolved constituents.

9.5.3 MECHANICAL EVAPORATION

Mechanical evaporation is available commercially to treat waters produced from shale gas production and can be used to recover water as well as a mixed salt product.[21] Mechanical evaporation is usually carried out using vertical tube, falling film, and vapor compression evaporation.[20] It is a popular choice since it is seen as robust and avoids the requirement for physical and chemical pretreatment and requires less maintenance than other technologies, while producing relatively little waste.[20]

9.5.4 CHEMICAL PRECIPITATION

Chemical precipitation is also used in the treatment of produced water with the major objective of treating the water for discharge. Recovery of either water or contaminants in the produced water is not a priority. This distinguishes it from more recent "unconventional" processes, where the emphasis is on recovering both the water and the contaminants (see Section 9.6.1).

For instance, in the modified hot lime process, the produced water is treated in two stages.[22] In the first stage of this process, the produced water is heated, thereby reducing the solubility of the dissolved scale-forming salts due to a negative temperature dependency, resulting in salt formation as a sludge. In the second stage, the chemical reaction stage, a hydroxide or carbonate reagent is added, and precipitation of hydroxide and carbonate salts occurs.[22] However, although water can be successfully recovered of feed-water quality for a steam generator by this process, the salts produced often need to be disposed because of the increased concentration of potentially toxic metals in the formed sludge.[20] An example of this drawback of chemical precipitation is given by Zhang and coworkers.[23] They showed that, when radium is removed from produced waters by coprecipitation with barium or other alkaline earth metals, the radium levels in the precipitate far exceeded the regulatory limits for disposal in municipal sanitary landfills.[23]

9.6 UNCONVENTIONAL CRYSTALLIZATION TECHNOLOGIES

While conventional methods (Section 9.4) exemplify the crystallization-based industrial workhorses for produced waters, other design strategies have also been employed with the aim of simultaneous recovery of water and dissolved compounds for near zero waste discharge. In particular, the recovered compounds (mostly salts) represent a potential stream of revenue to offset the cost for water treatment by selling them to the chemical industry. However, these promising examples of water treatment including precipitation,[24] eutectic freeze crystallization,[25–27] and membrane crystallization[28–30] remain mostly in the academic literature.

9.6.1 Recovery of Salts from Produced Water by Precipitation

Using chemical precipitation as a technology for the simultaneous treatment of produced water as well as recovery of salts is a relatively new area of application of conventional chemical precipitation. In this area, promising approaches have been reported on wastewater for the recovery of, among others, carbonates,[26,27,31] sulfides[32,33] as well as gypsum and magnesium hydroxide.[34] The same principles could also be applied to produced water. Some already promising work has been conducted by Hu and coworkers.[24] They have proven the simultaneous recovery of ammonium (NH_4^+), potassium (K^+), and magnesium (Mg_2^+) from produced water by struvite precipitation after a calcium pretreatment step with sodium carbonate (Na_2CO_3) addition or CO_2 stripping. The recovery efficiencies of the optimized laboratory conditions were reported as 85.9%, 24.8%, and 96.8% for NH_4^+, K^+ and Mg_2^+, respectively. They also found that no heavy metals and organic contaminants were accumulated in the recovered struvite, thus its quality was sufficient for further applications.[24]

9.6.2 Eutectic Freeze Crystallization

Eutectic freeze crystallization (EFC) is an innovative technology for the treatment of waste and saline waters.[35] It was first developed for the recovery of drinking water from seawater, for example, using natural cold sources in winter.[36–38] Early research showed that it was possible to precipitate two distinct solid phases from sodium chloride (NaCl), potassium chloride (KCl), and sodium bicarbonate ($NaHCO_3$) solutions.[39] Barduhn[38] gave a conceptual process design for concentrating dissolved inorganic and organic pollutants in the waste stream of a tertiary treatment plant. Since then, a significant amount of research has been carried out, mostly by the research groups at TU Delft[40–43] and at the University of Cape Town[44–50] to the extent that the process is currently at the commercialization stage.

EFC enables the recovery of both valuable dissolved compounds and water from aqueous streams.[10] Thus, EFC has the potential to treat complex, hypersaline coproduced brines as a sustainable technology for the treatment of produced water with the goal to achieve a near zero waste discharge because EFC changes the focus from the costs of a waste stream to a revenue generation stream (potable water and pure salts).[27,51]

A recent computational study has demonstrated the feasibility of EFC for the treatment of sodium- and chloride-rich produced water of the Marcellus Shale in the USA.[52] The modeling software OLI Stream Analyzer[53] was used to simulate the eutectic freeze crystallization of produced water. Various brine compositions generated from water quality reports were used as the starting point. It was shown that EFC could be capable of recovering selectively pure NaCl crystals while simultaneously producing pure ice crystals. The NaCl could be sold to the chemical industry and the recovered pure water (molten from the ice) would meet quality standards to be released into estuaries, used in agriculture, or reused for industrial purposes.[52]

The technical applicability of EFC for the removal of multiple salts from complex multicomponent, hypersaline brines has yet to be demonstrated, despite the fact that the thermodynamics of crystallization as separation technology are extensively

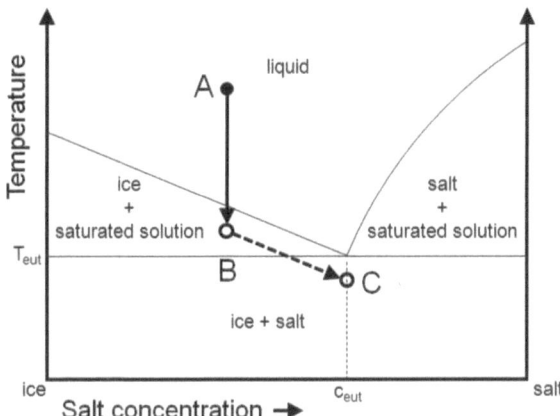

FIGURE 9.2 Principle of eutectic freeze crystallization (EFC) shown in a binary eutectic phase diagram of a brine. (Reprinted with permission from John Wiley and Sons, Inc.)[10]

known.[2–4,7,9] Briefly, the thermodynamic principle is illustrated in Figure 9.2 and is based on cooling the unsaturated solution at point A below its liquidus curve into the two-phase region to generate supercooling. For simplicity, the metastable zone is not shown in Figure 9.2.

Once the solution is supercooled, ice crystals will start to grow at point B. The ice crystals will continue to grow with further cooling below the eutectic temperature leading to an increase in the salt concentration in the solution along the BC line. At point C, the eutectic composition, salt crystals start to grow simultaneously besides the ice crystals. The EFC process can also be conducted if the initial salt concentration in the solution is larger than eutectic composition. In this scenario salt crystals will form first and the ice crystals start to grow once point C is reached.[26] The separation of the generated ice and salt is conducted based on their density differences leading to the settling of salt and floating of the ice crystals on the saturated solution. Further details on EFC can be found elsewhere.[25–27,40,44,51,52,54,55]

A case study by Lu[25] in 2014 on produced water from Kuwait found that ice and NaCl·2H$_2$O crystallizes simultaneously at the eutectic point of −23°C. After washing and melting the ice product, the recovered water contained about 100 mg kg^{-1} NaCl·2H$_2$O and 10 mg kg^{-1} of Ca and Mg. Inclusion of mother liquor in the ice and NaCl·2H$_2$O crystals was < 0.05wt% and < 0.1wt%, respectively. Moreover, after filtration and washing of the NaCl·2H$_2$O crystals, ≥99% of the impurities was left in the filtrate except for bromide, which was partially incorporated into the crystal lattice.[25]

Compared with known conventional separation technologies (e.g., evaporative or cooling crystallization, and freeze concentration) EFC can beneficially treat produced water in an ecologically and economically sustainable manner. Verbeek[55] showed that the overall efficiency of an EFC crystallizer is 59% and the energy requirement per unit feed is comparable to that of typical commercial evaporative crystallizers.[55] However, the cost might even be lower for optimized process conditions because the energy consumption of melt crystallization is seven times less compared to evaporation of the same amount of water in single stage evaporative crystallization.[25] However, to be competitive and of interest to the industry, EFC would need to be comparable to the injection disposal cost of \$1.00–\$6.50 per barrel.

It is clear that EFC is a very promising potential technology that can be used for the simultaneous treatment of produced water as well as the recovery of contaminants for their reuse. However, further investigations of the EFC process with respect to relevant process parameters and scaling up are needed to determine if EFC is a viable option to recover valuable products from oil coproduced water.

Research is actively being conducted into multiple aspects of EFC implementation. The research includes investigations into scientific and technical aspects as well as commercialization and economic analysis. Scientific and technical challenges include those of ice scaling[56] and multiple salt recovery.[57] Commercialization challenges are the next step.[57]

9.6.3 MEMBRANE CRYSTALLIZATION

The stand-alone unit operation membrane technology has a wide range of applications including the production of potable water, pharmaceuticals, chemicals, and electronics.[58–61] This technology is based on a thin barrier of functionalized/nanostructured material (membrane). It controls the exchange between two phases by external forces under the effect of fluid properties and the intrinsic characteristics of the membrane material.[58] For further details on the application of membrane technology for produced water, the interested reader is referred to the excellent review by Chang et al.[59] The focus of this section is on the application of the membrane technology as assisted operation in a hybrid process, setting together with crystallization because it enables the treatment of solutions beyond the saturation level. Specifically, this emerging approach utilizes the membrane to generate supersaturation and/or retain molecules but the nucleation and crystal growth take place in the crystallizer.[62] Moreover, the hybrid membrane crystallization process (Figure 9.3a) is a process intensification strategy,[63] which provides the opportunity to extract simultaneously fresh water and valuable compounds dissolved in produced waters. An extension of this hybrid concept combines the membrane technology and crystallization in

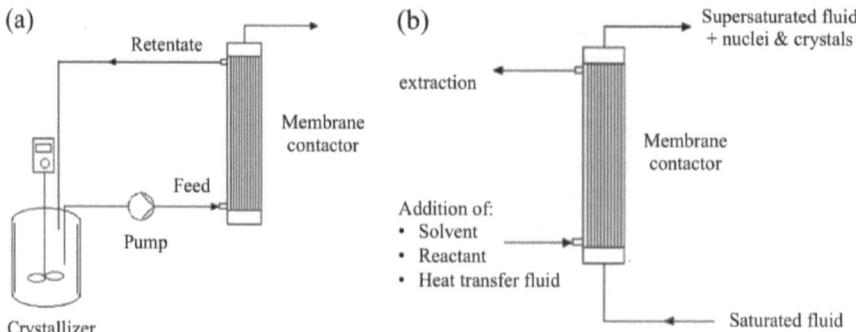

FIGURE 9.3 Schematic illustration of (a) hybrid membrane crystallization process and (b) integrated membrane crystallization (MCr) process. (Reprinted with permission from Elsevier.)[62]

one single step (Figure 9.3b), known as membrane crystallization (MCr). The basic idea is to utilize the membranes as heteronucleant to promote nucleation and thus crystallization directly inside the membrane module where the supersaturation is generated.[60,64,65]

Recent studies have demonstrated the potential of membrane crystallization as a separation strategy for produced waters. For instance, Kim et al.[28] treated synthetic shale gas produced water (SGPW) with high salinity employing a hybrid membrane crystallization process. The authors demonstrated a total recovery of water through this process intensification approach of up to 62.5% compared to the lower recovery of 37.5% obtained by a single membrane unit.[28] In a consecutive study, Kim et al.[29] applied the process to an actual SGPW from Eagle Ford Shale (USA) and proved its feasibility with a water recovery of up to 84% and a solid production rate of up to 2.72 kg m^{-2}day.[29]

Ali et al.[30] demonstrated that 16.4 kg of NaCl per cubic meter of produced water can be recovered by MCr with a purity of >99.9% at a recovery factor of 37%. The research team used an outside-in configuration to prevent clogging of the membrane fibers. After proving the technical feasibility in a laboratory scale, the process was successfully scaled up to a semi-pilot scale using commercially available membrane modules.

All studies showcase a promising pathway for the recovery of valuable compounds (dissolved) from produced water by applying membrane-assisted crystallization, which otherwise are considered as nuisance.[30] Moreover, Knutson et al.[66] reported in their work supported by the Department of Energy (USA) that a membrane/crystallizer configuration resulted in the lowest cost for the treatment of produced water, particularly when zero liquid discharge (ZLD) needs to be achieved.

9.7 CONCLUSION AND OUTLOOK

Looking at the current industrial approaches to treat produced water, it is obvious that the default approach is the disposal of the produced water as much as possible in a cheap manner. However, more recent advances in the academic literature have shown that it is possible to convert the produced water into a potential revenue stream and not just a waste stream. Knutson et al.[66] showed that the unconventional treatment methods for produced water with the goal of ZLD achieved by crystallization-involved separation technologies were found to be less expensive than conventional treatment methods without ZLD.[61] This is caused mainly by the high costs of disposal for the concentrated brines.[61]

Consequently, more progressive studies have recently demonstrated that not just pure water can be recovered but also other value-added products, mostly salts, from produced water. This is in line with the growing awareness of sustainability. For example, Arnold et al.[67] refer to improved management techniques that minimize the amount of water brought to the surface as well as consider the "excess" of water as a resource.

However, despite the highly promising work documented on innovative separation technologies based on and/or coupled with crystallization for produced water, further work is clearly required to solve both the technical and the commercialization challenges. Thus, current research efforts need to be intensified. The advancement of these research areas is paramount to ensure that the promising unconventional crystallization processes reported find their way into large-scale applications for produced water treatment and recovery of valuable resources.

REFERENCES

1. Lewis, A.; Mcmichael, L.; Glazewski, J. Water quality, fracking fluids and legal disclosure. In *Hydraulic Fracturing in the Karoo: Critical Legal and Environmental Perspectives*; Glazewski, J., Esterhuyse, S., Eds.; Juta: Claremont, South Africa, 2016; pp. 245–263.
2. Beckmann, W. *Crystallization: Basic Concepts and Industrial Applications*; Wiley-VCH Verlag GmbH & Co. KGaA: Weinheim, Germany, 2013.
3. Lewis, A.; Seckler, M.; Kramer, H.; Van Rosmalen, G. *Industrial Crystallization - Fundamentals and Applications*; Cambridge University Press: Cambridge, UK, 2015.
4. Myerson, A. S.; Erdemir, D.; Lee, A. Y. *Handbook of Industrial Crystallization*, 3rd ed.; Cambridge University Press: Cambridge, 2019.
5. Chianese, A.; Kramer, H. J. M. *Industrial Crystallization Process Monitoring and Control*; Wiley-VCH Verlag GmbH & Co. KGaA: Weinheim, Germany, 2012.
6. Fontalvo Gómez, M.; Johnson Restrepo, B.; Stelzer, T.; Romañach, R. J. Process analytical chemistry and nondestructive analytical methods: The green chemistry approach for reaction monitoring, control, and analysis. In *Handbook of Green Chemistry Volume 12: Green Chemical Engineering*; Lapkin, A., Ed.; Wiley-VCH Verlag GmbH & Co. KGaA: Weinheim, Germany, 2018; pp. 259–287.
7. Ulrich, J.; Glade, H. *Melt Crystallization - Fundamentals, Equipment and Applications*; Shaker Verlag: Aachen, 2003.
8. Ulrich, J.; Stepanski, M. Zur Begriffsklarung in Der Technischen Kristallisation. *Chemie Ingenieur Technik*, 60 (6), 481–483, 1988.
9. Arkenbout, G. F. *Melt Crystallization Technology*; Technomic Publishing Company, Inc.: Lancaster, PA, 1995.
10. Ulrich, J.; Stelzer, T. *Crystallization. Kirk-Othmer Encyclopedia of Chemical Technology*. John Wiley & Sons, Inc.: Hoboken, NJ, 2011, pp. 1–63.
11. Jones, A. G. *Crystallization Process Systems*; Butterworth-Heinemann: Oxford, 2002.
12. Garside, J.; Gibilaro, L. G.; Tavare, N. S. Evaluation of crystal growth kinetics from a desupersaturation curve using initial derivitives. *Chem. Eng. Sci.* **1982**, *37* (11), 1625–1628.
13. Tavare, N. S.; Chivate, M. R. Growth rate correlation for potassium sulphate crystals in a fluidized bed crystallizer. *Chem. Eng. Sci.* **1978**, *33* (9), 1290–1292.
14. Randolph, A. D.; Larson, M. A. *Theory of Particulate Processes*, 2nd ed.; Academic Press, Inc.: New York, 1988.
15. Ahmed, Q.; Nimer, H.; Hussien, M.; Zumrawi, M.; Babiker, E.; Fadul, I. Evaluation of seepage from evaporation ponds of produced water in some sudanese oilfields. In *Conference Proceedings Civil Engineering 2018*; 2018.
16. Lyman, S. N.; Mansfield, M. L.; Tran, H. N. Q.; Evans, J. D.; Jones, C.; O'Neil, T.; Bowers, R.; Smith, A.; Keslar, C. Emissions of organic compounds from produced water ponds I: Characteristics and speciation. *Sci. Total Environ.* **2018**, *619–620*, 896–905.
17. Igunnu, E. T.; Chen, G. Z. Produced water treatment technologies. *Int. J. Low-Carbon Technol.* **2014**, *9* (3), 157–177.
18. Velmurugan, V.; Srithar, K. Prospects and scopes of solar pond: A detailed review. *Renew. Sustain. Energy Rev.* **2008**, *12* (8), 2253–2263.

19. Boysen, J. E.; Harju, J. A.; Shaw, B.; Fosdick, M.; Grisanti, A.; Sorensen, J. A. The Current Status of Commercial Deployment of the Freeze Thaw Evaporation Treatment of Produced Water. March 1, 1999.

20. Jiménez, S.; Micó, M. M.; Arnaldos, M.; Medina, F.; Contreras, S. State of the art of produced water treatment. *Chemosphere* **2018**, *192*, 186–208.

21. Veolia Water Technologies. Brine Concentrator System: HPD® Evaporation and Crystallization.

22. Garbutt, C. F. Water Treatment Process for Reducing the Hardness of an Oilfield Produced Water. US5879562A, 1999.

23. Zhang, T.; Gregory, K.; Hammack, R. W.; Vidic, R. D. Co-precipitation of radium with barium and strontium sulfate and its impact on the fate of radium during treatment of produced water from unconventional gas extraction. *Environ. Sci. Technol.* **2014**, *48* (8), 4596–4603.

24. Hu, L.; Yu, J.; Luo, H.; Wang, H.; Xu, P.; Zhang, Y. Simultaneous recovery of ammonium, potassium and magnesium from produced water by struvite precipitation. *Chem. Eng. J.* **2020**, *382*, 123001.

25. Lu, X. Novel Applications of Eutectic Freeze Crystallization, Delft University of Technology, The Netherlands, 2014.

26. Chivavava, J.; Jooste, D.; Aspeling, B.; Peters, E.; Ndoro, D.; Heydenrych, H.; Pascual, M. R.; Lewis, A. Continuous eutectic freeze crystallization. In *The Handbook of Continuous Crystallization*; Yazdanpanah, N., Nagy, Z. K., Eds.; The Royal Society of Chemistry, London, 2020; pp. 508–541.

27. Lewis, A. E.; Nathoo, J.; Thomsen, K.; Kramer, H. J.; Witkamp, G. J.; Reddy, S. T.; Randall, D. G. Design of a eutectic freeze crystallization process for multicomponent waste water stream. *Chem. Eng. Res. Des.* **2010**, *88* (9), 1290–1296.

28. Kim, J.; Kwon, H.; Lee, S.; Lee, S.; Hong, S. Membrane distillation (MD) integrated with crystallization (MDC) for shale gas produced water (SGPW) treatment. *Desalination* **2017**, *403*, 172–178.

29. Kim, J.; Kim, J.; Hong, S. Recovery of water and minerals from shale gas produced water by membrane distillation crystallization. *Water Res.* **2018**, *129*, 447–459.

30. Ali, A.; Quist-Jensen, C. A.; Macedonio, F.; Drioli, E. Application of membrane crystallization for minerals' recovery from produced water. *Membranes (Basel).* **2015**, *5* (4), 772–792.

31. Costodes, V. C. T.; Lewis, A. E. Reactive crystallization of nickel hydroxy-carbonate in fluidized-bed reactor: Fines production and column design. *Chem. Eng. Sci.* **2006**, *61* (5), 1377–1385.

32. Guillard, D.; Lewis, A. E. Optimization of nickel hydroxycarbonate precipitation using a laboratory pellet reactor. *Ind. Eng. Chem. Res.* **2002**, *41* (13), 3110–3114.

33. Guillard, D.; Lewis, A. E. Nickel carbonate precipitation in a fluidized-bed reactor. *Ind. Eng. Chem. Res.* **2001**, *40* (23), 5564–5569.

34. Mokone, T. P.; van Hille, R. P.; Lewis, A. E. Effect of solution chemistry on particle characteristics during metal sulfide precipitation. *J. Colloid Interface Sci.* **2010**, *351* (1), 10–18.

35. Chen, G. Q.; Gras, S. L.; Kentish, S. E. Eutectic freeze crystallization of saline dairy effluent. *Desalination* **2020**, *480*, 114349.

36. Mokone, T. P.; van Hille, R. P.; Lewis, A. E. Metal sulphides from wastewater: Assessing the impact of supersaturation control strategies. *Water Res.* **2012**, *46* (7), 2088–2100.

37. Maharaj, C.; Chivavava, J.; Lewis, A. Treatment of a highly-concentrated sulphate-rich synthetic wastewater using calcium hydroxide in a fluidised bed crystallizer. *J. Environ. Manage.* **2018**, *207*, 378–386.

38. Barduhn, A. J.; Manudhane, A. Temperatures required for eutectic freezing of natural waters. *Desalination* **1979**, *28* (3), 233–241.

39. Powell, R. L. A Study of the Concentration of Waste Water by Eutectic Freezing, Syracuse University, 1964.

40. Stepakoff, G. L.; Siegelman, D.; Johnson, R.; Gibson, W. Development of a eutectic freezing process for brine disposal. *Desalination* **1974**, *15* (1), 25–38.

41. Pangborn, J. B. Observations of the eutectic freezing of salt solutions. *Rep. to AJ Barduhn, Dep. Chem. Eng. Syracuse Univ.* 1963.

42. Van der Ham, F. Eutectic Freeze Crystallization, Delft Technical University, The Netherlands, 1999.

43. Vaessen, R. J. C. Development of Scraped Eutectic Crystallizers, Delft Technical University, The Netherlands, 2003.

44. Himawan, C.; Vaessen, R. J. C.; Kramer, H. J. M.; Seckler, M. M.; Witkamp, G. J. Dynamic modeling and simulation of eutectic freeze crystallization. *J. Cryst. Growth* **2002**, *237–239*, 2257–2263.

45. Rodriguez Pascual, M. Physical Aspects of Scraped Heat Exchanger Crystallizers: An Application in Eutectic Freeze Crystallization, Delft Technical University, The Netherlands, 2009.

46. Randall, D. G.; Nathoo, J.; Lewis, A. E. A case study for treating a reverse osmosis brine using eutectic freeze crystallization—Approaching a zero waste process. *Desalination* **2011**, *266* (1–3), 256–262.

47. Hubbe, M. A.; Becheleni, E. M. A.; Lewis, A. E.; Peters, E. M.; Gan, W.; Nong, G.; Mandal, S.; Shi, S. Q. Recovery of inorganic compounds from spent alkaline pulping liquor by eutectic freeze crystallization and supporting unit operations: A review. *BioResources* **2018**, *13* (4), 9180–9219.

48. Heydenrych, H. R.; Rodriguez Pascual, M.; Lewis, A. E. economic and environmental evaluation of eutectic freeze crystallization vs. reverse osmosis for brine water treatment at industrial scale. In *19th International Symposium on Industrial Crystallization (ISIC19)*; Biscans, B., Ed.; Toulouse, France, 2014.

49. Jivanji, R.; Nathoo, J.; Merwe, W. V. D.; Human, A.; Lewis, A. Application of eutectic freeze crystallization to the treatment of mining wastewaters. In *22nd World Mining Congress*; Istanbul, 2011.

50. Lewis, A.; Randall, D. Using eutectic freeze crystallization to treat a range of brines. In *Desalination and Environment: A Water Summit Elsevier*; Rotana Beach, Abu Dhabi, 2011.

51. Genceli, F. E.; Gärtner, R.; Witkamp, G. J. Eutectic freeze crystallization in a 2nd generation cooled disk column crystallizer for $MgSO_4 \cdot H_2O$ system. *J. Cryst. Growth* **2005**, *275* (1–2), e1369–e1372.

52. Milligan, D. P. Investigation Into Oilfield Waste Water Disposal Through Eutectic Freeze Crystallization, University of Oklahoma, 2018.

53. OLI Systems Inc. OLI Studio: Stream Analyzer. https://www.olisystems.com/oli-studio-stream-analyzer.

54. Vaessen, R. J. C. Eutectic freeze crystallization using CO_2 clathrates. *Ann. N. Y. Acad. Sci.* **2000**, *912*, 483–495.

55. Verbeek, B. J. J. Eutectic Freeze Crystallization on Sodium Chloride, Delft Technical University, The Netherlands, 2011.

56. Leyland, D.; Chivavava, J.; Lewis, A. E. Investigations into ice scaling during eutectic freeze crystallization of brine streams at low scraper speeds and high supersaturation. *Sep. Purif. Technol.* **2019**, *220*, 33–41.

57. Panagopoulos, A.; Haralambous, K.-J.; Loizidou, M. Desalination brine disposal methods and treatment technologies - A review. *Sci. Total Environ.* **2019**, *693*, 133545.

58. Drioli, E.; Stankiewicz, A. I.; Macedonio, F. Membrane engineering in process intensification—An overview. *J. Memb. Sci.* **2011**, *380* (1), 1–8.

59. Chang, H.; Li, T.; Liu, B.; Vidic, R. D.; Elimelech, M.; Crittenden, J. C. Potential and implemented membrane-based technologies for the treatment and reuse of flowback and produced water from shale gas and oil plays: A review. *Desalination* **2019**, *455*, 34–57.

60. Di Profio, G.; Salehi, S. M.; Curcio, E.; Drioli, E. 3.11 membrane crystallization technology. In *Comprehensive Membrane Science and Engineering*; Drioli, E., Giorno, L., Eds.; Elsevier: Oxford, 2017; pp. 21–46.

61. Drioli, E.; Ali, A.; Lee, Y. M.; Al-Sharif, S. F.; Al-Beirutty, M.; Macedonio, F. Membrane operations for produced water treatment. *Desalin. Water Treat.* **2016**, *57* (31), 14317–14335.

62. Chabanon, E.; Mangin, D.; Charcosset, C. Membranes and crystallization processes: State of the art and prospects. *J. Memb. Sci.* **2016**, *509*, 57–67.

63. Stelzer, T.; Lakerveld, R.; Myerson, A. S. Process intensification in continuous crystallization. In *The Handbook of Continuous Crystallization*; Yazdanpanah, N., Nagy, Z. K., Eds.; The Royal Society of Chemistry: London, 2020; pp. 266–320.

64. Charcosset, C.; Kieffer, R.; Mangin, D.; Puel, F. Coupling between membrane processes and crystallization operations. *Ind. Eng. Chem. Res.* **2010**, *49* (12), 5489–5495.

65. Curcio, E.; Di Profio, G. Continuous membrane crystallization. In *The Handbook of Continuous Crystallization*; Yazdanpanah, N., Nagy, Z. K., Eds.; The Royal Society of Chemistry: London, 2020; pp. 321–352.

66. Knutson, C.; Dastgheib, S.A.; Yang, Y.; Ashraf, A.; Duckworth, C.; Sinata, P.; Sugiyono, I.; Shannon, M.A.; Werth, C.J. *Reuse of Produced Water from CO₂ Enhanced Oil Recovery, Coal-Bed Methane, and Mine Pool Water by Coal-Based Power Plants*; United States, 2012.

67. Arnold, R.; Burnett, D.; Elphick, J.; Feeley, T.; Galbrun, M.; Hightower, M.; Jiang, Z.; Khan, M.; Lavery, M.; Luffey, F.; Verbeek, P. Managing water—from waste to resource. *Oilf. Rev.* **2004**, *16*, 26–41.

10 Magnetic Separation of Micro- and Nanoparticles for Water Treatment Processes

*Jenifer Gómez-Pastora, Xian Wu,
and Jeffrey J. Chalmers*
The Ohio State University

CONTENTS

10.1 MAGNETIC NANOMATERIALS: SYNTHESIS, PROPERTIES, AND APPLICATIONS IN WATER TREATMENT

In the last two decades, nanotechnology has attracted significant attention and research has focused on miniaturized materials and systems, with the corresponding investment of research funding. In 2017, nanotechnology had a market value amounting to nearly 49 billion (U.S. dollars) worldwide and is projected to increase to nearly 76 billion by 2020 [1]. This is partly driven by advances in particle synthesis, which have allowed the fabrication of materials at the nanoscale with outstanding properties. There are some inherent qualities of nanoparticles that strictly depend on their small size. For example, decreasing the particle size increases the surface-area-to-volume ratio, which positively affects the performance of the material (i.e., it has more reactivity because the fraction of potentially reactive atoms at the surface is increased compared to microparticles or bulk materials) [2]. Nevertheless, the composition of these materials and the synthesis process employed in their fabrication, which in turn affects their properties, usually depend on the final application.

DOI: 10.1201/9781003091011-10

Although nanomaterials have been tested and employed in multiple applications, the most notable progress has been produced in the biomedical and chemical engineering fields. Particularly, nanotechnology has gained much attention in water remediation and purification. Water pollution is a direct consequence of industrialization and population growth, and unfortunately, currently a wide variety of pollutants can be found in water, including heavy metals, organic matter, and radioactive compounds. In this regard, nanomaterials have been employed for adsorption of contaminants, for destruction of pollutants through advanced oxidation processes or for the production of enhanced materials to be used in membrane processes (either reverse or forward osmosis) [3,4]. In comparison to traditional technologies, nanoparticle-based water treatment processes can improve the efficiency, reduce the energy consumption and avoid secondary pollution. Nevertheless, one of the main challenges of this technology is the recovery of the nanomaterials after the process is complete because of their small size and the high volumes of water that are required to be treated. One solution to this challenge is the design of micro- and nanoparticles with magnetic properties through the incorporation of nanometer-sized magnetic materials, magnetic nanoparticles, or MNPs, within the formulation of the specific nanoparticle to treat the water. In addition to providing a means to remove the reactive micro/nanoparticles from the water to be treated, since these micro/nanoparticles can be manipulated with an external field, the potential for recycling them for further use with corresponding economic advantages exists [5].

Regarding the composition of MNPs, bare magnetic nanomaterials have been applied for water treatment, such as zero-valent iron or iron oxides, but composites, especially spherical, core-shell particles, are the most widely used and studied for water treatment applications [5]. With composites, the core is usually composed of a magnetic material, such as Fe_3O_4; however, other magnetic cores containing cobalt or nickel ferrites are reported [5]. In these cases, the composition of the magnetic core does not strictly depend on the MNP application; it only provides a means to magnetically recover the particles. When the magnetic core is sufficiently small (on the order of 50 nm or less, depending on the compound), it can achieve superparamagnetic properties that allow the particle to only be magnetized when in the presence of an applied field. The shell or coating material that covers the magnetic core, or "islands within the particle", is generally determined by the specific application. For water treatment applications, the coating is usually composed of a material that protects the inner core and provides a functionalizable surface. For some applications, this coating makes the MNP stable against oxidation, corrosion, and spontaneous aggregation and increases the physicochemical stability of the material. The coating materials that can be used are multiple: organic materials such as surfactants, activated carbon or polymers (including dextran and polyethylene glycol), or inorganic components, such as metallic elements (e.g., gold or platinum), metal oxides (aluminum oxide and cobalt oxide), and silica. [5,6]. Finally, multiple surface groups (active sites or binding sites) can be deposited/attached to the coating surface depending on the water treatment process.

The amount of particles necessary to treat a unit of volume of water depends on many factors. One of the most important factors is the surface-area-to-volume ratio, which primarily depends on the particle size. One of the most important advantages of using nanometer-sized particles is the large specific surface area that they provide, requiring lower adsorbent doses and achieving faster adsorption kinetics in comparison with bulk adsorbent materials [5]. Generally, by adding coatings the size of the material increases, which affects the efficiency of the water treatment process. However, the surface-area-to-volume ratio can be kept constant by using porous materials, such as silica or carbon, which are among the most preferred shell materials [6]. Finally, it should be also noted that, by decreasing the particle size, the magnetic force that facilitates the separation of the material also decreases, as it will be reviewed in the next section. Hence, these factors must be balanced to achieve the required goals in both processes: water treatment process and magnetic recovery.

In the past several decades, a variety of methods have been developed to synthesize MNPs, which are typically divided into physical and chemical methods. The physical methods are typically "top-down" processes where a bulk magnetic material is decomposed into MNPs, i.e., mechanical milling [7]. In some cases, the milling process has been used to produce new materials from different starting compounds (mechanical alloying) and has been frequently used to produce metallic granular alloys, in which the main aim is to create ensembles of MNPs embedded in a diamagnetic matrix [8]. Alternatively, two other physical synthesis methods are vapor deposition and electrical explosion of wires, which can produce a great variety of particle compositions [8]. Although physical methods have higher production yield, the physical approaches have relatively high power consumption, and controlling the shape and size distributions is difficult [7]. In contrast, chemical approaches can provide control over the composition, size, shape, morphology, crystallinity, colloidal stability, and magnetic properties of the MNPs by tuning different parameters, such as the nature and concentration of the reacting agents and stabilizing surfactants, the pH and mixing of the solution, the reaction temperature, and time [8]. One specific type of chemical approach is the coprecipitation method, which is one of the oldest and most widely employed synthesis process, based on the coprecipitation of the precursor materials in a solution after the addition of a base under inert atmosphere, at room or high temperature, and generally employed to produce Fe_3O_4 MNPs [7,8]. The size and shape control of the magnetic nanoparticles using this technique is very challenging, and to overcome this problem, thermal methods are preferred; these include high-temperature thermal decomposition and hydrothermal and solvothermal synthesis processes [7]. Other chemical synthesis methods widely reported in the literature for the fabrication of MNPs include microemulsion synthesis, electrochemical deposition, microwave-assisted processes, sonochemical methods, and sol-gel or biosynthesis processes [7,8]. Figure 10.1 summarizes the applications of MNPs in water treatment processes, along with their special properties and the main synthesis methods that can be employed in their fabrication.

Finally, it should be noted that the ability to recover the nanomaterials with externally applied magnetic fields is one of the main advantages of using these particles and paramount for all the wastewater treatment processes in which they are involved. The environmental and economic viability of a water treatment process, such as adsorption, depends on the recovery of the solids and their further reusability. MNPs are perfect sorbents because of the low dose required and their full recovery using magnetic means. Furthermore, the reusability of these materials after several cycles

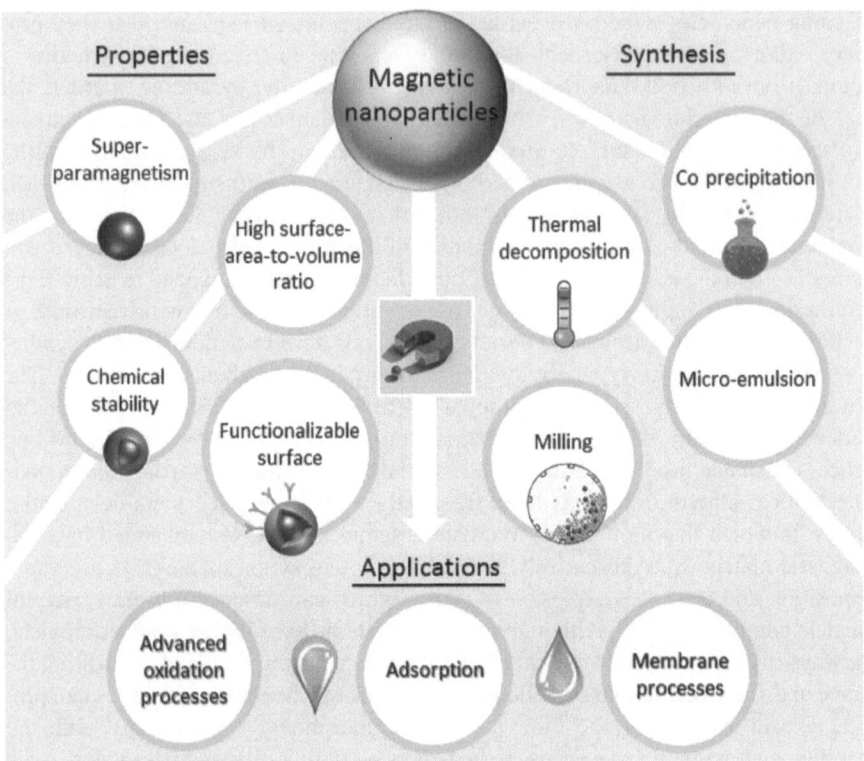

FIGURE 10.1 Properties and synthesis methods to produce MNPs, including the most developed MNP-based water treatment processes.

has been proven successful in several studies [5]. Nevertheless, while the application of MNPs on different water treatment processes is rather well studied, the separation of these small solids from liquid samples is relatively less explored, and the physics behind the nanoparticle behavior under magnetic fields is not totally understood [9]. Therefore, the principles of magnetic separations are reviewed in the following section as they represent the basic information needed to design the recovery process of MNPs. Moreover, Sections 10.3 and 10.4 will review the recent advances in the development of batch and continuous-flow separators, including the systems based on high and low gradient magnetic separation. Finally, several guidelines and future directions will be presented in the last section, among the main conclusions of this analysis.

10.2 PRINCIPLES OF MAGNETIC SEPARATIONS

Compared with conventional methods used to separate small solids from liquids, such as sedimentation and filtration, magnetic separations have multiple advantages [10]: (1) it is potentially more efficient than conventional processes, (2) it allows for the separation of magnetic materials from nonmagnetic solids, (3) it is generally faster than conventional methods, (4) the applied flow rates can be two orders of magnitude higher than what can be processed by ordinary filtration, (5) the separator requires less space in a full-scale conventional treatment plant, (6) noninvasive technology, e.g., the use of magnetic fields, does not modify the solution parameters, and (7) the separation is not sensitive to factors such as pH, temperature, or the ionic concentration of the solution, allowing operation in a wide range of these parameters [10]. Therefore, magnetic separations are preferable than other conventional technologies for separating small micro- and nanoparticles from fluids, such as water.

Magnetic separation is a process involving several forces acting on the material in the presence of an external magnetic field. These include the magnetic force (F_m), viscous drag (F_d), gravity, buoyancy, inertia, particle–fluid interactions (perturbations to the flow field), thermal kinetics (Brownian motion) and several interparticle effects (magnetic dipole interactions, Helmholtz double-layer interactions, and Van der Waals forces). Taking all these forces into account is not only complex, but unnecessary, as it often suffices to take into account the dominant F_m and F_d and neglect the other forces, which are generally of second order in effect [10]. Thus, to describe the separation/motion of micro- and nanoparticles due to a magnetic field (magnetophoresis), equations of the dominant forces acting on the process must be developed.

The magnetic force vector acting on a particle is proportional to its size, magnetization, and applied field gradient as follows [9]:

$$F_m = \mu_0 \frac{4}{3}\pi r^3 M_p \nabla H \tag{10.1}$$

where μ_0 represents the permeability of free space ($4\pi \times 10^{-7}$ H·m^{-1}), r is the particle radius, M_p is the magnetization, and H the external magnetic field. The magnetic behavior of MNPs employed in water treatment is usually superparamagnetic, as long as the particle size of the magnetic material remains below a material-dependent threshold, generally around 30 nm [11]. Superparamagnetic nanoparticles provide a strong response to an external field and exhibit very large saturation magnetization values, several orders of magnitude higher than what is observed in paramagnetic materials. Moreover, their magnetic moment vectors relax to random directions (i.e. an unmagnetized state) in the absence of an applied magnetic field, in which case they have no attraction for each other, thereby reducing the occurrence of particle aggregation. To predict the magnetization of superparamagnetic particles, which represents the magnetic response of the MNP, several models can be applied. For example, linear models are very popular in the literature [10]. With these models, the particle magnetization below saturation can be assumed as a linear function of the applied magnetic field (expressed as $M_p = \chi H$, being χ the magnetic susceptibility of the material) up to saturation, where it reaches the maximum value $M_{p,s}$. Generally, for fields above 0.5 T, the particle can be assumed saturated with $M_p = M_{p,s}$ [9]. The $M_{p,s}$ value depends on the material used as core, the presence of coatings, and their volume ratio with regard to the magnetic core, as well as the MNP size and shape. A recent publication reported that, for MNPs composed of iron oxide particles (nine different types of iron oxide nanoparticles were analyzed), the saturation

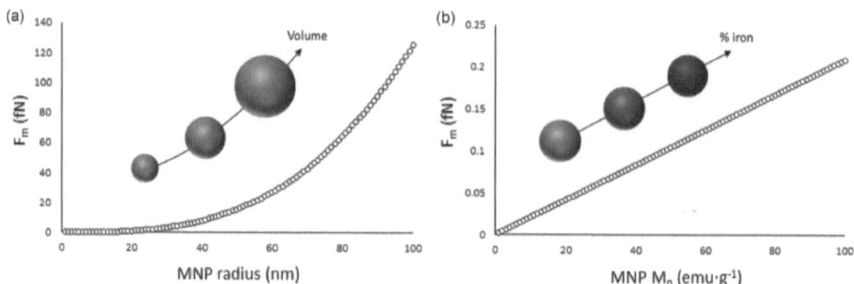

FIGURE 10.2   (a) Magnetic force exerted on a saturated MNP as a function of the particle radius (particle magnetization of 60 emu·g^{-1} and field gradient of 100 T·m^{-1}); (b) Magnetic force exerted on a 20-nm MNP as a function of the particle magnetization (field gradient of 100 T·m^{-1}).

magnetization ranges from 40 to 85 emu·g^{-1} with most particles being in the 50–70 emu·g^{-1} range [12]. In addition to the magnetization of the MNP, beyond whether the MNP is saturated, the magnetic force operating on the MNP depends on its volume and the applied magnetic field gradient.

Figure 10.2 presents the magnetic force (in fN) as a function of (1) particle radius and (2) magnetization (in this case saturation magnetization expressed in emu·g^{-1}), both for a constant magnetic field gradient of 100 T·m^{-1}. The magnetic force scales with r^3 and underscores the strong effect that particle size has on the potential to magnetically move MNP; in fact, it is generally recognized that below a specific size, depending on the magnetic material, it is not possible to magnetically remove the particles in a flowing or stationary fluid (unless the MNP agglomerate first) [9]. It is this trade off, decrease in MNP surface area with increasing size (and thereby easier magnetic separation), that further complicates the practical applications of MNP.

The primary force vector opposing the applied magnetic force on an MNP is the fluid drag force, generally assumed to be approximated using Stokes' approximation for the drag on a sphere:

$$F_d = 6\pi\eta r\left(v - v_p\right) \tag{10.2}$$

where η and v are the viscosity and the velocity vector of the fluid and v_p denotes the particle velocity. To solve this force, the fluid velocity field governed by the Navier–Stokes equations is usually included within the magnetophoresis model:

$$\rho\left(\frac{\partial v}{\partial t} + v\nabla v\right) = -\nabla P + \nabla\left(\eta\nabla v\right) \tag{10.3}$$

$$\nabla v = 0 \tag{10.4}$$

where ρ represents the density of the fluid and P the pressure.

Finally, these forces and velocity equations are included into models comprising a set of equations that relates the particle movement and its separation to multiple

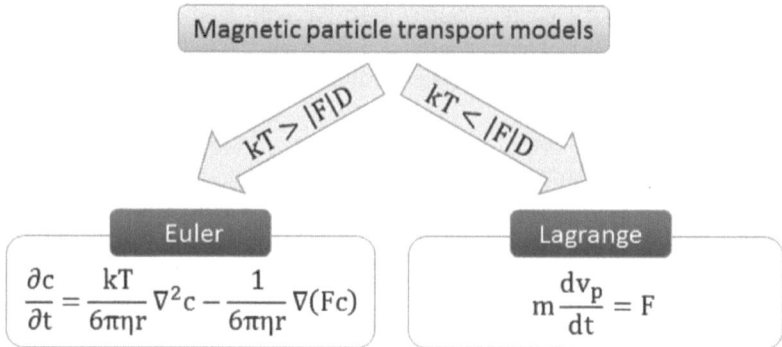

FIGURE 10.3 Magnetic particle transport models, explaining the criterion to solve either an Euler approach or a Lagrangian model: K, Boltzmann constant; T, temperature; |F|, magnitude of the total force acting on the particle, D, particle diameter; c, particle concentration; t, time; η, fluid viscosity; r, particle radius; F, resultant force acting on the particle; m, particle mass; and v_p, particle velocity.

input parameters and operating variables. Figure 10.3 presents different models for predicting the transport of magnetic particles in a carrier fluid. Each model has a different regime of transport. The Euler equation approach should be solved when the thermal energy kT (k representing Boltzmann's constant and T the temperature) is higher than the product between the magnitude of the total force acting on the particle and its size (|F|D). In this model, Brownian motion is accounted and a drift-diffusion equation is solved to predict the behavior of a concentration of noninteracting magnetic nanoparticles [13]. On the contrary, for particles of submicron size or larger, kT is usually smaller than |F|D and Brownian motion is neglected; under this condition, a Lagrangian approach solving the classical Newtonian physics is useful to predict the motion of individual particles. More information above these models can be found in specific literature [10,13].

As presented above, the basic elements of any magnetic separation are the magnetic particles themselves, employed in a specific application, and the magnetic source (field and gradient created), which, in turn, depends on the design of the magnetic separator [10]. In the following sections, we discuss recent developments in the field as well as useful guidelines on the applications of existing magnetic separator systems.

10.3 BATCH MAGNETIC SEPARATORS

Magnetic separators can be classified into different categories depending on the operating mode, the scale, and the source of the magnetic field and gradient. According to the operating mode, they can be classified into batch (sometimes called "catch and release") or continuous systems. The difference lies in the method by which the particles are collected. Batch separators are based on columns or vessels that accumulate the particles on trapping elements using an external field, allowing their subsequent recovery when the field is removed. In continuous systems, the forces acting on the particles are balanced so that they are concentrated in different outlets due to the deflection of their flow trajectories. For both modes, several designs can be applied. In this section, batch separators, both high and low gradient systems, are analyzed whereas continuous-flow systems, at large and microscale, are presented in the following section.

To date, most of the conventional water treatment technologies utilizing magnetophoretic-driven separations operating at relatively high flow rates achieve the separation in batch operation mode using high gradient magnetic separation (HGMS) columns. These columns are filters that trap the particles on packing within the columns, a typical a packing is ferromagnetic (steel) filaments. The filaments, with a saturation of 0.8 T, diameters around 50 μm, and gaps of 10–100 μm between them, create high gradients on their surface (on the order of $10^4 – 10^5$ T·m^{-1}) when the column is magnetized externally with an electromagnet [10]. Once magnetized, they trap the MNPs on their surface, and the fluid, free of MNPs, can be collected at the outlet. The process can be repeated until the required efficiency is achieved or until the column becomes saturated (no available area to deposit more particles on the matrix); then, the MNPs can be recovered by switching off the electromagnet. Figure 10.4a presents a scheme of the HGMS technology for the filtration of MNPs.

HGMS columns have multiple advantages. Among them, the generation of extremely high magnetic gradients should be highlighted, which increases the magnetic force acting on the particles. As the magnetic force increases, the use of higher flow rates is possible, even though this increased fluid velocity results in increased drag force. Nevertheless, these filters have proven to be able to recover MNP of small sizes, as well as weakly paramagnetic particles, at relatively high flow rates. For these reasons, HGMS has become a well stablished technology for the separation of magnetic solids in different industrial fields and has been successfully operating for more than 40 years [10].

Table 10.1 presents studies in which HGMS technology has been applied for the recovery of magnetic micro- and nanoparticles [14–23]. It can be observed that, even with particles sizes below 30 nm, successful capture is achieved with these filters

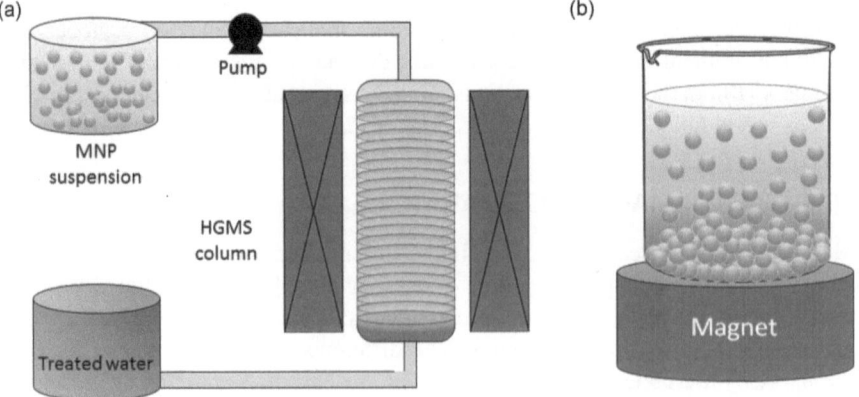

FIGURE 10.4 Schematic representation of (a) high gradient magnetic separation technology and (b) low gradient magnetic separation.

TABLE 10.1

Magnetic Separation of Micro- and Nanoparticles via High Gradient Magnetic Separation (HGMS) Technology

External Field	Matrix	Particle Composition	Particle Size (μm)	Flow Rate (L·min⁻¹)	Efficiency	Comments[a]	Ref.
Superconducting magnets (2–4 T)	Steel wool	Fe_2O_3 and small quantities of CaO, MgO, and SiO_2	0.1–8 (6.431 average)	30–70	Up to 99%	Two separators in series. Particles of 40–100 nm can be trapped at 3.75 T	14
NdFeB permanent magnets (0.8 T)	Fe_3O_4-SiO_2 or steel wool	Fe_2O_3	1–30 (9.340 average)	0.05–0.15	>90% for 10 minutes	Fe_3O_4-SiO_2 matrix was about 30%–40% as effective as steel	15
Electromagnet (10 kG)	Net of permalloy strips	Cu_2O and CuO	0.4–60 (average 6–20)	0.04–0.45 depending on the scale	88%–100%	Copper ion was reduced to metallic copper and simultaneously separated	16
SmCo permanent magnets (240 kA·m⁻¹)	Ferromagnetic membrane	Fe_3O_4 (26 nm) adsorbed on hydroxylapatite	0.05–2 (1 average?)	0.3–1.2	≈100% for particle size > 1 μm	The magnetic moment of the suspension decreased by 20%–25% after the separation	17
Superconducting magnet (2 T)	Steel wool	Magnetic zeolite	5	-	99.80%	Particles used to remove ammonia nitrogen. Fluid velocity: 0.5–1 m·s⁻¹	18
Superconducting magnet (1–6 T)	Steel wool	Zirconium ferrite	10	-	≈100% for fields > 1 T	Particles previously incubated with water for the removal of phosphorus. Fluid velocity: 1 m·s⁻¹	19
Superconducting magnets (1.59 T)	Steel wool	Fe_2O_3	1	1–3	71%–95%	After removing the matrix, capture of Fe_3O_4 (1 μm) is possible at 1 L·min⁻¹	20
SmCo permanent magnets (0.5 T)	Steel wool	Fe_2O_3	0.8	6.6–10.2	100% after 10 minutes	HGMS operating in recycling mode	21
Electromagnet (1.3 T)	Steel wool	Polymer-coated Fe_3O_4	0.026	-	70%–90%	Fluid velocity: 0.02–2 cm·s⁻¹	22
Electromagnet (up to 0.5 T)	Steel wool	Fe_3O_4	0.004–0.02	0.02	> 90% for sizes > 9 nm	Particles in hexane used for arsenic removal	23

[a] Particles suspended in water unless stated otherwise.

at relatively high fluid velocities. Moeser et al. used HGMS to separate polymer-coated MNPs employed for the extraction of organic contaminants from water [22]. They synthesized superparamagnetic Fe_3O_4 particles (7.5 nm in size and a saturation magnetization of 63 emu·g^{-1}) and coated them with a graft copolymer (MNP final size of 26 nm). By using a HGMS column equipped with steel wool matrix and an electromagnet operating at 1.3 T, these authors could recover between 70% and 90% of the particles at flow rates between 0.02 and 2 cm·s^{-1}. The study by Yavuz et al. deserves special attention; they employed 12 nm Fe_3O_4 nanocrystals for removing arsenic from water samples, achieving an efficiency higher than 98% in the removal of As (III) and As (V). Such a reduction resulted in the arsenic level well below the current U.S. standards for drinking water [23]. These authors found that, when comparing the arsenic adsorption capabilities of their small MNPs to a commercially available material of 300 nm in size, their material provided extremely concentrated wastes (they achieved a 99% reduction in solid waste). Finally, after the wastewater treatment process was completed, they used HGMS for recovering the particles. Nearly 100% of the 12-nm-diameter MNPs were retained in the HGMS column at applied fields of only 0.2 T generated by an electromagnet and operating at a flow rate of 20 mL·min^{-1}.

It should be also noted that HGMS filters can be applied for the recovery of any particulate matter with magnetic properties. In fact, this technology was originally designed for recovering magnetic particles in the mineral industry and iron oxide particles in steel and power plants. For example, Li et al. [14] employed a superconducting HGMS system to remove the turbidity of wastewater from a steelmaking plant. They found that by using two filters in series and working at relatively high magnetic fields (3 T), not only the ferromagnetic but also weak magnetic and non-magnetic substances such as CaO, SiO_2, and MgO could be separated in the system (either adsorbed, filtered, or flocculated). In that study, even small sized particles of several tens of nanometers could be efficiently separated at a flow rate of 50 L·min^{-1}, and their particle capture efficiency was close to 100%, with a turbidity reduction of 99%, which allowed further reutilization of the water.

As can be observed in Table 10.1, the filters are generally magnetized with electromagnets and superconducting magnets. The disadvantage of electromagnets is the energy loss due to Joule effect, which can be overcome by the use of superconducting magnets [21]; these can produce strong magnetic fields, as seen in Table 10.1. Conversely, the use of superconducting magnets, while effectively having no heat loss, require prohibitively expensive cryogenic cooling systems [10]. Because of this challenge, a number of studies have investigated the feasibility of HGMS using permanent magnets as the source field, obtaining promising efficiency results while reducing the electrical usage and improving the system portability [17,21]. However, the design of switchable permanent magnets (necessary for the rinsing of the matrix and to recover the particles) is challenging [10]. Regarding the matrix elements, the most popular material is stainless steel wool, although recent studies have addressed the possibility of using ferromagnetic membranes [17] or supported magnetite composites [15].

Besides the high capture efficiency of HGMS technology, it poses several disadvantages. For example, the installation and operation cost of the HGMS columns is extremely high as huge amounts of energy are required for the electromagnet to generate a magnetic field that is sufficiently intense to induce a successful collection of the particles [10,21,24]. Furthermore, the presence of the matrix generates inhomogeneous magnetic fields and forces inside the column, which are difficult to describe or simulate due to the different length scales involved in the process (nanometer to micrometer scale to solve the particle–wire interactions and macroscopic scale to solve the entire filter) [5,10,21,24]. Also, the magnetic matrix may promote the trapping of solids that are not magnetic (non-specificity) [10,17,20] or the permanent trapping of magnetic particles [5,23]. Finally, the steel filaments are expensive, especially those having a diameter below 100 μm, and it should be noted that no retention takes place on a stainless steel wire when the axis of the wire is parallel to the applied magnetic field, which causes approximately one-third of the wires in a stainless steel wool to be magnetically inactive [15]. Thus, intense research is being conducted on the development of micro-ordered matrices able to generate much higher magnetic gradients [24].

Due to the previous disadvantages, a number of recent studies in magnetic separation have been focused on the development of low gradient magnetic separation (LGMS). LGMS is a cost-effective alternative that can achieve separation under externally applied field gradients on the order of 100 T·m^{-1}. LGMS involves a simple setup; the MNP suspension is contacted with a magnetic field in batch mode without the presence of a ferromagnetic packing inside the vessel (Figure 10.4b). Generally, the field is provided by a permanent magnet, and although the field produced with current rare earth magnets cannot exceed 2 T, higher fields are not necessary as the particles become saturated below 1 T. Under nonhomogeneous magnetic fields provided by the magnet, MNPs migrate toward the location where the gradient is highest. However, the separation of MNPs under low gradients is very challenging, as the magnetic field gradient, and therefore the magnetic force, is much lower than what can be achieved by HGMS. For this reason, the separation time could reach several hours or even days. Jin and Kim [25] studied the separation of magnetite nanoparticles in aqueous solution at low magnetic field gradients (23–34 T·m^{-1}). They found that, by monitoring the magnetic weight at the bottom of the vessel, the separation could take more than 1–2 days. Bakhteeva et al. [26] studied the separation of magnetite particles coated with silica ($Fe_3O_4@SiO_2$) of 20–30 nm in size in pure water under low fields (0.3 T) and gradients (13 T·m^{-1}). The separation in water could be greatly speeded up (separation time decreased by 3 orders of magnitude) by promoting steric and electrostatic particle–particle interactions achieved when several salts were added to the solution. These long operation times are due to the fact that, the gradient decay in a strong, non-linear manner with the distance. By using a simple magnet, the gradient is lower than 20 T·m^{-1} for distances of about 1 cm [9]. Thus, to speed up the magnetophoresis of small MNPs with this technology and to reach the performance of HGMS, different cooperative effects have been introduced into the separation process.

The first effect, called "cooperative magnetophoresis", is based on the aggregation of particles under a magnetic field, and their subsequent collection once the aggregate reaches a sufficient volume to induce a large magnetic force which can overcome the randomizing thermal energy. To observe cooperative magnetophoresis, the parameter ψ, which represents the ratio between the dipole–dipole contact energy and the thermal energy, must be greater than 1 [9]:

$$\Psi = \frac{\text{dipole} - \text{dipole} \ \text{contact energy}}{\text{thermal energy}} = \frac{\mu_0 \pi r^3 M_p^2}{9\,kT} \qquad (10.5)$$

To observe aggregation and cooperative magnetophoresis in the case of diluted suspensions of MNPs, the concentration should be above a critical value (see Ref. 27 for more information).

It should be noted that the phenomenon "cooperative magnetophoresis" can speed up the separation of small particles under magnetic fields generated by a handheld magnet. For example, Andreu et al. [27] observed that when cooperative magnetophoresis takes place, the separation time decreases from days to minutes. In their study, the single-particle separation of 12 nm iron oxide nanoparticles under field gradients between 30 and 60 T·m^{-1} took 30 hours, but when the particle size increased to 200 nm, cooperative magnetophoresis was observed and the separation took place in 2 minutes.

The second effect studied in the literature to favor the magnetophoresis of small MNPs under LGMS is the hydrodynamic effect or the induction of fluid flow due to particle motion. A few studies have addressed that magnetic field gradients can induce substantial convective currents during the magnetophoresis of superparamagnetic nanoparticles, which enhances the efficiency of the LGMS process. For example, Leong et al. [28,29] have studied the magnetophoresis of MNPs under LGMS with a system setup similar to the one presented in Figure 10.4b. These authors found that the induced convection boosts the magnetophoretic capture of MNPs by approximately 30 times compared to the situation with no convection [28]. Moreover, they demonstrated that the magnetophoretic collection rate was determined by the magnetic field gradient at the surface of contact of the vessel with the magnet and the area of this surface, but the kinetics of the process was independent of the magnetic field distribution across the solution subjected to magnetophoresis (as long as the convective current was magnetically induced) [29]. The two-way momentum transfer between the magnetic materials and the surrounding fluids has also been reported for microscale particle magnetophoresis [30,31]. These studies, although very useful for the design of magnetophoretic separators with permanent magnets, are limited and more research is still needed to make the most of LGMS.

The conventional HGMS filters and LGMS systems described in this section have several advantages for the separation of small MNPs from water samples. Nevertheless, the batch operation mode could be seen as a disadvantage because the disruption of the operation is needed for MNP elution. Moreover, in the case of HGMS, the accumulation of the particles on the trapping elements could decrease the efficiency and capacity of the separator with time and also creates strong particle–particle interactions on the matrix which often create stable aggregates, hindering the future reutilization of the material. Finally, the use of batch systems facilitates the nonspecific entrapment of sample impurities in the capturing regions, generating contamination. These limitations could be mitigated by the use of continuous-flow separators, which are reviewed in the following section.

10.4 CONTINUOUS-FLOW SEPARATORS

To overcome the issues reported by batch separators, continuous-flow systems, where the particles are deflected from their initial path and recovered in different outlets, could be employed. Open gradient magnetic separation (OGMS) columns represent a type of large-scale, continuous-flow, magnetic separator. In these systems, the magnetic fields and gradients are provided by external magnets specifically arranged around the column walls (Figure 10.5a). Contrary to HGMS, the column has no matrix on the inside to produce high gradients; the gradient is limited to the magnet and column design. The separator has multiple inlets and outlets, and the particle suspension is injected near the center of the column, and by the generation of a magnetic force in the direction toward the outer wall (radial direction) of the column, particles are deflected radially and exit the device through outer outlets (Figure 10.5b). The magnetic characteristics of the particles (susceptibility and magnetization) determine the direction and angle of this deflection [32]. To achieve deflection across the fluid streams, laminar flow should be assured; thus, fluid velocities are generally low. In OGMS, magnets can be arranged so that the magnetic field gradient inside the system is constant, with fields ranging from zero at the center to the maximum at the walls (Figure 10.5a). This isodynamic configuration, which means that the magnetic force is nearly constant throughout the working volume, greatly simplifies the design, modeling, optimization, and scale-up of the process [32]. Nevertheless, the presence of zero fields at the center of the column makes this area magnetically inactive, and thus, nonmagnetic cylinders are placed at this position [10].

Table 10.2 presents several studies where OGMS has been employed for the recovery of magnetic particles from aqueous streams in continuous flow [32–37]. With these separators, the processed flow rates might be lower than the ones achieved

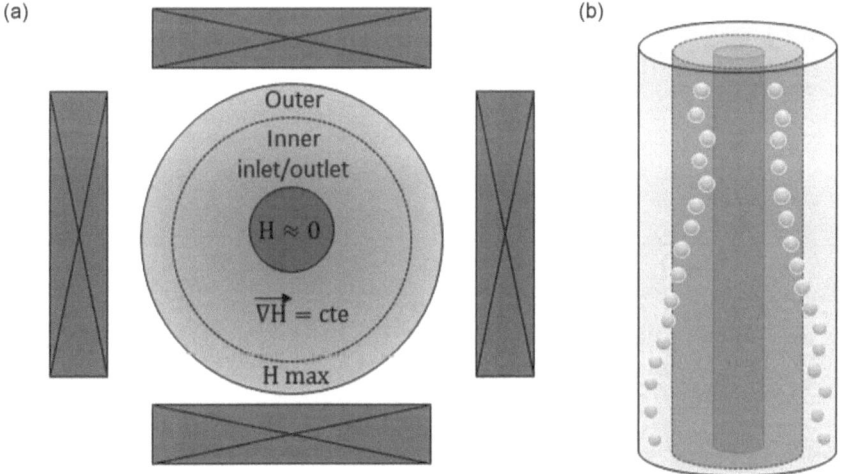

FIGURE 10.5 Open gradient magnetic separation: (a) top view of the system, showing the magnet arrangement (quadrupole magnet configuration) and the inlets/outlets and (b) schematic representation of particle deflection inside the column.

TABLE 10.2

Magnetic Separation of Microparticles from Aqueous Solutions via Open Gradient Magnetic Separation (OGMS) Technology

External Field	Field and Gradient	Particle Composition	Particle Size (μm)	Flow Rate (L·min⁻¹)	Efficiency	Comments	Ref.
Superconducting (Nb3Sn) magnet	0.5–2.5 T	Fe_3O_4	-	3–6	40%–60% in the separation of Fe, Cr, Ni attached to the particles	Particles used for the adsorption of metals (emulated wastewater). Flow rate did not significantly affect the efficiency	32
Permanent magnets (NdFeB)	0.3 T	Fe_3O_4	-	4–12	45%–60% in the separation of Ni	Device tested with real wastewater for Ni recovery. Efficiency dropped with increased flow rates	32
High-temperature superconducting (Bi2223) magnet	2 T at the center of the bore	Fe_3O_4	-	1	More than 99% of particles removed continuously	Efficiency measured as Fe content. Optimization studies revealed that to attain practical processing speed: field > 5 T and $r > 5$ μm	33
High (Bi2223) and low (NbTi) temperature superconducting magnet	2 T (Bi2223), 5 T (NbTi)	Fe_3O_4	10	1	More than 99% removed continuously	Fe content in the outer outlet is about 2.5 times larger than the initial Fe content with the Bi2223 magnet and 11 times more with the NbTi magnet	34
Superconducting (NbTi) magnet	0.7–3 T	Fe_3O_4/silicate	< 5 (Fe_3O_4)	4.5–7	82% and 55% in the removal of Cr and Mo from synthetic and real wastewaters	The results from real wastewater in the separation of other metals were not satisfactory because of problems with pH stabilization	35
Quadrupole electromagnet	0.5 T (gradient: 10 T·m⁻¹)	Fe_3O_4	-	5	95% in the recovery of iron particles	Fe content in the outer outlet is about 2 times higher than the initial. Continuous separation of paramagnetic particles of micrometer scale is feasible with higher gradients	36
Quadrupole superconducting magnet	11 T (gradient: 20–200 T·m⁻¹)	2 particle types	2	Maximum velocity 0.01–0.1 m·s⁻¹	Complete at high velocity when the susceptibility ratio is ≈ 100	Two kinds of particles with different magnetic susceptibility can be selectively separated	37

by HGMS due to a decrease in the magnetic field gradient as a consequence of the matrix absence and the need to avoid turbulent flows. Nevertheless, the separation of particles as small as 5 μm can be achieved at flow rates of several liters per minute. Furthermore, since the susceptibility of the particles determines the deflection angle, different types of magnetic particles can be separated with this device. For example, Takahashi et al. [37] numerically studied the separation of two types of particles with different susceptibilities. Their results confirmed that the separation of these particles (both 1 μm in radius with a susceptibility ratio of ≈10) is possible at a reasonable fluid velocity and field gradient.

Reviewing the examples in Table 10.2, it can be seen that the magnetic field inside the column can be generated by simple and inexpensive permanent magnets, or sophisticated and costly electromagnets and superconducting magnets. For example, Fukui et al. [34] compared the performance of an OGMS column to separate 10 μm Fe_3O_4 by using high (Bi2223)- and low (NbTi)-temperature superconducting magnets, operating at 2 and 5 T, respectively. The low-temperature magnet achieved a better separation than what was observed with the high-temperature superconducting magnet (particle concentration in the outer outlet was 11 times larger than the initial value with the low-temperature magnet, in comparison to the system operated with a high-temperature magnet that achieved only 2.5 times the initial value). However, the enhanced performance of the low-temperature magnet is achieved at a cost of operating at 4.2 K, in comparison to the high-temperature magnet, which operated at 20 K. Even though superconducting separators are smaller and consume less power than their normal conducting counterparts, the material price is expensive and they require extremely low operation temperatures [38]. For example, the price of NbTi is much lower than that of Nb_3Sn, but Nb_3Sn wire has higher critical temperature and flux density; on the contrary, Bi223 magnets can operate at higher temperatures, but the cost of the material increases by more than 10 times in comparison to the previous materials. Thus, the material cost as well as the required cooling power are factors to consider in the design of OGMS when superconducting magnets are necessary.

In fact, it has been suggested that in OGMS, the gradient that can be created by optimizing the magnet design is limited to 50 T·m⁻¹ [35]. That might be correct for large-scale separators, where the column diameter has several centimeters in length. For these large-scale systems, the field on the wall should be maximized as much as possible to achieve high gradients (note that the field at the center of the column is zero for a quadrupole magnet configuration, thus the higher the field on the wall, the higher the gradient for a specific column diameter). To achieve this high field at the system wall, the scientific community has been focused on using superconducting magnets, with can achieve fields larger than 10 T. However, high gradients can be also generated by reducing the device diameter. For example, we have estimated the gradient generated inside the columns when quadrupole magnetic fields are applied for different column diameters, shown in Figure 10.6. It can be seen that for diameters above 1 mm, the gradient drastically diminishes regardless the value of the magnetic field applied on the walls. Nevertheless, by using columns in the micrometer scale and medium magnetic fields on the wall provided by simple and nonexpensive permanent magnets (1–2 T), gradients as high as the ones provided

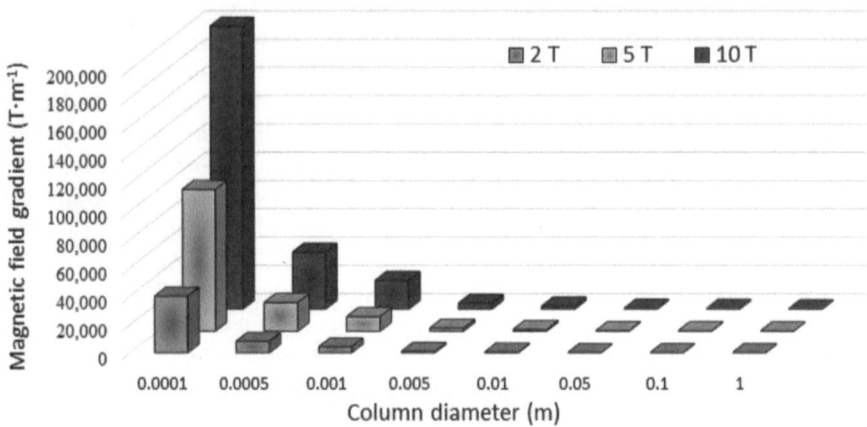

FIGURE 10.6 Magnetic field gradient generated inside an OGMS column as a function of the column diameter when quadrupole magnetic fields are employed. The colors represent different maximum magnetic fields applied on the separator walls.

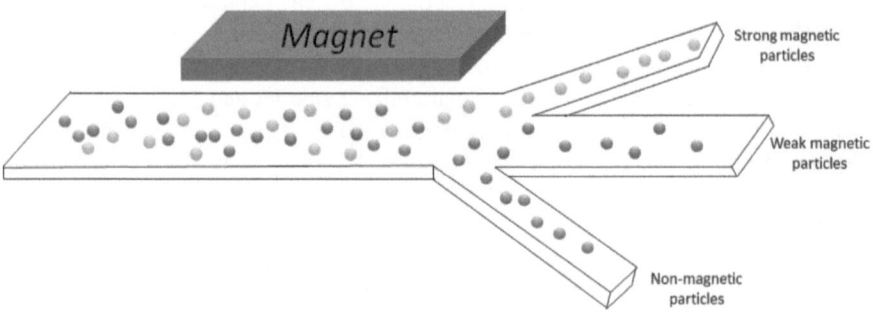

FIGURE 10.7 Separation of multiple particle types inside a continuous-flow microdevice.

by HGMS technology can be generated. This is one of the multiple advantages that microfluidics can offer.

Microfluidics is a technology that emerged in the 2000s, focused on manipulating small volumes of liquids within an area with dimensions of tens to hundreds of μm. Batch and continuous-flow magnetophoretic microdevices have attained great attention in the last 10–15 years because these devices can integrate multiple fluids and particles, with multiple functionalities. Figure 10.7 represents the operation of free-flow particle magnetophoresis of materials with different magnetic characteristics. Through a combination of specifically designed magnetic field gradients, particles with different magnetic susceptibility can be separated within the same device by magnetically deflecting them into different outlets. Note that the particles are randomly distributed at the separator inlet, and their trajectories change as a function of their size and composition. So far, devices with up to 25 outlets have been reported [39]. Furthermore, due to the laminar flow developed in these devices, fluids containing different reagents can be employed, and magnetic materials can be deflected across them [40]. This enables the reaction between the particles and the fluids in a sequential manner, and thus, chemical reactions and washing steps can be implemented within the same device (i.e., lab-on-a-chip (LOC) and "micro total analysis systems" (µTAS)) [30].

Owing to the features of these devices, the potential applications are multiple, including assays in limited resource settings (point-of-care devices), environmental monitoring, food-contaminant analysis, etc. [10]. While microdevices have been primarily employed within biomedical fields due to the small volumes that can be processed [41–44], applications in water treatment such as water quality assessment can be found as well. In fact, for water monitoring applications, batch and continuous microfluidic devices could be an affordable option for recovering magnetic particles employed in the analysis and can provide a rapid readout while using small sample volumes and particle concentrations. This is important as the magnetic particles employed in these processes are expensive, and the use of small volumes might reduce the risks due to exposures to hazardous materials. For example, several works have addressed the advantages of combining magnetic particles and microfluidics in the detection and analysis of several waterborne pathogens, such as E. Coli [45–48]. These studies have reported that the magnetophoretic microdevices could detect the analyte with high sensitivity and high selectivity in small volumes in a simple manner [45] and are also able to treat medium volumes (≈ 100 mL) in short times ($\approx$15 minutes) [46], suggesting the great potential of this technology for point-of-need water quality monitoring. Although most of these studies employed batch devices, the detection and analysis of multiple analytes could simultaneously be performed within the same device in continuous mode by employing various particles (with different functionalizations and magnetic properties) and recovering them in different outlets of the device for further analysis, as schematized in Figure 10.7.

Within microfluidics, free-flow magnetophoresis can be easily achieved with permanent magnets at relatively high flow rates by using enhanced magnet designs. For example, recent works [9, 49–51] have employed microfluidic channels equipped with strong quadrupole magnetic fields provided by permanent magnets to separate magnetic particles and cells under continuous-flow. Due to the small dimension of the channels, the gradients generated varied from 286 to 1750 T·m^{-1}, which is much more than what LGMS or OGMS can achieve [9]. The targeted particles and cells could be easily separated at high flow rates and short operation times under these conditions [49–51]. Thus, although the main disadvantage of microfluidics is the low throughput, the ability to use multiple devices in parallel could overcome this issue [52]. Due to the use of inexpensive materials for chip fabrication and the negligible operation costs when operating with permanent magnets, parallelization of hundreds of microfluidic chips could potentially achieve in the future the flow rates that are nowadays processed by larger separators that require the use of complex electromagnets to process high volumes.

It should be noted that both OGMS columns as well as magnetophoretic microdevices can also operate in batch mode, with impressive efficiency in the separation of nanometer-sized particles. In comparison with HGMS, these devices are more flexible as they can operate in batch or continuous mode depending on the process requirements. Moreover, they can be employed for the separation of multiple particle

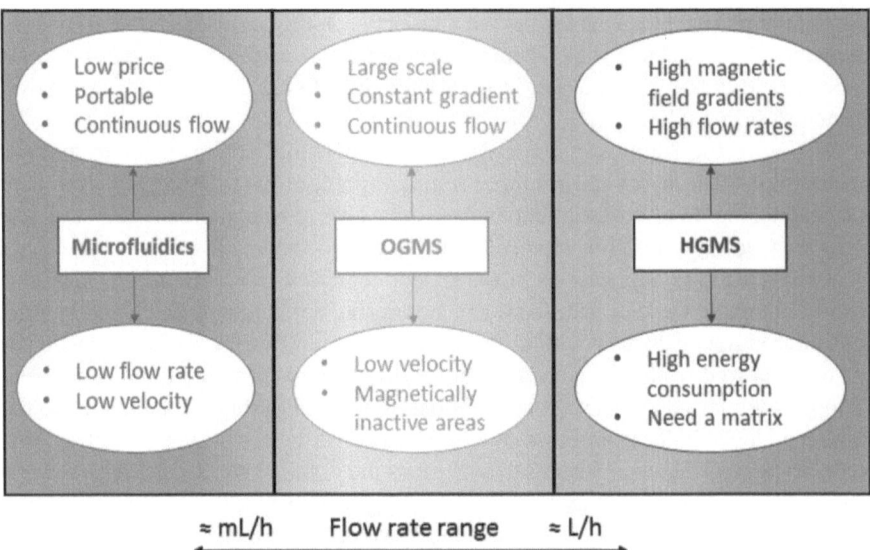

FIGURE 10.8 Advantages and disadvantages of microfluidic, OGMS and HGMS systems, including the flow rate range that they can process.

types (fractionation of magnetic and nonmagnetic particles) as long as they have different magnetic susceptibilities or sizes, which is not easily achieved by batch devices. Finally, Figure 10.8 presents the most important advantages and disadvantages of the separators reviewed in this chapter, showing the range of flow rates that can be processed by each system.

10.5 CONCLUSIONS

In recent years, there has been a growing interest in the application of micro- and nanoparticles within water treatment processes. These small particles present superior physical properties in comparison to bulk materials. Moreover, their easy functionalization to accommodate custom surface chemistry allows the design of multiple composites to separate or eliminate hazardous contaminants from water matrices. When incorporating a magnetic material in their formulation, the separation of the particles from the treated water could be achieved by including a magnetic separation stage. However, the fundamentals of magnetic separations are poorly understood and should be further described and reviewed to facilitate the successful integration of these technologies.

In this chapter, the principles of magnetic separation have been described as they are the theoretical background necessary for designing the separation process. The magnetic and drag forces have been presented as the dominant forces acting on the particles, and the numerous variables and parameters that affect the separation have been analyzed. Different alternatives for the separation of magnetic particles have been analyzed and classified depending on the operation mode. Systems operating only in batch mode have been first presented. Among them, HGMS, which is a well-established technology, has been introduced as the best available option when the treatment of high flow rates is necessary. Due to the high gradients achieved with this technology, the successful separation of nanoparticles under high fluid velocities is possible. However, these systems are only used when the consumption of high amounts of energy is justified by the need of treating high volumes and flow rates. A more economical batch separation technique, LGMS, has also been presented. However, much theoretical and experimental research seems necessary to increase the understanding of the process and to achieve higher separation efficacies in short times with this technology.

Finally, continuous-flow devices were introduced. Large-scale, OGMS columns are good candidates when the magnetic separation needs to be carried out in continuous operation mode. The use of OGMS with high fields provided by superconducting magnets enables the separation of magnetic microparticles at flow rates of several liters per minute. Moreover, the design and optimization of these devices is very simple when quadrupole magnetic fields are employed, because of the constant magnetic field gradient generated inside the working volume. Lastly, continuous-flow microfluidic devices are presented as an interesting option for analytical applications. These LOC devices could detect different analytes with high sensitivity and selectivity by using small sample volumes in a rapid, simple, and economical manner. Therefore, their application for point-of-need water quality monitoring is encouraged.

Finally, we can conclude that the future of magnetic separation is very promising due to the exceptional features of magnetic nanomaterials. It has been demonstrated that complex reactions or analysis can be efficiently performed with the assistance of magnetic fields. Therefore, the implementation of magnetic technologies and materials in an increasing number of fields is expected in the future. In this regard, less expensive magnetic sources (new materials, novel arrangements, etc.) and magnetic particles with stronger magnetic response (higher magnetic moments) might be the key for further development.

ACKNOWLEDGMENTS

We thank the National Heart, Lung, and Blood Institute (1R01HL131720-01A1) for their financial assistance.

REFERENCES

1. M. Garside, 2019. Global nanotechnology market value 2010–2020. Available at: https://www.statista.com/statistics/1073886/global-market-value-nanotechnology/. Accessed: March 26, 2020.
2. C. Buzea, I.I. Pacheco, K. Robbie, Nanomaterials and nanoparticles: Sources and toxicity, *Biointerphases* 2 (2007) MR17–MR71.
3. P. Xu, G.M. Zeng, D.L. Huang, C.L. Feng, S. Hu, M.H. Zhao, C. Lai, Z. Wei, C. Huang, G.X. Xie, Z.F. Liu, Use of iron oxide nanomaterials in wastewater treatment: A review, *Sci. Total Environ.* 424 (2012) 1–10.
4. B. Al-Najar, C.D. Peters, H. Albuflasa, N.P. Hankins, Pressure and osmotically driven membrane processes: A review of the benefits and production of nano-enhanced membranes for desalination, *Desalination* 479 (2020) 114323.

5. J. Gómez-Pastora, E. Bringas, I. Ortiz, Recent progress and future challenges on the use of high performance magnetic nano-adsorbents in environmental applications, *Chem. Eng. J.* 256 (2014) 187–204.
6. J. Gómez-Pastora, E. Bringas, I. Ortiz, Design of novel adsorption processes for the removal of arsenic from polluted groundwater employing functionalized magnetic nanoparticles, *Chem. Eng. Trans.* 47 (2016) 241–246.
7. M.R.Z. Kouhpanji, B.J. Stadler, A guideline for effectively synthesizing and characterizing magnetic nanoparticles for advancing nanobiotechnology: A review, *Sensors* 20(9) (2020) 2554.
8. J. Alonso, J.M. Barandiarán, L. Fernández Barquín, A. García-Arribas, Magnetic nanoparticles, synthesis, properties, and applications. In A.A. El-Gendy, J.M. Barandiarán, R.L. Hadimani (Eds). *Magnetic Nanostructured Materials*; Elsevier: Amsterdam, The Netherlands, 2018; pp. 1–40.
9. J. Gómez-Pastora, X. Wu, N. Sundar, J. Alawi, G. Nabar, J.O. Winter, M. Zborowski, J.J. Chalmers, Self-assembly and sedimentation of 5 nm SPIONs using horizontal, high magnetic fields and gradients, *Sep. Purif. Technol.* 248 (2020) 117012.
10. J. Gómez-Pastora, X. Xue, I.H. Karampelas, E. Bringas, E.P. Furlani, I. Ortiz, Analysis of separators for magnetic beads recovery: From large systems to multifunctional microdevices, *Sep. Purif. Technol.* 172 (2017) 16–31.
11. A.G. Kolhatkar, A.C. Jamison, D. Litvinov, R.C. Willson, T.R. Lee, Tuning the magnetic properties of nanoparticles, *Int. J. Mol. Sci.* 14 (2013) 15997–16009.
12. D. Cao, H. Li, L. Pan, J. Li, X. Wang, P. Jing, X. Cheng, W. Wang, J. Wang, Q. Liu, High saturation magnetization of γ-Fe_2O_3 nano-particles by a facile one-step synthesis approach, *Sci. Rep.* 6 (2016) 32360.
13. E.P. Furlani, Magnetic biotransport: Analysis and applications, *Materials* 3 (2010) 2412–2446.
14. S.Q. Li, M.F. Wang, Z.A. Zhu, Q. Wang, X. Zhang, H.Q. Song, D.Q. Cang, Application of superconducting HGMS technology on turbid wastewater treatment from converter, *Sep. Purif. Technol.* 84 (2012) 56–62.
15. A.D. Ebner, J.A. Ritter, Retention of iron oxide particles by stainless steel and magnetite magnetic matrix elements in high-gradient magnetic separation, *Sep. Sci. Technol.* 39 (2004) 2863–2890.
16. W.I. Wu, C.H. Wu, P.K.A. Hong, C.F. Lin, Capture of metallic copper by high gradient magnetic separation system, *Environ. Technol.* 32 (2011) 1427–1433.
17. S.N. Podoynitsyn, O.N. Sorokina, A.L. Kovarski, High-gradient magnetic separation using ferromagnetic membrane, *J. Magn. Magn. Mater.* 397 (2016) 51–56.
18. T. Sugawara, Y. Matsuura, T. Anzai, O. Miura, Removal of ammonia nitrogen from water by magnetic zeolite and high-gradient magnetic separation, *IEEE Trans. Appl. Supercond.* 26 (2016) 7395296.
19. T. Ishiwata, O. Miura, K. Hosomi, K. Shimizu, D. Ito, Y. Yoda, Removal and recovery of phosphorus in wastewater by superconducting high gradient magnetic separation with ferromagnetic adsorbent, *Physica C* 470 (2010) 1818–1821.
20. K. Yokoyama, T. Oka, H. Okada, K. Noto, High gradient magnetic separation using superconducting bulk magnets, *Physica C* 392–396 (2003) 739–744.
21. G. Mariani, M. Fabbri, F. Negrini, P.L. Ribani, High-gradient magnetic separation of pollutant from wastewaters using permanent magnets, *Sep. Purif. Technol.* 72 (2010) 147–155.
22. G.D. Moeser, K.A. Roach, W.H. Green, T.A. Hatton, High-gradient magnetic separation of coated magnetic nanoparticles, *AIChE J.* 50 (11) (2004) 2835–2848.
23. C.T. Yavuz, J.T. Mayo, W.W. Yu, A. Prakash, J.C. Falkner, S. Yean, L. Cong, H.J. Shipley, A. Kan, M. Tomson, D. Natelson, V.L. Colvin, Low-field magnetic separation of monodisperse Fe_3O_4 nanocrystals, *Science* 314 (2006) 964–967.

24. W. Ge, A. Encinas, E. Araujo, S. Song, Magnetic matrices used in high gradient magnetic separation (HGMS): A review, *Results Phys.* 7 (2017) 4278–4286.

25. D. Jin, H. Kim, Agglomeration dynamics of magnetite nanoparticles at low magnetic field gradient, *Bull. Korean Chem. Soc.* 39 (2018) 729–735.

26. I.A. Bakhteeva, I.V. Medvedeva, M.A. Uimin, I.V. Byzov, S.V. Zhakov, A.E. Yermakov, N.N. Shchegoleva, Magnetic sedimentation and aggregation of $Fe_3O_4@SiO_2$ nanoparticles in water medium, *Sep. Purif. Tech.* 159 (2016) 35–42.

27. J.S. Andreu, J. Camacho, J. Faraudo, M. Benelmekki, C. Rebollo, L.M. Martınez, Simple analytical model for the magnetophoretic separation of superparamagnetic dispersions in a uniform magnetic gradient, *Phys. Rev. E Stat. Nonlin. Soft Matter Phys.* 84 (2) (2011) 021402.

28. S.S. Leong, Z. Ahmad, J. Lim, Magnetophoresis of superparamagnetic nanoparticles at low field gradient: hydrodynamic effect, *Soft Matter* 11 (2015) 6968.

29. S.S. Leong, Z. Ahmad, J. Camacho, J. Faraudo, J. Lim, Kinetics of low field gradient magnetophoresis in the presence of magnetically induced convection, *J. Phys. Chem. C* 121 (2017) 5389–5407.

30. J. Gómez-Pastora, C. González-Fernández, E. Real, A. Iles, E. Bringas, E.P. Furlani, I. Ortiz, Computational modeling and fluorescence microscopy characterization of a two-phase magnetophoretic microsystem for continuous-flow blood detoxification, *Lab Chip* 18 (2018) 1593–1606.

31. J. Gómez-Pastora, I.H. Karampelas, E. Bringas, E.P. Furlani, I. Ortiz, Numerical analysis of bead magnetophoresis from flowing blood in a continuous-flow microchannel: Implications to the bead-fluid interactions, *Sci. Rep.* 9 (1) (2019) 7265.

32. T. Hartikainen, R. Mikkonen, Open-gradient magnetic separator with racetrack coils suitable for cleaning aqueous solutions, *IEEE Trans. Appl. Supercond.* 16 (2) (2006) 1130–1133.

33. H. Nakajima, H. Kaneko, M. Oizumi, S. Fukui, M. Yamaguchi, T. Sato, M. Oizumi, H. Imaizumi, S. Nishijima, T. Watanabe, Separation characteristics of open gradient magnetic separation using high-temperature superconducting magnet, *Physica C* 392–396 (2003) 1214–1218.

34. S. Fukui, Y. Takahashi, M. Yamaguchi, T. Sato, H. Imaizumi, M. Oizumi, S. Nishijima, and T. Watanabe, Study on open gradient magnetic separation using superconducting solenoid magnet, *IEEE Trans. Appl. Supercond.* 14 (2) (2004) 1568–1571.

35. T. Hartikainen, J.P. Nikkanen, R. Mikkonen, Magnetic separation of industrial waste waters as an environmental application of superconductivity, *IEEE Trans. Appl. Supercond.* 15 (2005) 2336–2339.

36. S. Fukui, H. Nakajima, A. Ozone, M. Hayatsu, M. Yamaguchi, T. Sato, H. Imaizumi, S. Nishijima, T. Watanabe, Study on open gradient magnetic separation using multiple magnetic field sources, *IEEE Trans. Appl. Supercond.* 12(1) (2002) 959–962.

37. M. Takahashi, T. Umeki, S. Fukui, J. Ogawa, M. Yamaguchi, T. Sato, H. Imaizumi, M. Oizumi, S. Nishijima, T. Watanabe, Numerical evaluation of separation characteristics of open gradient magnetic separation using quadrupole magnetic field, *IEEE Trans. Appl. Supercond.* 15 (2005) 2340–2343.

38. M. Ahoranta, J. Lehtonen, R. Mikkonen, Open gradient magnetic separation utilizing NbTi, Nb_3Sn and Bi-2223 materials, *Supercond. Sci. Technol.* 15 (2002) 1421–1426.

39. C. Carr, M. Espy, P. Nath, S.L. Martin, M.D. Ward, J. Martin, Design, fabrication and demonstration of a magnetophoresis chamber with 25 output fractions, *J. Magn. Magn. Mater.* 321 (2009) 1440–1445.

40. J. Gómez-Pastora, V.A. Roodan, I.H. Karampelas, A.Q. Alorabi, M.D. Tarn, A. Iles, E. Bringas, V.N. Paunov, N. Pamme, E.P. Furlani, I. Ortiz, Two-step numerical approach to predict ferrofluid droplet generation and manipulation inside multilaminar flow chambers, *J. Phys. Chem. C* 123(15) (2019) 10065–10080.

41. M.D. Tarn, M.J. Lopez-Martinez, N. Pamme, On-chip processing of particles and cells via multilaminar flow streams, *Anal. Bioanal. Chem.* 406 (2014) 139–161.
42. C. Phurimsak, M.D. Tarn, N. Pamme, Magnetic particle plug-based assays for biomarker analysis, *Micromachines* 7(5) (2016) 77.
43. N. Pamme, On-chip bioanalysis with magnetic particles, *Curr. Opin. Chem. Biol.* 16 (2012) 436–443.
44. J. Gómez-Pastora, E. Bringas, M. Lázaro-Díez, J. Ramos-Vivas, I. Ortiz, *Drug Delivery Systems*, World Scientific, Hackensack, NJ, 2017, pp. 207–244.
45. B. Ngamsom, A. Truyts, L. Fourie, S. Kumar, M.D. Tarn, A. Iles, K. Moodley, K.J. Land, N. Pamme, A microfluidic device for rapid screening of E.coli O157:H7 based on IFAST and ATP bioluminescence assay for water analysis, *Chem. Eur. J.* 23(52) (2017) 12754–12757.
46. K.Y. Castillo-Torres, E.S. McLamore, D.P. Arnold, A high-throughput microfluidic magnetic separation (μFMS) platform for water quality monitoring, *Micromachines* 11(1) (2020) 16.
47. E.N. Kasap, Ü. Doğan, F. Çoğun, E. Yıldırım, İ.H. Boyacı, D. Çetin, Z. Suludere, U. Tamer, N. Ertaş, Fast fluorometric enumeration of E. coli using passive chip, *J. Microbiol. Methods* 164 (2019) 105680.
48. V. Kamat, S. Pandey, K. Paknikar, D. Bodas, A facile one-step method for cell lysis and DNA extraction of waterborne pathogens using a microchip, *Biosens. Bioelectron.* 99 (2018) 62–69.
49. M. Zborowski, L. Sun, L.R. Moore, P.S. Williams, J.J. Chalmers, Continuous cell separation using novel magnetic quadrupole flow sorter, *J. Magn. Magn. Mater.* 194(1–3) (1999) 224–230.
50. L. Sun, M. Zborowski, L.R. Moore, J.J. Chalmers, Continuous, flow-through immunomagnetic cell sorting in a quadrupole field, *Cytometry: J. Int. Soc. Anal. Cytol.* 33(4) (1998) 469–475.
51. T. Schneider, L.R. Moore, Y. Jing, S. Haam, P.S. Williams, A.J. Fleischman, S. Roy, J.J. Chalmers, M. Zborowski, Continuous flow magnetic cell fractionation based on antigen expression level, *J. Biochem. Biophys. Methods* 68(1) (2006) 1–21.
52. L.R. Moore, P.S. Williams, J.J. Chalmers, M. Zborowski, Tessellated permanent magnet circuits for flow-through, open gradient separations of weakly magnetic materials, *J. Magn. Magn. Mater.* 427 (2017) 325–330.

11 Influence of Colloids on Mineralization in Unconventional Oil and Gas Reservoirs and Wellbores
A Case Study with the Marcellus Shale

J. Alexandra Hakala, Wei Xiong, Justin Mackey,
Meghan Brandi, James Gardiner, Nicholas Siefert,
Christina Lopano, and Barbara Kutchko
National Energy Technology Laboratory

B.J. Carney
Northeast Natural Energy

CONTENTS

DOI: 10.1201/9781003091011-11

11.1 USE OF PRODUCED WATER FOR HYDRAULIC FRACTURING OF UNCONVENTIONAL OIL AND GAS RESERVOIRS

Development of unconventional tight hydrocarbon reservoirs is a major driver for increases in United States-based oil and gas production over the past decade (EIA 2020). Significant volumes of water are injected during hydraulic fracturing of these reservoirs and can range from less than 10,000 to ~150,000 m³ per well depending on lateral length (Gallegos et al. 2015, Kondash, Lauer, and Vengosh 2018). The volume of water produced during flowback and production, which can range from less than 10,000 to ~85,000 m³ per well (Kondash, Lauer, and Vengosh 2018), ultimately requires management and treatment either for disposal or beneficial use, especially in regions where more water is produced than used for completion activities.

Use of low total dissolved solids (TDS; <1,000 mg L⁻¹) freshwater for hydraulic fracturing fluid (HFF) base water is limited in certain regions due to its availability, such as for shale gas production in Texas, USA, where it comprises less than 20% of the base water (Nicot and Scanlon 2012, Rodriguez and Soeder 2015). Blending water sources, such as brackish groundwater (TDS 1,000–10,000 mg L⁻¹), reclaimed municipal waste water, and recycled produced water, is applied by some Permian operators to ensure a sustainable water source for unconventional operations (e.g., Pioneer 2020). In other regions, freshwater is abundant; however, limited produced water treatment and disposal options exist due to technological, regulatory, or disposal well access challenges, such as in the Appalachian Basin, USA (Rodriguez and Soeder 2015, He et al. 2014, Shaffer et al. 2013, Rassenfoss 2011). Use of produced water as the base fluid for hydraulic fracturing may provide a cost benefit for an operator's overall operations (LeBas et al. 2013, Monroe et al. 2013) even if the produced water is a blend of waters from multiple formations (e.g., both conventional and unconventional) or a blend of high- and low-salinity waters.

Colloids are defined as being a dispersed phase (discontinuous phase) distributed in a dispersion medium (continuous phase) (Everett 1972). Mineral colloids are present in natural waters and serve as reactive surfaces for mineral nucleation and growth (Stumm 1993), and they influence many of the potential geochemical reactions that occur on steel surfaces and within the stimulated unconventional reservoir (Hakala et al. 2021, Mackey et al. 2019). These materials include sulfate-bearing phases, metal oxides, and clay phases that are present in waters injected into and produced from unconventional oil and gas reservoirs (Flynn et al. 2019, Phan, Hakala, and Bain 2018) and have been detected in HFF generated at Marcellus Shale well pads (Hakala et al. 2021). Mineral colloids may be present prior to blending water streams (e.g., Phan, Hakala, and Bain 2018) and can nucleate in solution due to blending of water streams with distinct chemical composition (e.g., Jun, Kim, and Neil 2016). They can aggregate and undergo sedimentation on rough surfaces such as corroded well tubing (Mackey et al. 2021) and fractured rock (Zhang et al. 2012). This results in mineral scale growth due to reaction between dissolved aqueous species and reactive surfaces (e.g., Stumm 1993) that may be present in the accumulated colloidal material or on the steel and rock surfaces.

This book chapter is focused on how mineral colloids in injected HFF can impact mineral scale development in unconventional oil and gas well tubing and reservoir rock, with a focus on systems where produced water is used as a base fluid for hydraulic fracturing operations. The produced water may require treatment by multiple techniques to control both reactive colloids and dissolved ions that are precursors to mineral scale deposition in the subsurface. The Marcellus Shale case study focuses on two scenarios that involve blending of Marcellus produced water with distinct surface waters that resulted in two different operational challenges. The experimental and modeling techniques demonstrate methods that can be applied toward other operational scenarios to inform the treatment goals and strategies for produced waters designated for beneficial use as HFF base fluids.

11.2 FLOW RESTRICTION IN UNCONVENTIONAL OIL AND GAS WELLS AND RESERVOIRS

Development of an onshore unconventional oil and gas well involves multiple activities associated with completions and production: wellbore drilling and well placement; opening of perforations and cleaning of perforations with concentrated acid; injection of HFF; a period of shut-in for the well; flowback of fracturing fluids; and production of hydrocarbons and water from the reservoir (Figure 11.1) (GWPC 2020). Mineral colloids can be transported into the well and reservoir as a polydisperse system or can nucleate due to reaction between mixing of fluids with different compositions (Jun, Kim, and Neil 2016). Colloids injected with fracturing fluid may consist of gelling agent from the HFF, natural organic matter, and scale-forming mineral particles. This chapter is focused on mineral colloids as precursors for development of mineral scale deposits within well tubing and unconventional reservoir fractures. Primary scale-forming mineral colloids in oil and gas reservoirs include barium sulfate ($BaSO_4$), strontium sulfate ($SrSO_4$), divalent metal carbonates, and metal oxides (Kumar, Naiya, and Kumar 2018, He et al. 2014).

11.2.1 ABOVEGROUND PIPING, PUMPS, AND FILTRATION SYSTEMS IMPACTED BY MINERAL SCALE PRECIPITATION

Mineral colloids may aggregate, undergo sedimentation, and cause mineral scale buildup in aboveground flow lines, pumps, and filtration systems. Corrosion of steel components within these systems can generate rough surfaces that promote colloid

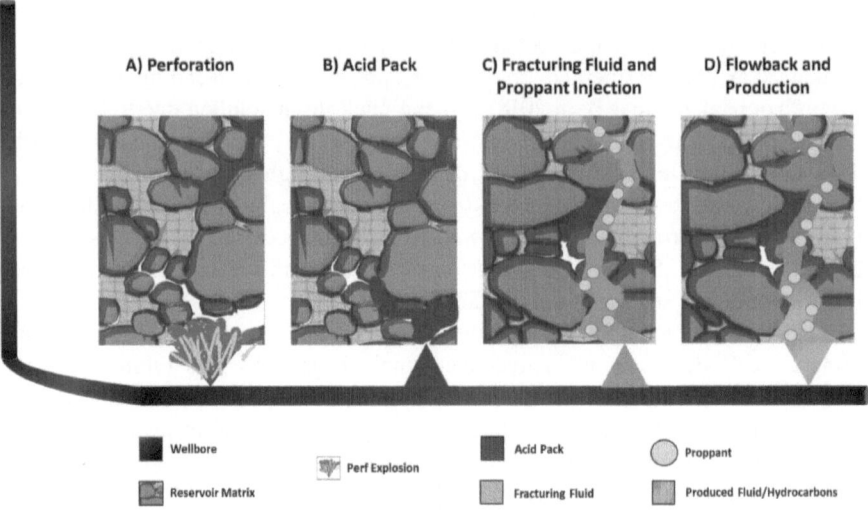

FIGURE 11.1 Schematic showing horizontal well for onshore unconventional oil and gas resources, different completion steps (after well placement), and production.

sedimentation, and corroded surfaces can react with dissolved solutes resulting in mineral scale deposition. Management of mineral scale buildup commonly involves filtration and application of scale and corrosion inhibitors (Olajire 2015).

11.2.2 WELLBORE ZONES IMPACTED BY MINERAL SCALE PRECIPITATION

An unconventional well configuration consists of multiple components, designed to ensure environmental and wellbore integrity during the lifetime of the well (API 2009). Locations where mineral scale development could impact flow within the well include the interiors of the production casing and the production tubing (Figure 11.2). For the purposes of discussion in this chapter, "well tubing" will be used as the primary term indicating fluid–steel contact points during injection and production. Recommended practices exist for drilling, casing, cementing, and perforation of horizontal wells in unconventional oil and gas reservoirs (API 2009).

Multiple interactions between HFF and steel influence mineral scale development in well tubing. Corrosion occurs during the acid pack stage (Figure 11.1) due to contact between the low-pH acid and steel (GWPC 2020, Mackey et al. 2021). Fluids injected during hydraulic fracturing are aerated and often contain dissolved O_2 that induces additional corrosion on surfaces of the well tubing (Craig et al. 2019). Addition of oxidizing biocides can result in high oxidation-reduction potentials of the HFF (e.g., Kahrilas et al. 2015). Mineral colloids present in the injected fluid can sediment onto corroded surfaces, and dissolved ions in the injected fluid can react with both the corroded surface and deposited mineral colloids to form mineral scale (Mackey et al. 2019, Olajire 2015). These processes can be compounded during the flowback and production periods, where casing or tubing steel is exposed to reservoir-reacted fracturing fluids, brine, and hydrocarbons. In this scenario, mineral

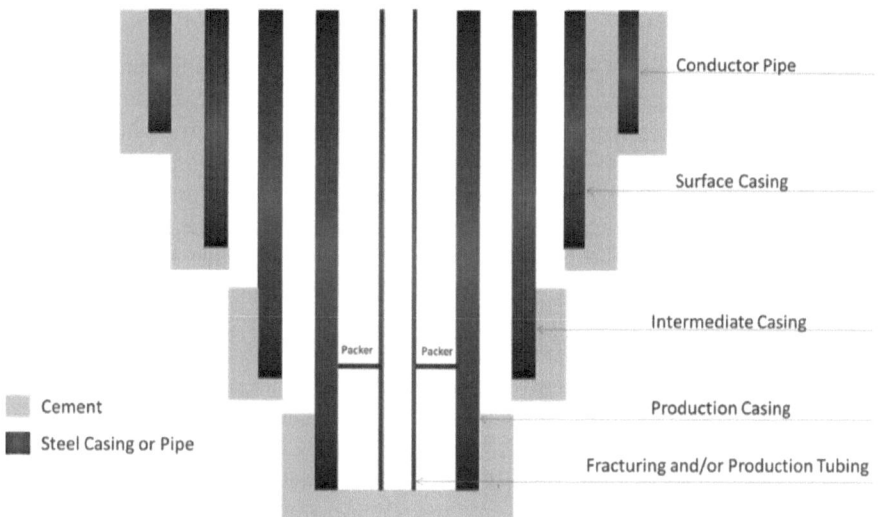

Conductor Pipe

Surface Casing

Intermediate Casing

Packer   Packer

Production Casing

Cement

Steel Casing or Pipe

Fracturing and/or Production Tubing

FIGURE 11.2   Schematic showing basic components of the vertical section of a completed unconventional oil and gas well based on a schematic developed by the American Petroleum Institute (API 2009). Spaces between casing strings are likely to be filled with cement.

colloids present in the produced fluids (Flynn et al. 2019, Phan, Hakala, and Bain 2018) can sediment on rough corroded surfaces, promoting development of mineral scale deposits at these sites (Olajire 2015). The corrosion products can serve as nucleation points for mineral scale deposition, in addition to the accumulated mineral colloids (Mackey et al. 2019, Olajire 2015). Corrosion inhibitors are usually included with fracturing fluids to limit the extent of corrosion that develops (GWPC 2020).

Different types of mineral scale can develop at different locations in the well tubing due to changes in temperature and pressure from surface conditions (as low as ~14 psi and ~25°C expected at the separator) to typical reservoir depths (~300 psi and ~65°C for the Marcellus Shale in northern West Virginia and southwestern Pennsylvania) (Xiong et al. 2020a). Carbonate (except for witherite and siderite), iron oxide (hematite), aluminum oxide (gibbsite), silicate, and clay phases were expected to develop (either as a fluid-dispersed mineral colloid or as a mineral precipitate on steel and shale surfaces) during both early flowback and late flowback and production, for a Marcellus Shale well (Hakala et al. 2021). Screening and geochemical modeling of produced water compositions from residual waste reports submitted to the Pennsylvania Department of Environmental Protection revealed a spatial and temporal evolution of the saturation and type of mineral scale expected to develop within Marcellus Shale wells (or reservoir) (Mackey et al. 2020). The type of scale that forms within wells is expected to vary spatially across an unconventional basin and occurs during different operational periods due to variable fluid compositions in contact with the well tubing. Many chemical or physical treatments designed to treat mineral scaling in aboveground flow lines may be less effective in the subsurface. For example, scale inhibitors efficacy can change with increased pressure and temperature at depth and in the presence of complex subsurface fluids (He and Vidic 2016).

11.2.3 RESERVOIR ZONES IMPACTED BY MINERAL SCALE PRECIPITATION

Results from recent studies show that chemical reactions between injected fluids and shale can affect fracture and matrix permeability, potentially impacting long-term hydrocarbon production from unconventional oil and gas reservoirs. Fluid composition, reservoir mineral assemblage, fracture network geometry, and kinetics of fluid–rock interactions will affect whether chemistry-induced formation damage will occur. Within hydraulically fractured reservoirs, mineral scale precipitation in primary fracture apertures (up to ~0.5 μm in laboratory-measured Marcellus Shale under confining pressure) (Frash and Carey 2017), in the shale matrix (pore diameters from ~10 nm to ~1 μm) (Milliken et al. 2013), and across the fracture-to-matrix interface will affect flow in different ways. Injected and in situ generated mineral colloids can obscure micron-scale fractures and matrix pore throats and result in development of near-wellbore mineral scale due to chemical and thermodynamic gradients at the wellbore-to-reservoir contact (Vankeuren et al. 2017, Jew et al. 2019).

For propped fractures, mineral colloids may accumulate in dense proppant packs and serve as nucleation sites for more significant mineral precipitation (Vankeuren et al. 2017). Mineral colloids such as barite are present in HFF-injected fluids (Hakala et al. 2021) and were observed to stick to the proppant pack in the absence of scale inhibitor in laboratory-scale experiments (He and Vidic 2016). Certain mineral reactions result in mechanical weakening of the fracture face, causing proppant embedment and fracture closure (Du et al. 2018, Vankeuren et al. 2017).

Increasing long-term production from unconventional reservoirs depends on production of hydrocarbons from the reservoir matrix via induced fracture networks (Karra et al. 2015). Chemical and physical processes that occur along primary fracture faces impact which matrix geochemical reactions occur (Vankeuren et al. 2017, Xiong et al. 2020a). Fluid residence time also affects the mineral reactions that occur at the fracture–matrix interface (Xiong et al. 2020a). The extent of mineral reaction depends on the composition of injected fluid and minerals present both at the fracture surface and within the shale matrix.

Calcite dissolution co-located with barite precipitation in laboratory core flood experiments with Marcellus Shale, which resulted in a net fracture–matrix aperture increase for fracturing fluids with varied TDS (Vankeuren et al. 2017). Experiments focused on changes to matrix porosity in Marcellus Shale showed that pyrite oxidation caused a thick sulfur oxidation zone with a thinner iron oxidation zone, co-located with carbonate dissolution and secondary porosity development at the shale–fluid interface up to 200 μm from the fracture face (Li et al. 2019). The net impact of chemical-mechanical processes on permeability across the fracture–matrix interface, and impacts to production from unconventional reservoirs, is an active area of current research (e.g., Birkholzer et al. 2021).

Scale inhibitors are commonly employed to prevent growth of scale-forming minerals that nucleate in oilfield fluids (Olajire 2015). Scale inhibitor efficacy is impacted by concentrations of dissolved metal ions (such as ferrous iron per Zhang et al. 2019), and laminar versus turbulent flow in fractured shale reservoirs (Yan et al. 2017). Clay stabilizers are applied to control clay colloids and "fines" from the formation (Nguyen et al. 2007). Manipulation of fluid zeta potentials was proposed as a means for controlling colloidal stability within HFF, where colloids with higher zeta-potential are less likely to result in agglomeration, sedimentation, and subsequent blockage of flow pathways (Ali and Hascakir 2017). Knowledge of colloidal stability can define which HFF parameters to adjust to mitigate these processes (Ali and Hascakir 2017). Treatment of produced water for use as HFF base fluid requires processes for removing reactive components from the produced water, and addition of chemicals that can mitigate mineral colloid development, aggregation, sedimentation, and reactivity within the reservoir.

11.3 CASE STUDY: EFFECT OF SULFATE AND SCALE INHIBITOR ON MINERAL SCALE DEVELOPMENT – MARCELLUS SHALE, NORTHWESTERN WEST VIRGINIA, USA

This section focuses on two scenarios where minimally treated Marcellus produced water was blended with two distinct surface fresh water sources in West Virginia and resulted in different operational challenges: (1) scaling damage to aboveground water transfer pumps and manifolds (Scenario 1 – Monongahela River) and; (2) excessive injection filter blockage from barite-rich mineral precipitates (Scenario 2 – Paw Paw Creek). Addition of scale inhibitors to the base water mitigated flow restriction issues in the aboveground lines, which were attributed to mineral colloid aggregation and sedimentation and subsequent mineral scale development. However, some question remained in both scenarios regarding potential subsurface mineral scale accumulations in well tubing and reservoir rock due to interactions with HFF.

Scenario 1 focuses on a HFF base fluid composed of Monongahela River water blended with Marcellus Shale produced water and Scenario 2 focuses on a HFF base fluid composed of Paw Paw Creek water blended with Marcellus Shale produced water. The chemical interactions controlling mineral scale were evaluated using benchtop laboratory experiments and modeling for Scenario 1 and Scenario 2. These scenarios provide examples for experimental and modeling approaches that may be used to characterize the types and extent of mineral scale that can develop in unconventional oil and gas well tubing and reservoir rock. The results may be used to assist in design of HFF base water pretreatment approaches that limit the development of damaging mineral scale deposits in well tubing and the reservoir.

11.3.1 Experimental Evaluation for Scenario 1: Blending Monongahela River Water with Marcellus Produced Water

The Scenario 1 problem involved excessive buildup of mineral scale within pump manifolds that shut down the fresh and produced water blending pumps, after Monongahela River water and Marcellus Shale produced water were blended in a 5:1 ratio at a northern West Virginia well pad (Figure 11.3). Both water sources were piped to and stored in aboveground tanks prior to piping water from the upper section of the tanks, with no additional treatment. The problematic mineral scale primarily contained barite. To solve the problem, scale inhibitor including ethylene glycol, amine triphosphate, and sodium phosphate was added to the freshwater stream, prior

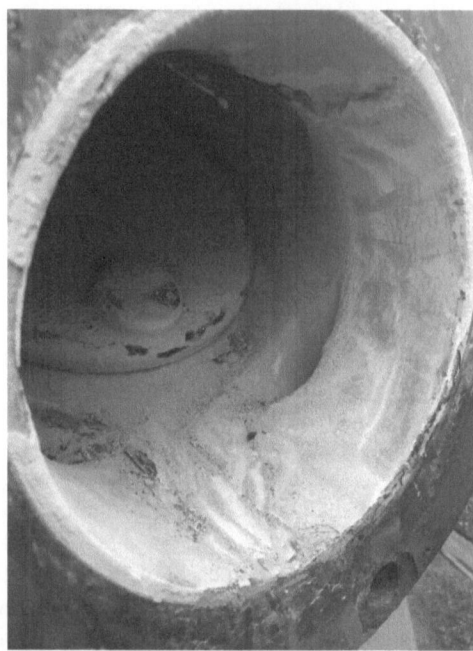

FIGURE 11.3 Photograph of blending pump disabled by mineral scale buildup, prior to addition of scale inhibitor to the freshwater stream at the well pad site for Scenario 1.

to blending with the produced water. The blended fluid was then filtered (25 μm) prior to use as a base fluid for hydraulic fracturing of Marcellus Shale.

Although inclusion of the scale inhibitor solved the problem with the aboveground lines, data were not available to identify whether mineral scale could develop in the well and reservoir. Laboratory experiments were designed to determine whether well tubing corrosion and scaling, or reservoir damage, could occur due to chemical reactions with the blended water as a HFF base fluid.

11.3.1.1 Field Sampling and Experimental Methods

Sample Collection and Characterization

Recycled Marcellus produced water and Monongahela River freshwater samples were collected from a release valve at the bottom of an array of production tanks on location during active hydraulic fracture operations. A separate blended fluid containing a 5:1 ratio of fresh water (including scale inhibitor) and produced water was sampled from a pump manifold. All samples were filtered in the laboratory (0.2 μm) and analyzed by inductively coupled plasma optical emission spectroscopy (ICP-OES) at NETL's Pittsburgh Analytical Laboratory and ion chromatography (IC) for dissolved cation, anion, and metal concentrations at NETL's Brine Chemistry Laboratory. The ICP-OES detection limits were in mg/L: Ba (0.03), Ca (0.15), Fe (0.15), K (10.88), Mg (0.94), Ni (0.02), Sr (1.89), and Zn (0.02). The detection limit for analytes measured by IC (Li^+, Na^+, NH_4^+, K^+, Mg^{2+}, Ca^{2+}; Cl^-, SO_4^{2-}, Br^-, NO_3^-) was

0.5 mg L^{-1}. For analytes measured by both methods, the larger concentration value was used. Grade 4 Whatman filter paper (20–25 µm) was used to filter blended water used in the experiments.

Experiments with Blended Water and Low Carbon Steel

Experiments conducted to determine whether added scale inhibitor and filtration prior to injection impacted steel surfaces due to HFF exposure were performed using blended water from the well pad and steel coupons in laboratory autoclaves. Low carbon steel coupons procured from McMaster Carr were used as a proxy for J-55 production tubing (Fe (90.06%–99.42%), C (0.13%–0.20%), Mn (0.30%–0.90%), P (0.04% Max), Si (0.15%–0.30%), and S (0.50% Max)). Steel was machined to ~1 in × 1 in × 1/8 in coupons, rinsed in ethanol, and then placed in a 1 N HCl bath for 1 hour to remove residual oils, scales, and corrosion byproducts. Samples were held in polytetrafluoroethylene (PTFE) beakers and exposed to both 25 µm-filtered and unfiltered blended fluid at a 1:8 solid/liquid ratio in a Parr 4680 Autoclave under a nitrogen atmosphere at 23°C and 14.5 psi to simulate shallow well conditions, and 50°C and 2,000 psi to simulate deep well conditions. Mineral scale occurs at a range of depths within unconventional wells, and these experiments represent well depths within the vertically cased component of a Marcellus well. Standard hydrostatic pressure (0.460 psi ft^{-1}) and geothermal (25°C km^{-1}) gradients were used to replicate the in-situ conditions of a Marcellus Shale well, given a depth range between ~1.4 and 2.7 km. Surface conditions representative of the near-surface injection point and tubing (14.5 psi, 23°C) are represented by the Steel-1 (with unfiltered blended water; location A in Figure 11.4) and Steel-2 (with filtered blended water; location B in Figure 11.4) experiments. Subsurface temperature and pressure ranges from 35°C to 67°C and ~2,000–4,000 psi (location C in Figure 11.4) are represented by the Steel-3 (with unfiltered blended water) and Steel-4 (with filtered blended water) experiments.

Post-reaction steel samples were rinsed with the reacted fluids to remove any settled particulate, removed from batch reactors, dried in a nitrogen supplied desiccator and characterized via scanning electron microscopy with energy dispersive x-ray spectroscopy (SEM-EDS). Reacted effluent was evaluated by ICP-OES and IC for major cations, metals, and anions, at NETL's Brine Chemistry Laboratory and Pittsburgh Analytical Laboratory, with the same detection limits reported for the field-collected water. The steel experiments were performed without replicates due to reactor availability, and therefore the reported measurement error is within the range of analytical standards applied during the ICP-OES and IC analyses reported above.

Experiments with Blended Water and Marcellus Shale[1]

Experiments conducted to determine whether added scale inhibitor and filtration prior to injection resulted in new mineral nucleation were performed using blended well pad fluid and crushed Marcellus Shale in laboratory benchtop experiments. Aliquots of the blended fluid from the well pad and powdered Marcellus Shale were reacted in amber glass bottles with a cap and magnetic stir bar, and heated on a

---

[1] Results for fluid-shale experiments were reported previously in Xiong et al. 2020b.

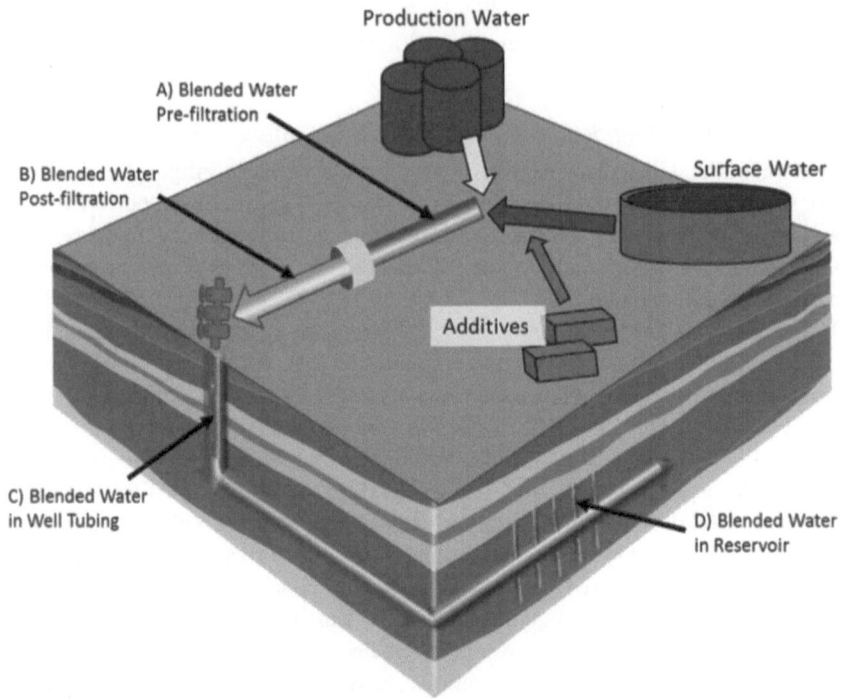

FIGURE 11.4 Representative locations in a Marcellus Shale well-to-reservoir system for which bench-scale experiments were designed to evaluate the potential for corrosion (in the case of steel) and mineral scale precipitation (in the cases of both steel and shale). The schematic is a modification of that presented in Mackey et al. (2021).

stirred hot plate. The blended water was filtered with 20–25 µm filter paper prior to the experiment to simulate the on-site filtration process at the Scenario 1 well pad, and to characterize colloids present in the unfiltered blended pad fluid. Marcellus Shale from the Marcellus Shale Energy and Environmental Laboratory (MSEEL) site in Monongalia County, WV (7491.11 ft depth) (Phan et al. 2019), was powdered with a tungsten ball mill to an average size of less than 10 µm. The shale mineralogy consisted of (in volume %) quartz (20.4), illite (36.2), muscovite (24.6), clinochlore (2.3), pyrite (2.3), calcite (1.0), dolomite (0.7), and organic matter (12.4) (Phan et al. 2019).

Experiments were conducted with 1 g of Marcellus Shale powder and 100 mL of the filtered blended water for 2 days (Shale-1) and 14 days (Shale-2) at 66°C. One control experiment was conducted without shale at 66°C for 2 days (Shale-3) to characterize changes in fluid chemistry in the absence of shale, and another control experiment was conducted with shale powder at 20°C for 2 days (Shale-4) to test the effect of lowering the temperature on HFF-shale interactions. Reactions observed at elevated temperature and low pressure are expected to occur under reservoir temperature and pressure conditions, with increased mineral reaction kinetics (e.g., BinMerdah, Yassin, and Muherei 2010). Post-reaction solids analysis was performed on 0.2 μm filter paper solid retentates with SEM-EDS, fluids were analyzed by ICP-OES and IC, and Geochemists Workbench (GWB) was used to calculate mineral saturation indices (SI).

11.3.1.2 Water Analysis and Experimental Results

Aqueous Chemistry of Water Sources and
Identification of Colloids in Blended Water[2]

Chemical analysis of the Monongahela River, Marcellus produced water, and blended water samples are presented in Table 11.1. Results show expected low concentrations of dissolved solids in the Monongahela River water, high concentrations in the Marcellus produced water, and an intermediate composition for the blended water. The Monongahela River water has higher concentrations of SO_4^{2-} and NO_3^-, and otherwise the remaining measured constituents are present in higher concentrations in the produced water (Table 11.1). Filtration of the blended fluid (25 μm) in the laboratory resulted in filter retention of round barite particles, as confirmed by SEM-EDS (Figure 11.5). Most of these particles are within 30–50 μm diameter, with some smaller than 20 μm and others as large as 100 μm. Barite appears as rounded particles while the other darker phases are mixtures of oxides, halides, phosphates, and silicates/SiO_2 (Figure 11.5).

Results from Experiments with Steel Coupons

Interaction between the blended fluid (pre-filtration) and steel components at surface conditions (location A in Figure 11.4; Steel-1 in Table 11.1) resulted in pitting, precipitation of iron (oxy)hydroxides, and colloid accumulation on the steel surface (Figure 11.6). Analysis of the EDS from the areas with significant corrosion-related precipitation revealed high Ca, Fe, and Cl, with detectable levels of Ba, Sr, Mg, Al, P, and Na (Figure 11.6). Interaction between blended fluid (post-filtration) and steel components at surface conditions (location B in Figure 11.4; Steel-2 in Table 11.1) resulted in some pitting and iron-(oxy) hydroxide precipitation, where Fe, O, and Cl were major components identified in the precipitates (Figure 11.7).

Interaction between blended fluid (post-filtration) with steel components at reservoir conditions (location C in Figure 11.4; Steel-3 in Table 11.1) also resulted in some pitting and iron (oxy)hydroxide formation, where Fe and Cl were the major components identified in the precipitates (Figure 11.8). Comparatively, unfiltered blended fluid (representing a situation where blended fluids with scale inhibitor are injected without a filtration step; Steel-4 in Table 11.1) reacted significantly with steel, where significant corrosion and pitting were observed, along with significant iron-(oxy) hydroxide precipitation and colloid accumulation (Figure 11.9). In this case, Fe and Cl were the major components and increased weight percentages of other metals, including Ca, Na, Ba, Sr, and Mg, were observed compared to the filtered experiment at elevated pressure and temperature shown in Figure 11.8. Additionally, a Ba-rich precipitate accumulated near locations where Fe and O are present at high levels (upper left quadrant of Figure 11.9).

---

[2] Results for colloids filtered from the blended water were reported previously in Xiong et al. (2020b).

TABLE 11.1

Composition of Fluids Collected from the Scenario 1 Field Site and Laboratory Experimental Results

Concentrations in mg/L	Blended Water[a]	Fresh Water (Mon. River)[a]	Produced Water (Marcellus Shale)[a]	Steel-1	Steel-2	Steel-3	Steel-4	Shale-1[b]	Shale-2[b]	Shale-3[b]	Shale-4[b]
Fluid preparation for experiment	N/A[a]	N/A[a]	N/A[a]	Unfiltered	Filtered	Filtered	Unfiltered	Filtered	Filtered	Filtered	Filtered
Solid preparation for experiment	N/A[a]	N/A[a]	N/A[a]	Polished steel coupon	Polished steel coupon	Polished steel coupon	Polished steel coupon	Milled shale average < 10 μm	Milled shale average < 10 μm	No Shale	Milled shale average < 10 μm
Designation in Fig 20.4	Blended Water Pre-Filtration	Fresh Water	Production Water	A	B	C	C; "Unfiltered Scenario"	D; 2-day reaction	D; 14-day reaction	D; shale-free control	D; Low temp.
Experimental Conditions				14.5 psi; 23°C	14.5 psi; 23°C	2,000 psi; 50°C	2,000 psi; 50°C	14.5 psi; 66°C	14.5 psi; 66°C	14.5 psi; 66°C	14.5 psi; 20°C
Ba	551.0	0.9	2,938.0	405.5	419.9	458.9	456.7	420.5 ± 13.6	287.3 ± 24.6	550.8 ± 0.1	499.7 ± 4.8
Ca	3,910.0	31.5	18,370.0	3,711.5	3,696.5	3,821.9	3,857.0	3,930.5 ± 50.5	3,804.4 ± 20	3,861.5 ± 12.5	3,843.5 ± 4.5
Fe	2.2	0.0	65.8	24.0	8.0	31.9	169.6	BDL[c]	8.2 ± 6.3	BDL[c]	BDL[c]
Mg	343.1	6.4	1,553.9	325.2	327.9	335.4	343.6	351.7 ± 3.3	347.3 ± 1.1	340.2 ± 0.8	337.4 ± 4.4
Na	8,107.4	0.0	38,408.4	7,258.6	7,309.6	7,483.0	7,666.5	8,242.4 ± 47	7,745.9 ± 135.5	8,088.9 ± 3.5	8,112.9 ± 49.5
SiO$_2$	32.4	15.3	15.0	n.a.[d]	n.a.[d]	n.a.[d]	n.a.[d]	26.7 ± 2.3	49.8 ± 5.5	10.1 ± 0.1	12.1 ± 5.4
Sr	643.9	1.0	3,149.0	436.1	457.5	501.4	489.9	643.7 ± 3.4	612.2 ± 8.8	638.2 ± 2.5	634.1 ± 6.2
Li	17.4	0.0	49.9	16.5	16.6	17.0	17.4	n.a.[d]	n.a.[d]	n.a.[d]	n.a.[d]
NH$_4^+$	58.2	0.0	265.9	58.9	60.2	61.4	62.9	59.9 ± 0.4	57.8 ± 0.8	57.4 ± 0.1	59.2 ± 0.2
K	104.2	0.0	116.0	81.5	82.1	83.7	85.8	131.0 ± 2.6	139.3 ± 5.4	91.3 ± 1.6	136.3 ± 6.2
Cl$^-$	22,474.6	76.6	122,074.2	18,865.8	19,233.2	18,682.8	18,784.8	21,978.0 ± 121.5	22,088.4 ± 272.8	22,808.0 ± 178.1	21,851.9 ± 172.8
SO$_4^{2-}$	29.1	84.7	14.7	16.1	15.7	9.0	15.1	32.6 ± 0.5	21.3 ± 0.6	32.8 ± 2.8	36.4 ± 3
Br$^-$	201.8	114.2	918.8	134.5	133.5	128.5	121.2	216.7 ± 1.2	151.6 ± 3.3	213.0 ± 5.9	203.3 ± 8.3
NO$_3^-$	BDL[c]	103.2	BDL[c]	n.a.[d]	n.a.[d]	n.a.[d]	n.a.[d]	n.a.[d]	n.a.[d]	n.a.[d]	n.a.[d]
pH	6.3	7.5	6.1	6.7	6.8	6.6	6.7	7.0	4.6	7.2	7.4

[a] Not applicable – samples were filtered according to analytical chemistry protocols for ICP-OES, ICP-MS, and IC.

[b] Data reproduced from tables in (Xiong et al. 2020b). Error values reported are the standard deviation for the mean of experiments performed in duplicate.

[c] BDL, below detection limit.

[d] n.a., not analyzed.

FIGURE 11.5 Photograph of material retained on the 25 μm filter paper after filtering 650 mL of blended water with a 125 mm scale bar at the bottom (a). SEM image of solids retained on the filter with a 500 μm scale bar shown in the lower right corner (b). SEM image of solids retained on the filter with a 100 μm scale bar shown in the lower right corner (c). The white boxes (yellow in the online version) highlight the bright white spheres of barite retained on the filters, as confirmed by EDS spectra (d).

Results from Experiments with Shale

Unreacted shale powders contain mostly Al-Si-O clay minerals with some pyrite grains or framboids (Figure 11.10a). Powdered shale reacted with filtered blended fluid at high temperature showed the development of new euhedral barite crystals (Figure 11.10b); these grew larger in the longer term experiment with shale at high temperature (Xiong et al. 2020b). Filtered blended fluid reacted at high temperature, without shale, resulted in limited new colloid development (Figure 11.10c). The precipitates shown in Figure 11.10c are primarily iron oxides and spherical barite, and crystallized NaCl from drying of the 0.2 μm filter paper used to separate suspended solids after the experiment. The filtered blended fluid retained elevated SO_4^{2-} in the 2-day HFF-shale experiment, which decreased in the 14-day experiment (Shale-2; Table 11.1). Barium concentrations in the fluid phase decreased in experiments containing shale (Shale-1, Shale-2, and Shale-4), relative to the concentration in both the unreacted blended well pad water and the shale-free experiment (Shale-3) (Table 11.1). SI calculated using the experimental fluid chemistries show that layered silicates (antigorite, chrysotile), quartz, and barite are saturated or supersaturated

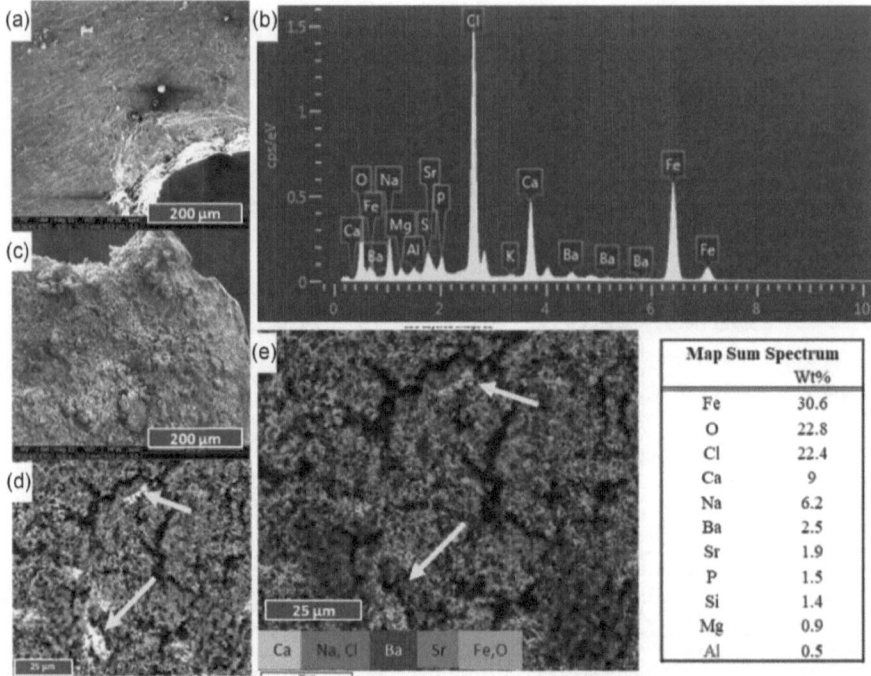

FIGURE 11.6 Results from the Steel-1 experiment. Comparison of pre-exposure (a) and post-exposure (c) with a 200 µm scale bar SEM images for the steel coupon reacted with unfiltered blended fluid at surface P/T (14.5 psi; 23°C; location A in Figure 11.4). EDS spectra of reacted steel surface (b). A zoomed-in scan of the post-exposure precipitate with a 25 µm scale bar shows a corrosion deposits with a high z-contrast (d) that are associated with high Ba concentrations (e). Table listing map elemental concentrations in the bottom right.

in the 2-day HFF-shale experiment (Shale-1); however, only quartz and barite are saturated in the 14-day experiment (Shale-2), and barite was supersaturated in the unfiltered blended pad fluid (Table 11.2).

11.3.1.2.1   Discussion – Influence of Dissolved Chemical Species and Colloids on Reactions with Steel and Shale

The roundness of barite particles in the blended fluid is likely due to the scale inhibitor added to the freshwater stream. Filtration of the blended fluid prior to injection should remove most of the barite particles; however, some of the smaller-sized particles may pass through the filter. The filtered fluid is still supersaturated with barite.

The major impact to steel components of the system, both in aboveground flow lines and within the reservoir, is corrosion and pitting with development of Fe-rich precipitates on the surface. Unfiltered fluids result in greater accumulation of surface deposits on the steel due to corrosion and deposition of secondary precipitates containing other divalent cations. The colocation of Ba-rich zones near regions with evidence for high Fe oxidation suggests that oxidized components of well tubing steel may provide adsorption sites for sedimentation of existing mineral colloids from the

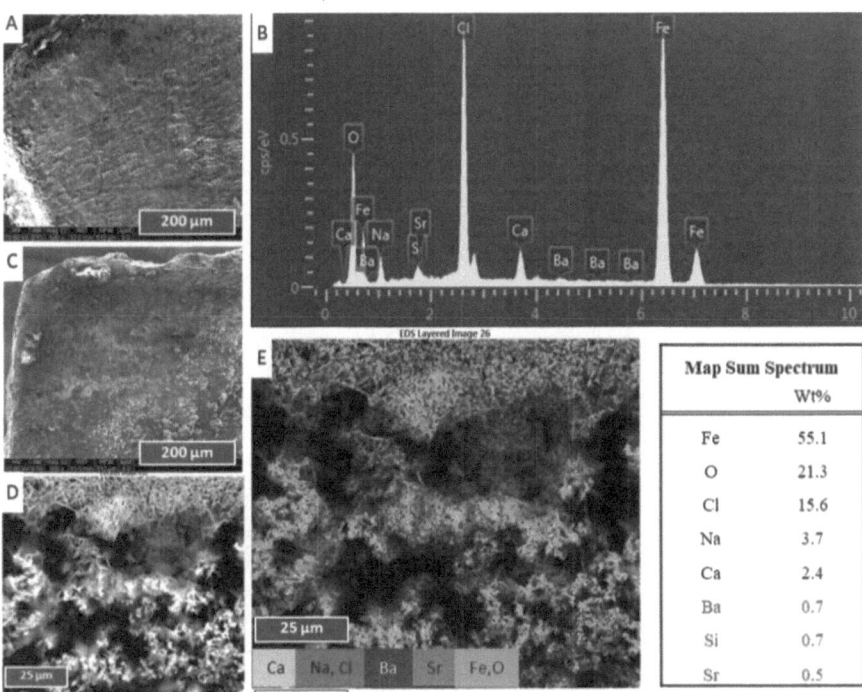

FIGURE 11.7 Results from the Steel-2 experiment. Comparison of pre-exposure (a) and post-exposure (c) with a 200-μm scale bar SEM images for the steel coupon reacted with filtered blended fluid at surface P/T (14.5 psi; 23°C; Location B in Figure 11.4). EDS spectra of reacted steel surface (b). A zoomed-in scan of the post-exposure shows iron (oxy)hydroxide precipitates with a 25 μm scale bar (d) with their associated element maps (e). Table listing map elemental concentrations in the bottom right. Note element concentrations in bold text at the bottom of the list (highlighted in red online) are below the confidence threshold of the software.

unfiltered fluids (Mackey et al. 2021). These sites could facilitate further nucleation and mineral scale growth in an active well. The results also demonstrate that removal of mineral colloids through filtration at least partially mitigated surface corrosion and mineral colloid sedimentation on the steel surfaces.

New barite crystals developed in the presence of shale due to continued geochemical reactions between the blended fluid and the shale. The shale is expected to play a role in the development of new barite crystals, as changes in experimental barium concentrations only occurred in experiments containing shale (Shale-1, Shale-2, and Shale-4) with little change in the shale-free control experiment (Shale-3; Table 11.1). Although SO_4^{2-} remains elevated during short-term (2 day) experiments containing shale (Shale-1 and Shale-4), concentrations decrease during the long-term (14 day) experiment (Shale-2; Table 11.1). Sulfate may be released to solution from the shale, as indicated by the slight increase in the short-term experiments in this study (Table 11.1), and has been observed in core flood experimental studies (Xiong et al. 2020a). Longer fluid–shale exposures can result in barite precipitation and removal

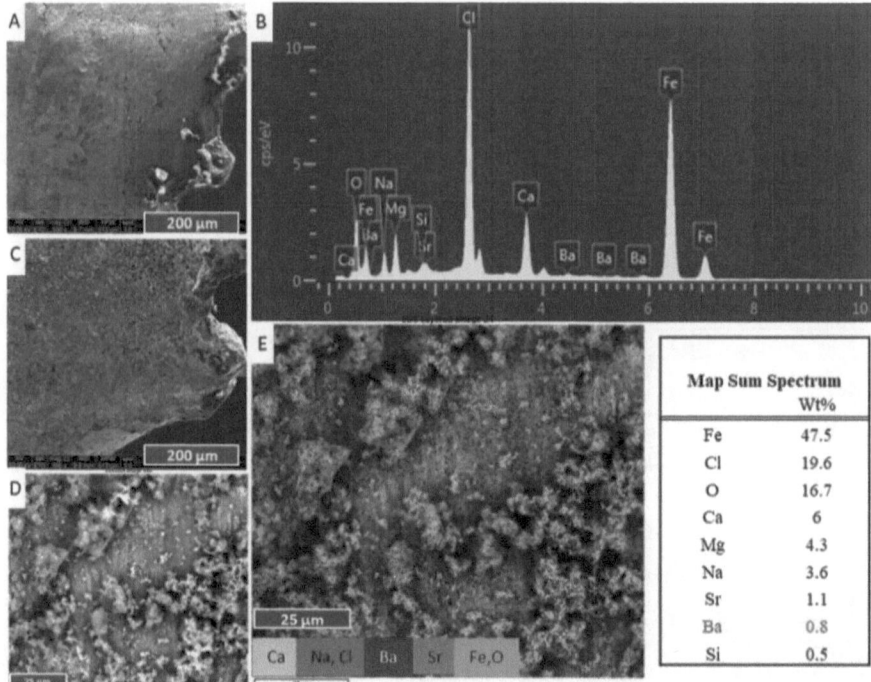

FIGURE 11.8 Results from the Steel-3 experiment. Comparison of pre-exposure (a) and post-exposure (c) with a 200-μm scale bar SEM images for the steel coupon reacted with filtered blended fluid at wellbore P/T (2,000 psi; 50°C; Location C in Figure 11.4). EDS spectra of reacted steel surface (b). A zoomed-in scan of the post-exposure shows some iron (oxy) hydroxide precipitates with a 25-μm scale bar (d) that are associated with high Ba concentrations (e). Table listing maps elemental concentrations in the bottom right. Note element concentrations in bold text at the bottom of the list (highlighted in red online) are below the confidence threshold of the software.

of dissolved SO_4^{2-} (Xiong et al. 2020a), and similar reactions likely occur in the experiments presented here. Reactions occurring over longer time frames within the reservoir could result in continuous barite colloid formation. Application of predictive modeling to identify the chemical conditions promoting corrosion and mineral supersaturation can aid in selecting appropriate water treatment and chemical additive systems for mitigating mineral scale development.

11.3.2 Application of Chemical Process and Reaction Path Modeling to Characterize the Potential for Mineral Scale Development in Unconventional Oil and Gas Systems

Blending Marcellus Shale produced water with freshwater piped in from surface water sources in West Virginia is a common practice for development of hydraulic fracturing base fluids in the Appalachian Basin. An important variation in fresh

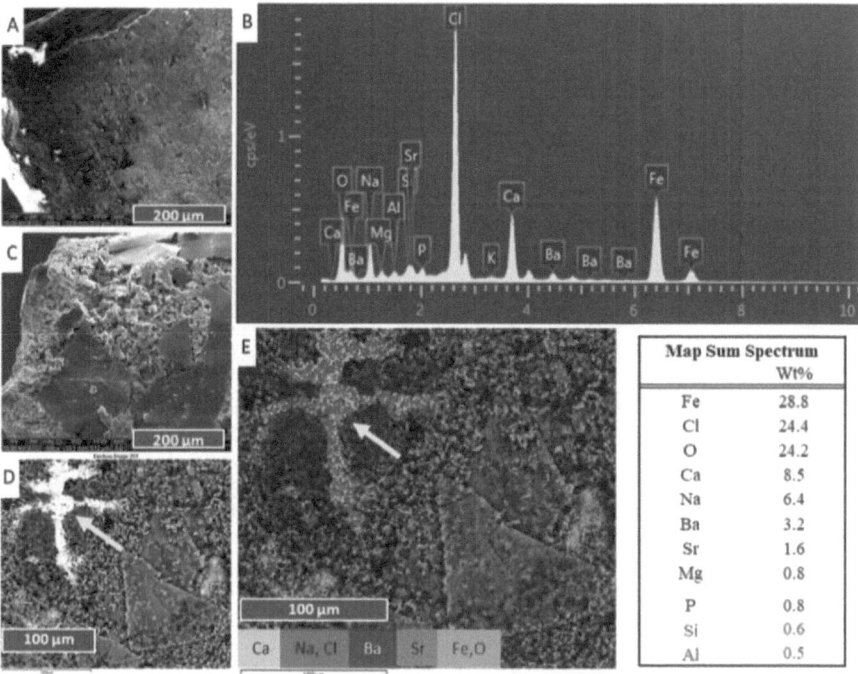

FIGURE 11.9 Results from the Steel-4 experiment. Comparison of pre-exposure (a) and post-exposure (c) with a 200-μm scale bar SEM images for the steel coupon reacted with unfiltered blended fluid at wellbore P/T (2,000 psi; 50°C). EDS spectra of reacted steel surface (b). A zoomed-in scan of the post-exposure shows mineralization with high z-contrast with a 100-μm scale bar (d) that is associated with high Ba concentrations (e). Table listing maps elemental concentrations in the bottom right. Note elements highlighted in bold text at the bottom of the list (highlighted in red online) are below the confidence threshold of the software.

waters relevant for barite colloid formation is fluctuation in SO_4^{2-}. Marcellus Shale produced water SO_4^{2-} concentrations are typically very low and often are below detection limit although barium often is elevated in produced water and increases over time (Tieman et al. 2020, Phan, Hakala, and Sharma 2020).

The Monongahela River and Paw Paw Creek both serve as freshwater sources for blending with produced water to develop HFF base fluid in northern West Virginia. Both freshwaters are surface tributaries in West Virginia, with chemistries influenced by extensive regional bituminous coal mining and elevated levels of SO_4^{2-} relative to most surface waters (Corbett 1977, Anderson et al. 2000). Challenges encountered with blending Monongahela River water with Marcellus Shale produced water (Scenario 1) were described in Section 11.3.1. For Scenario 2, challenges with significant temporal variability in Paw Paw Creek SO_4^{2-} concentrations resulted in severe clogging of 100 μm filters used to prepare HFF for injection at a different well pad. Scale inhibitor was added to both the freshwater and produced water streams prior to blending, which reduced filter clogging; however no data were available to characterize processes occurring in the well and reservoir. The following examples

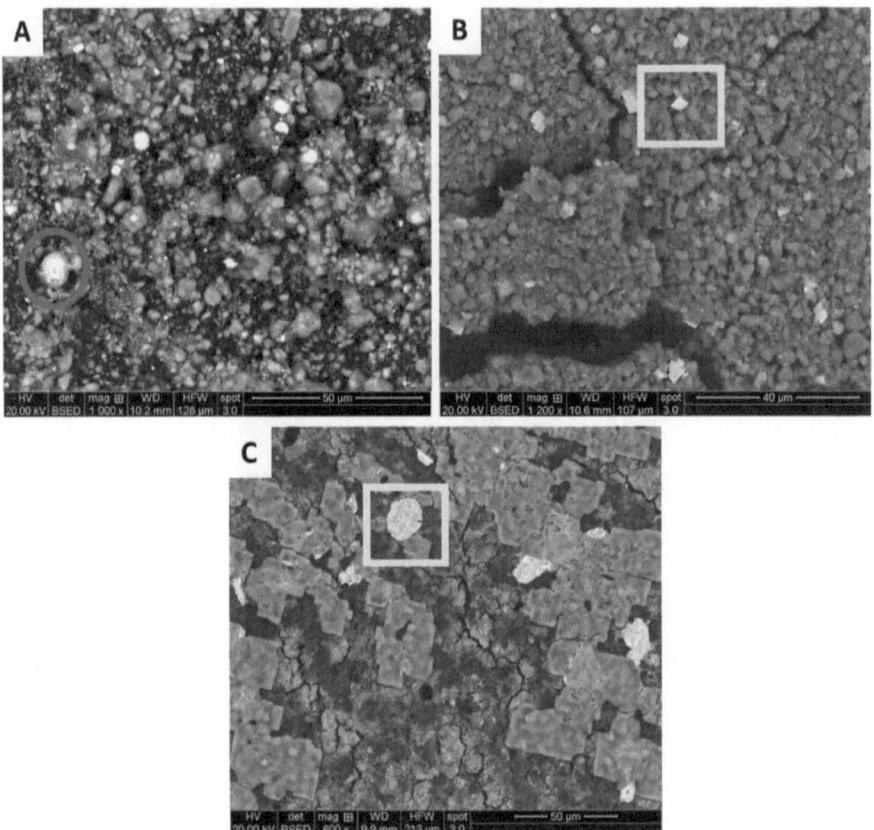

FIGURE 11.10 Unreacted Marcellus Shale powder (a) with a 50-μm scale bar in the lower right corner. Reacted powdered shale solids retained on a 0.2-μm filter showing new euhedral barite development (b) with a 40-μm scale bar in the lower right corner (representing Location D in Figure 11.4). Control experiment without shale solids retained on 0.2-μm filter showing spherical barite, iron oxides, and NaCl (c) with a 50-μm scale bar in the lower right corner. The circle (red online) highlights framboidal pyrite in the unreacted shale. The white boxes (yellow boxes in the online version) highlight the euhedral barite (b) and spherical barite (c).

of chemical process and reaction path modeling describe approaches that can be employed to identify potential mineral reactions occurring in well tubing and shale reservoirs.

11.3.2.1 Predictive Modeling Approach

An approach was designed using the OLI software package, with a focus on connecting processes that occur across the diverse pressure, temperature, and chemical conditions that exist in the well-to-reservoir system (Figure 11.11). OLI Flowsheet was used to charge balance the fresh and produced water streams (Figure 11.11a and b), obtain a composition for the blended water (Figure 11.11c), and identify precipitates formed in the produced water, fresh water, and blended water. OLI

TABLE 11.2

SI (logQ/K) from Geochemists Workbench Modeling of the Unfiltered Blended Pad Fluid, Shale-1 (Reservoir 2-Day), and Shale-2 (Reservoir 14-Day) Fluids[a,b]

SI (logQ/K)	Antigorite $Mg_3(Si_2O_5)(OH)_4$	Chrysotile $Mg_3Si_2H_4O_9$	Quartz SiO_2	Barite $BaSO_4$	Celestite $SrSO_4$	Gypsum $CaSO_4$	Brucite $Mg(OH)_2$	Halite NaCl	$BaCl_2$	$SrCl_2$
Unfiltered blended water (direct measurement)	−106.06	−7.55	0.86	2.45	−1.69	−1.68	−6.4	−2.63	−6.23	−11.77
Shale-1 experiment	39.97 ± 14.10	1.77 ± 0.89	0.02 ± 0.04	1.67 ± 0.00	−1.55 ± 0.01	−1.66 ± 0.00	−2.31 ± 0.32	−2.69 ± 0.00	−5.97 ± 0.02	−10.35 ± 0.01
Shale-2 experiment	−182.48 ± 25.70	−12.16 ± 1.61	0.3 ± 0.05	1.33 ± 0.03	−1.75 ± 0.02	−1.85 ± 0.01	−7.15 ± 0.57	−2.72 ± 0.01	−6.13 ± 0.03	−10.37 ± 0.01

[a] Input data are presented in Table 11.1 for the blended water, Shale-1, and Shale-2 experiments.
[b] Uncertainty values reported reflect the standard deviation of SI values calculated for duplicate experiments.

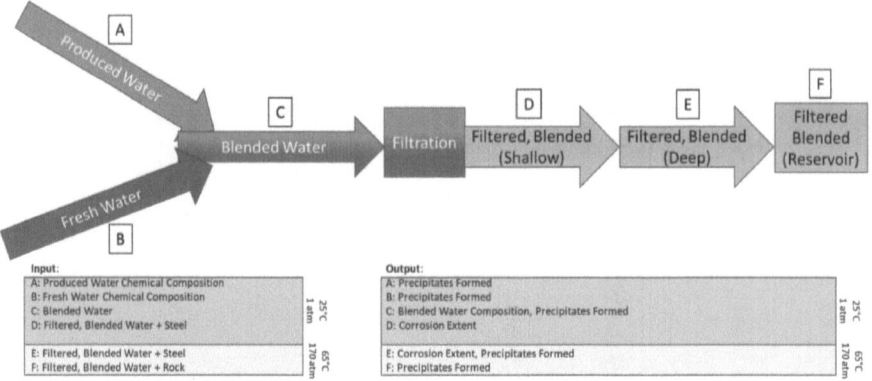

FIGURE 11.11 Design of the modeling scheme for evaluating how blended waters from Scenario 1 and Scenario 2 affect corrosion and mineral precipitate development in the well-bore tubing and reservoir.

Corrosion Analyzer was used to calculate corrosion rates for well tubing at shallow (Figure 11.11d) and deep (Figure 11.11e) pressure and temperature conditions, where the software's Low Carbon Steel G10100 option was used as a proxy for the well-bore tubing steel. Mineral precipitation in the well tubing was calculated using OLI Flowsheet (Figure 11.11e and f).

For Scenario 1, elevated $SO_4{}^{2-}$ in the Monongahela River water is hypothesized to control barite formation during blending of river water and Marcellus produced water containing high Ba^{2+} concentrations (Xiong et al. 2020a). Modeling focused on testing the influence of different starting $SO_4{}^{2-}$ concentrations in the Monongahela River water on predicted mineral saturation, where low $SO_4{}^{2-}$ would require less scale inhibitor due to lower expected SI for barite, and high $SO_4{}^{2-}$ would require more scale inhibitor due to increased expected SI for barite. The data used for developing the input streams for A and B (Figure 11.11) were generated based on field data collected (as described in the experimental section above), with some calculations and assumptions to adjust $SO_4{}^{2-}$ concentrations, account for alkalinity in the system, and to charge balance (Table 11.3). Sulfate was adjusted based on low and high values reported for the Monongahela River (TRQ 2020). Dissolved O_2 concentration for the fresh waters (6.6 mg L^{-1}) was based on expected O_2 solubility in surface water at standard atmospheric and pressure conditions (Marcon et al. 2017). Carbonate alkalinity was set to 100 mg L^{-1} $CaCO_3$ for all water streams, based on values commonly encountered in surface and produced waters in the Appalachian Basin and measured by our operator partner at one of their WV well pad sites. The pH values entered represent values common for the Monongahela River and Marcellus Shale produced waters in the region (TRQ 2020).

Scenario 2 explores how variations in both fresh and produced water composition, in the presence and absence of scale inhibitors, affect the potential for corrosion of steel components and mineral precipitate development both at surface and deep pressure and temperature conditions. Data provided by the operator for Scenario 2 were used to develop the input chemical streams for the produced water and freshwater.

TABLE 11.3

Input Water Compositions for OLI Flowsheet and Corrosion Modeling

Water:		Scenario#1 Mon River Low Sulfate	Scenario#1 Mon River High Sulfate	Scenario#1 Prod. Water	Scenario#2Paw April 24	Scenario#2 Paw Paw June 4	Scenario#2 Prod. Water April 24	Scenario#2 Prod. Water June 4
pH (median)		7.2	7.2	5.5	6.6	6.9	6.2	6.2
Alkalinity (as CaCO$_3$)	mg L^{-1}	100[d]	100[d]	100[d]	42	100	124	100
Total dissolved solids	mg L^{-1}	--	--	--	180	660	137,500	129,000
Specific elect. cond.	µmho cm^{-1}	--	--	--	225	825	172,000	155,000
SiO$_2$	mg L^{-1}	15.3	15.3	15.0	NA[a]	NA[a]	NA[a]	NA[a]
Al$_2$O$_3$	mg L^{-1}	NA[a]	NA[a]	NA[a]	1.0	0.4	BDL[b]	BDL[b]
O$_2$(aq)[c]	mg L^{-1}	6.6[d]	6.6[d]	--	6.6[d]	6.6[d]	--	--
Ethylene glycol[c]	mg L^{-1}	4.3	19.2	--	--	--	--	--
Sodium phosphate[c]	mg L^{-1}	4.3	19.2	--	--	--	--	--
Citric acid[c]	mg L^{-1}	--	--	--	4.8	4.8	4.8	4.8
Sodium (Na$^+$)	mg L^{-1}	68.1	147	38,400	9.4	96.4	37,300	32,000
Potassium (K$^+$)	mg L^{-1}	NA[a]	NA[a]	116	1.5	2.1	273	261
Ammonium (NH$_4^+$)	mg L^{-1}	NA[a]	NA[a]	266	NA[a]	NA[a]	NA[a]	NA[a]
Magnesium (Mg^{+2})	mg L^{-1}	6.4	6.4	1,550	6.7	12.3	1,680	1,480
Calcium (Ca^{+2})	mg L^{-1}	31.5	31.5	18,400	26.7	52.6	16,900	14,500
Strontium (Sr^{+2})	mg L^{-1}	1.0	1.0	3,150	0.2	0.4	2,980	2,750
Barium (Ba^{+2})	mg L^{-1}	0.9	0.9	2,940	0.1	BDL[b]	3,120	3,230
Manganese (Mn^{+2})	mg L^{-1}	0.1	0.1	8.1	BDL[b]	BDL[b]	20.6	BDL[b]
Iron (Fe^{+2})	mg L^{-1}	BDL[b]	BDL[b]	65.8	0.5	2.7	388	101
Chloride (Cl$^-$)	mg L^{-1}	124	76.7	122,000	7.0	23.0	93,800	772,00
Bromide (Br$^-$)	mg L^{-1}	114	114	91.9	BDL[b]	BDL[b]	1,100	1,340
Sulfate (SO$_4^{-2}$)	mg L^{-1}	20.0	250[d]	14.7	54.0	223	BDL[b]	BDL[b]

[a] NA, not analyzed in the original fluid sample used as a basis for the composition.
[b] BDL, below detection limit in the original fluid sample used as a basis for the composition.
[c] Parameter value set based on expected values. Dashed lines indicate that the parameter was not included in the input to the model.
[d] Value not measured, however assumed based on expected values for the natural water.

These data were input into the OLI model as received, except for dissolved O_2 and alkalinity which were input with the same values for the Scenario 1 case (described above). Solution chemistries were charge balanced in OLI prior to initiating the calculations (Table 11.3). Modeling for Scenario 2 focused on low-$SO_4{}^{2-}$ (April 24; "Paw Paw Low Sulfate") and high-$SO_4{}^{2-}$ (June 4; "Paw Paw High Sulfate") cases and only included calculation of points A – E in Figure 11.11 to focus on reactions that may occur in the well tubing.

Modeling Mineral Saturation under Reservoir Conditions

The advantage of using the OLI modeling approach allows for updating of fluid chemistry inputs for each portion of the system being modeled. For example, the fluid chemistry interacting with well tubing will differ from fluid chemistry after the interaction. HFF–rock interactions for Scenario 1 (high-$SO_4{}^{2-}$ surface water composition presented in Table 11.3 used as the starting input) were considered as an example of how OLI could be applied toward modeling reservoir geochemistry, after HFF–steel reactions. Because OLI Flowsheet is stream-based, where calculations involve mixing or reacting flowing streams, approximations were made to account for a "rock stream" to mimic the reservoir rock chemical conditions. This "rock stream" does not represent the geometry or time variability of processes that occur in fractured unconventional reservoirs, however, is designed to simulate a steady-state chemical condition. This portion of the calculation was performed to explore how a comprehensive software package (OLI) could handle the coupled well tubing-to-reservoir system.

The "rock stream" approximation modeled reservoir minerals as an influx stream with a target of ~80% fluid:rock mass ratio based on the expected volume of fluid in contact with fractures. The injected fluid was set at a volume of 1,351 m³ (8,500 bbl). The mass of rock contacted by the injected fluid was assumed to equal the volume of fluid produced during initial production from Marcellus wells in the same region. The value used for this modeling study was derived by converting the cumulative first month volume of water produced from MSEEL MIP 5H well (~640 m³ (4,000 bbl)(MSEEL 2017)) to a rock mass using the density of quartz (2.65 g cm⁻³) as the bulk rock density. The mineral composition used for the model is based on the minerals available in OLI, and the composition and expected reactive components in Marcellus Shale [mass %]: $CaCO_3$ [15], $Mg(CO_3)_2$ [3], FeS_2 [3], $BaCl_2$ [0.03], SiO_2 [78.97], and additional modeling parameters are shown in Table 11.3. All solids were set to precipitate completely. Future improvements in the accuracy of the model predictions may be obtained through development of conversions on how to relate what passes through certain sized filters in the field scenario into an entrainment value for the calculation.

Prior work showed that reaction path modeling provided reasonable approximations for HFF–rock reactions when compared to experimental results (e.g., Hakala et al. 2021). A reaction path model for Scenario 1 was designed based on the HFF-shale experiments described above, using the blended pad fluid composition (composition in Table 11.1 with 6.6 mg L⁻¹ O_2 and 100 mg L⁻¹ alkalinity as $CaCO_3$) and Marcellus Shale mineralogy for shale used in the experiments. GWB calculations were performed with the default thermodynamic database (thermo.dat), along with the V8.R6 database due to the greater representation of iron-bearing minerals in the latter database. Although the Pitzer database is favored for modeling high ionic strength solutions, it lacks inclusion of carbon species, which are important for modeling carbonate mineral chemistry in the HFF-shale system. The reaction path models were performed at 65°C, and NH_4^+ was not included in the V8.R6 reaction path calculation.

11.3.2.2   Predictive Modeling Results

Well Tubing Conditions

For Scenario 1, the mass of barite retained in the filter is over 10 times greater for Mon River High Sulfate versus the Mon River Low Sulfate, when assuming a 100% precipitate removal efficiency (439 and 39 kg hour^{-1}, respectively) (Figure 11.12). Similar rates of carbonate formation in the well were calculated for both (8 and 10 kg hour^{-1}), and the maximum thickness of carbonate deposition on well walls was calculated to be 26 mm year^{-1} for Mon River High Sulfate and 31 mm year^{-1} for Mon River Low Sulfate (Figure 11.12). Calcite was the primary carbonate mineral predicted to form in well tubing, and both conditions resulted in similar corrosion rates at the surface and at depth, with higher corrosion rates under deep well conditions (Figure 11.12).

For Scenario 2, the mass of barite retained by filtration is five times higher for Paw Paw High Sulfate compared to Paw Paw Low Sulfate, assuming a 100% filtration efficiency (423 and 84 kg hour^{-1}, respectively) (Figure 11.12). Similar rates of carbonate formation in the well and maximum carbonate deposition on the well

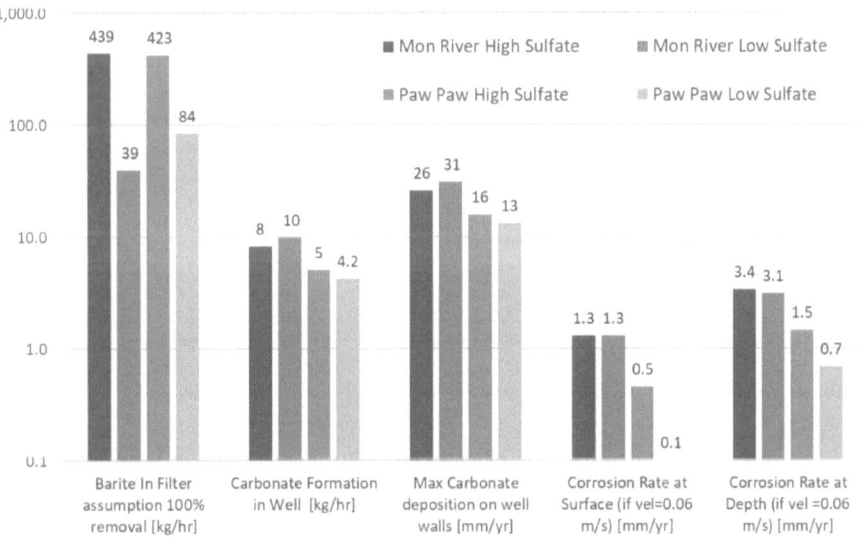

FIGURE 11.12   Comparison of OLI modeling results for different cases from Scenario 1 (Mon River water blended with Marcellus produced water) and Scenario 2 (Paw Paw Creek water blended with Marcellus produced water). Input parameters for the models are presented in Table 11.3.

walls were calculated for both conditions, with a slightly lower deposition for Paw Paw Low Sulfate (Figure 11.12). Iron carbonates were the primary carbonate mineral predicted to form for both. The corrosion rate is five times higher for Paw Paw High Sulfate under shallow well conditions and about two times higher under deep well conditions (Figure 11.12).

Reservoir Conditions

For Scenario 1, the OLI modeling approach showed that SiO_2 phases, FeS_2, and $MgCO_3$ are expected to dissolve, and $CaCO_3$, $SrCO_3$, $Mg(OH)_2$, FeS, and $MnCO_3$ are expected to precipitate (Table 11.4) The data presented in Table 11.4 show the percentage change of minerals present in the "solid stream" after the simulation, normalized as a percentage change of total solids in the system. The material balance groups (MBG) in the "liquid stream" identify which major chemical components change relative to the starting "liquid stream," where positive values indicate

---

TABLE 11.4

Expected Reservoir Geochemistry after Fluid–Shale Interaction based on Predictive Modeling for Scenario 1

OLI Reservoir Modeling Percentage Mass Change from Input Streams				Geochemists Workbench (GWB) Reaction Path Model Results Minerals with SI>0
Solids		**Liquid**		**Minerals predicted by both databases:**
Lechatelierite (SiO₂)	−0.0276%	O(−2)	0.0373%	Quartz (SiO_2)
				Witherite ($BaCO_3$)
Calcite ($CaCO_3$)	0.6087%	H(+1)	−0.0001%	Pyrite (FeS_2)
				Barite ($BaSO_4$)
Pyrite (FeS_2)	−0.0019%	Cl(−1)	0.0102%	Strontianite ($SrCO_3$)
Magnesite ($MgCO_3$)	−0.5950%	Ca(+2)	−0.2437%	Muscovite ($KAl_2(AlSi_3O_{10})(F,OH)_2$)
Strontianite ($SrCO_3$)	0.0695%	Mg(+2)	0.1703%	
$BaCl_2$	−0.0300%	Ba(+2)	0.0198%	
Brucite ($Mg(OH)_2$)	0.0029%	Sr(+2)	−0.0413%	thermo.dat database V8.R6 database
Pyrrhotite (FeS)	0.0027%	Si(+4)	0.0129%	Saponite-Ca Nontronite-Ca
Rhodochrosite ($MnCO_3$)	0.0001%	C(+4)	0.0060%	$(Ca_{0.25}(Mg,Fe)_3$ $((CaO_{0.5},Na)_{0.3}Fe^{3+}_2(Si,Al)_4O_{10}$
		N(-3)	−0.0001%	$(Si,Al)_4O_{10}$ $(OH)_2 \cdot nH_2O)$
				$(OH)_2 \cdot nH_2O)$
		Fe(+2)	−0.0009%	Kaolinite Nontronite-Mg ($Mg_{0.17}Fe_{1.67}Al_{0.}$
		N	0.0001%	$(Al_2Si_2O_5(OH)_4)$ $_{67}Si_{3.66}O_{10}(OH)_2)$
		Mn(+2)	−0.0001%	Pyrolusite (MnO)
SUM	0.0295%	SUM	−0.0295%	Mesolite
				$(Na_2Ca_2(Al_2Si_3O_{10})_3 \cdot 8H_2O)$
				Diaspore (α-AlO(OH))

an increase in the component in the "liquid stream" after mixing with the "solid stream" (Table 11.4). The "liquid stream" results show that there is a net dissolution of carbonate (positive percentage change for O(−2) and C(+4)) and precipitation of phases including Ca(+2), Sr(+2), and Mn(+2), which corresponds with the Ca, Sr, and Mn carbonates predicted to form in the "solid stream." Comparison of the solid and liquid streams suggests that pyrite is dissolving and forming pyrrhotite, as pyrite decreases and pyrrhotite increases in the solid stream, and Fe(+2) decreases in the liquid stream. Mass balance is retained as demonstrated by the same absolute value of summated percentages across both the liquid and solid streams (Table 11.4).

Reaction path models performed with GWB for Scenario 1 showed that the major saturated phases (SI>0) included witherite ($BaCO_3$), barite ($BaSO_4$), strontianite ($SrCO_3$), SiO_2 phases (quartz, tridymite, chalcedony, and cristobalite), dolomite ($CaMg(CO_3)_2$), and $CaCO_3$ phases (calcite, aragonite) (Table 11.4). Multiple clay and clay-like phases were also calculated with SI>0 (kaolinite, muscovite, saponite-Ca for the thermo.dat- based calculation; mesolite, muscovite, nontronite-Ca, and nontronite-Mg for the V8.R6-based calculation) as a result of the HFF-shale reaction (Table 11.4). Calculation with the V8.R6 database also showed SI>0 for manganese oxide (pyrolusite) and aluminum oxyhydroxide (diaspore) (Table 11.4).

11.3.2.3   Discussion – Mineral Precipitation Predicted in the Modeled Systems

Different carbonate minerals were predicted to precipitate in well tubing for the two scenarios, where calcite was the major phase predicted to precipitate for the Monongahela River calculation (Scenario 1) and siderite was the major phase predicted to precipitate for the Paw Paw Creek calculation (Scenario 2). These types of differences are important to consider when identifying which types of mineral scale need to be mitigated in the well tubing and result from differences in the chemistry of the freshwater streams.

Comparison of the reservoir results from the OLI Flowsheet and GWB reaction path models shows that barite is saturated in the reaction path model, however is not saturated in OLI. The OLI calculation uses the output from each calculation step (e.g., after fluid–steel interaction) and applies it to the next calculation step. Therefore, the reaction between the fluid and the reservoir in the OLI approach represents a reaction between the steel-reacted fluid and the reservoir. For the GWB reaction path model, mineral SI were calculated based on the HFF (blended pad fluid) chemistry at the surface and the reservoir minerals. Differences between the OLI and GWB reaction path model approaches highlight how the chemistry that occurs within well tubing may impact the mineral reactions expected to occur in the reservoir. Accounting for alkalinity in both models is important for predicting the potential for barium and strontium carbonate precipitation.

Experimental results for the well tubing reactions confirm the potential for iron oxyhydroxide precipitation, as predicted by OLI. Experimental results for the shale experiments reflect precipitation of barite predicted by the GWB reaction path models. Future experimental studies may require a design that accounts for HFF–steel reactions that influence the composition of the fluid interacting with the shale reservoir. The modeling performed as part of this study is based on thermodynamics and is unable to capture nuances of mineral reaction kinetics or surface reactions between colloids and scale inhibitor. The ability to include these nuances in modeling packages is a needed area for future research.

11.4 CONCLUSIONS – WHAT TO CONSIDER WHEN DESIGNING HFF BASE WATER PRETREATMENT

Application of produced water as a base fluid for HFF operations, especially when blending the produced water with fresh water, requires consideration of the potential mineral reactions that may occur within the well tubing and reservoir that can lead to mineral colloid formation and eventual flow-restricting mineral scale precipitation. Application of benchtop experimentation coupled with chemical process flow modeling and geochemical reaction path modeling can provide insight on which chemical reactions are most important to target for controlling mineral precipitation (e.g., Hakala et al. 2021, He et al. 2014, McDevitt et al. 2020). Mineral colloid size control and filtration of colloids likely will prevent sedimentation issues in the well tubing. However, dissolved species present in the injected fluids, such as SO_4^{2-}, may continue to interact with dissolved cations and generate new colloids that sediment in the well and reservoir, and new mineral scale that can develop on steel and shale surfaces. Water treatment to control the ionic composition of injected fluids, such as SO_4^{2-} concentrations, in combination with targeted scale inhibitor application can reduce mineral colloid-related scaling and corrosion challenges that affect unconventional hydrocarbon production.

ACKNOWLEDGMENTS AND DISCLAIMER

Research was supported through the U.S. Department of Energy, Office of Fossil Energy, Oil and Natural Gas, through the National Energy Technology Laboratory's Research and Innovation Center research portfolio in Onshore Unconventional Resources. Analytical support was provided by the NETL Pittsburgh Analytical Laboratory team. J.M., M.B., and W.X. were supported either fully or in part by research appointments at NETL through the Oak Ridge Institute for Science and Education. J.M. and J.G. are part of the Leidos/Battelle RSS Support Team to the NETL RIC.

Neither the United States Government nor any agency thereof, nor any of their employees, nor Leidos Research Support Team (LRST), nor any of their employees make any warranty, expressed or implied, or assume any legal liability or responsibility for the accuracy, completeness, or usefulness of any information, apparatus, product, or process disclosed, or represents that its use would not infringe privately owned rights. Reference herein to any specific commercial product, process, or service by trade name, trademark, manufacturer, or otherwise does not necessarily constitute or imply its endorsement, recommendation, or favoring by the United States Government or any agency thereof. The views and opinions of authors expressed herein do not necessarily state or reflect those of the United States Government or any agency thereof.

REFERENCES

Ali, M., Hascakir, B. 2017. "Water/rock interaction for Eagle Ford, Marcellus, Green River, and Barnett Shale samples and implications for hydraulic-fracturing-fluid engineering." *SPE Journal* 22:162–171.

Anderson, R. M., Beer, K. M., Buckwalter, T. F., Clark, M. E., McAuley, S. D., Sams III, J. I., Williams, D. R. 2000. Water Quality in the Allegheny and Monongahela River Basins: Pennsylvania, West Virginia, New York, and Maryland, 1996–1998. edited by U.S. Department of the Interior. Denver, CO, USA: U.S. Geological Survey.

API. 2009. Hydraulic Fracturing Operations - Well Construction and Integrity Guidelines. In Upstream Segment. Washington, D.C.: API Publishing Services.

BinMerdah, A. B., Yassin, A. A. M., Muherei, M. A. 2010. "Laboratory and prediction of barium sulfate scaling at high-barium formation water." *Journal of Petroleum Science and Engineering* 70 (1–2):79–88. doi: 10.1016/j.petrol.2009.10.001.

Birkholzer, J. T., Morris, J., Bargar, J. R., Brondolo, F., Cihan, A., Crandall, D., Deng, H., Fan, W., Fu, W., Fu, P., Hakala, A., Hao, Y., Huang, J., Jew, A. D., Kneafsey, T., Li, A., Lopano, C., Moore, J., Moridis, G., Nakagawa, S., Noel, V., Reagan, M., Sherman, C. S., Settgast, R., Steefel, C., Voltolini, M., Xiong, W., Ciezobka, J. 2021. "A new modeling framework for multi-scale simulation of hydraulic fracturing and production from unconventional reservoirs." *Energies* 14 (641). doi: 10.3390/en14030641.

Corbett, R. G. 1977. "Effects of coal mining on ground and surface water quality, Monongalia County, West Virginia." *Science of the Total Environment* 8 (1):21–38. doi: 10.1016/0048-9697(77)90059-6.

Craig, B., Blumer, D., Huizinga, S., Young, D., Singer, M. 2019. Management of corrosion in shale development. In *NACE Corrosion Conference and Exo 2019, Nashville, TN*.

Du, J., Hu, L., Meegoda, J. N., Zhang, G. 2018. "Shale softening: Observations, phenomenological behavior, and mechanisms." *Applied Clay Science* 161:290–300 doi: 10.1016/j.clay.2018.04.033

Everett, D. H. 1972. "Manual of symbols and terminology for physicochemical quantities and units." *Pure and Applied Chemistry* 31 (4):577–638.

Flynn, S. L., von Gunten, K., Warchola, T., Snihur, K., Forbes, T. Z., Goss, G. G., Gingras, M. K., Konhauser, K. O., Alessi, D. S. 2019. "Characterization and implications of solids associated with hydraulic fracturing flowback and produced water from the Duvernay Formation, Alberta, Canada." *Environmental Science Processes & Impacts* 21 (2):242–255. doi: 10.1039/c8em00404h.

Frash, L. P., Carey, J. W. 2017. Experimental measurement of fracture permeability at reservoir conditions in Utica and Marcellus Shale. In *Unconventional Resources Technology Conference, Austin, TX*.

Gallegos, T. J., Varela, B. A., Haines, S. S., Engle, M. A. 2015. "Hydraulic fracturing water use variability in the United States and potential environmental implications." *Water Resources Research* 51:5839–5845. doi: 10.1002/2015WR017278.

GWPC. Hydraulic Fracturing 2020. Available from https://fracfocus.org/hydraulic-fracturing-process.

Hakala, J. A., Vankeuren, A. N. P., Scheuermann, P. P., Lopano, C., Guthrie, G. D. 2021. "Predicting the potential for mineral scale precipitation in unconventional reservoirs due to fluid-rock and fluid mixing geochemical reactions." *Fuel* 284. doi: 10.1016/j.fuel.2020.118883.

He, C., Vidic, R. D. 2016. "Impact of antiscalants on the fate of barite in the unconventional gas wells." *Environmental Engineering Science* (Special Issue: The Science and Innovation of Emerging Subsurface Energy Technologies). doi: 10.1089/ees.2015.0603.

He, C., Zhang, T., Zheng, X., Li, Y., Vidic, R. D. 2014. "Management of Marcellus Shale produced water in Pennsylvania: A review of current strategies and practices." *Energy Technology* no. 2 (12):968–976.

Jew, A. D., Li, Q., Cercone, D., Brown Jr., G. E., Bargar, J. R. 2019. A new approach to controlling barite scaling in unconventional systems. In *Unconventional Resources Technology Conference, Denver, CO.*

Jun, Y.-S., Kim, D., Neil, C. W. 2016. "Heterogeneous nucleation and growth of nanoparticles at environmental interfaces." *Accounts of Chemical Research* 49 (9):1681–1690. doi: 10.1021/acs.accounts.6b00208.

Kahrilas, G. A., Blotevogel, J., Stewart, P. S., Borch, T. 2015. "Biocides in hydraulic fracturing fluids: A critical review of their usage, mobility, degradation, and toxicity." *Environmental Science & Technology* 49:16–32. doi: dx.doi.org/10.1021/es503724k.

Karra, S., Makedonska, N., Viswanathan, H. S., Painter, S. L., Hyman, J. D. 2015. "Effect of advective flow in fractures and matrix diffusion on natural gas production." *Water Resources Research* 51:8646–8657. doi: 10.1002/2014WR016829.

Kondash, A. J., Lauer, N. E., Vengosh, A. 2018. "The intensification of the water footprint of hydraulic fracturing." *Science Advances* 4:8. doi: 10.1126/sciadv.aar5982.

Kumar, S., Naiya, T. K., Kumar, T. 2018. "Developments in oilfield scale handling towards green technology - A review." *Journal of Petroleum Science and Engineering* 169: 428–444. doi: 10.1016/j.petrol.2018.05.068.

LeBas, R., Lord, P., Luna, D., Shahan, T. 2013. Development and use of high-TDS recycled produced water for crosslinked-gel-based hydraulic fracturing. In *SPE Hydraulic Fracturing Technology Conference*. The Woodlands, TX: Society of Petroleum Engineers.

Li, Q., Jew, A. D., Kohli, A., Maher, K., Brown Jr., G. E., Bargar, J. R. 2019. "Thickness of chemically altered zones in shale matrices resulting from interactions with hydraulic fracturing fluid." *Energy & Fuels* 33:6878–6889 doi: 10.1021/acs.energyfuels. 8b04527.

Mackey, J., Gardiner, J., Kutchko, B., Brandi, M., Fazio, J., Hakala, J. A. 2019. Is it in the water? Elucidating mineral scale precipitation mechanisms on unconventional production string components. In *Unconventional Resources Technology Conference*, Denver, CO, USA.

Mackey, J., Gardiner, J., Kutchko, B., Brandi, M., Fazio, J., Hakala, J. A. 2021. "Characterizing mineralization on low carbon steel exposed to aerated and degassed synthetic hydraulic fracture fluids." *Journal of Petroleum Science and Engineering* 202:108514.

Mackey, J., Gardiner, J., Lackey, G., Kutchko, B., Hakala, J. A. 2020. From waste to insight: Generating high resolution geochemical models from publibly available residual waste profiles. In *Unconventional Resources Technology Conference, Austin, TX.*

Marcon, V., Joseph, C., Carter, K. E., Hedges, S. W., Lopano, C. L., Guthrie, G. D., Hakala, J. A. 2017. "Experimental insights into geochemical changes in hydraulically fractured Marcellus Shale." *Applied Geochemistry* 76:36–50 doi: 10.1016/j. apgeochem.2016.11.005.

McDevitt, B., Cavazza, M., Beam, R., Cavazza, E., Burgos, W. D., Li, L., Warner, N. R. 2020. "Maximum removal efficiency of barium, strontium, radium, and sulfate with optimum AMD-Marcellus flowback mixing ratios for beneficial use in the Northern Appalachian Basin." *Environmental Science & Technology* 54 (8):4829–4839.

Milliken, K. L., Rudnicki, M., Awwiller, D. N., Zhang, T. 2013. "Organic matter-hosted pore system, Marcellus Formation (Devonian), Pennsylvania." *AAPG Bulletin* 97 (2):177–200. doi: 10.1306/07231212048.

Monroe, S., McCracken, D., Dawson, K., Mouallem, S. 2013. Production gains through reuse of fraac fluid for hydraulic fracturing: A 10 year study. In *SPE Annual Technical Conference and Exhibition, New Orleans, LA.*

MSEEL, Marcellus Shale Energy and Environmental Laboratory. 2017. Production Data edited by West Virginia University. mseel.org/research/research_MIP.html.

Nguyen, P.D., Weaver, J.D., Rickman, R.D., Dusterhoft, R.G., Parker, M.A. 2007. "Controlling formation fines at their sources to maintain well productivity." *SPE Production & Operation.* doi: 10.2118/97659-PA.

Nicot, J.-P., Scanlon, B. R. 2012. "Water use for shale-gas production in Texas, U.S." *Environmental Science & Technology* 46 (6):3580–3586. doi: 10.1021/es204602t.

Olajire, A. A. 2015. "A review of oilfield scale management technology for oil and gas production." *Journal of Petroleum Science and Engineering* 135:723–737 doi: 10.1016/j.petrol.2015.09.011.

Phan, T. T., Hakala, J.A., Bain, D. J. 2018. "Influence of colloids on metal concentrations and radiogenic strontium isotopes in groundwater and oil and gas-produced waters." *Applied Geochemistry* 95:85–96 doi: 10.1016/j.apgeochem.2018.05.018.

Phan, T. T., Hakala, J. A., Lopano, C. L., Sharma, S. 2019. "Rare earth elements and radiogenic strontium isotopes in carbonate minerals reveal diagenetic influence in shales and limestones in the Appalachian Basin." *Chemical Geology* 509:194–212 doi: 10.1016/j.chemgeo.2019.01.018.

Phan, T. T., Hakala, J. A., Sharma, S. 2020. "Application of isotopic and geochemical signals in unconventional oil and gas reservoir produced waters toward characterizing in situ geochemical fluid-shale reactions." *Science of the Total Environment* 714. doi: 10.1016/j.scitotenv.2020.136867.

Pioneer. Resilient: 2020 Sustainability Report. Pioneer Natural Resources 2020. Available from https://www.pxd.com/sites/default/files/files/2020-Sustainability-Report.pdf.

Rassenfoss, S. 2011. "From flowback to fracturing: Water recycling grows in the Marcellus Shale." *Journal of Petroleum Technology* 63 (7):48–51.

Rodriguez, R. S., Soeder, D. J. 2015. "Evolving water management practices in shale oil & gas development." *Journal of Unconventional Oil and Gas Resources* 10:18–24 doi: 10.1016/j.juogr.2015.03.002.

Shaffer, D. L., Chavez, L. H. A., Ben-Sasson, M., Castrillon, S. R.-V., Yip, N. Y., Elimelech, M. 2013. "Desalination and reuse of high-salinity shale gas produced water: Drivers, technologies, and future directions." *Environmental Science & Technology* 47 (17):9569–9583.

Stumm, W. 1993. "Aquatic colloids as chemical reactants: surface structure and reactivity." In *Colloids in the Aquatic Environment*, edited by Th. F. Tadros and J. Gregory, 1–18. Elsevier, Amsterdam.

Tieman, Z. G., Stewart, B. W., Capo, R. C., Phan, T. T., Lopano, C. L., Hakala, J. A. 2020. "Barium isotopes track the source of dissolved solids in produced water from the unconventional Marcellus Shale gas play." *Environmental Science & Technology* 54 (7):4275–4285. doi: 10.1021/acs.est.0c00102?ref=pdf.

TRQ. 2020. 2009 to 2019 Yearly Averages Mon Ohio Allegheny 2020 [cited June 2020]. Available from https://3riversquest.wvu.edu/data/3rq-maps.

U.S. Energy Information Administration's (EIA) 2020. Annual Energy Outlook 2020. Edited by U.S. Department of Energy. Washington, D.C., USA: U.S. Department of Energy.

Vankeuren, A. N. P., Hakala, J. A., Jarvis, K., Moore, J. E. 2017. "Mineral reactions in shale gas reservoirs: Barite scale formation from reusing produced water as a hydraulic fracturing fluid." *Environmental Science & Technology* 51 (16):9391–9402. doi: 10.1021/acs.est.7b01979.

Xiong, W., Gill, M., Moore, J., Crandall, D., Hakala, J.A., Lopano, C. 2020a. "Influence of reactive flow conditions on barite scaling in Marcellus Shale during stimulation and shut-in period of hydraulic fracturing." *Energy & Fuels* 34 (11):13625–13635. doi: 10.1021/acs.energyfuels.0c02156.

Xiong, W., Lopano, C., Hakala, A., Carney, B.J. 2020b. Investigation of barite scaling during reaction between pre-treated hydraulic fracturing fluid from the field and Marcellus Shale. In *Unconventional Resources Technology Conference, Austin, TX.*

Yan, F., Dai, Z., Ruan, G., Alsaiari, H., Bhandari, N., Zhang, F., Liu, Y., Zhang, Z., Kan, A., Tomson, M. 2017. "Barite scale formation and inhibition in laminar and turbulent flow: A rotating cylinder approach." *Journal of Petroleum Science and Engineering* 149:183–192 doi: 10.1016/j.petrol.2016.10.030.

Zhang, P., Zhang, Z., Liu, Y., Kan, A. T., Tomson, M.B. 2019. "Investigation of the impact of ferrous species on the performance of common oilfield scale inhibitors for mineral scale control." *Journal of Petroleum Science and Engineering* 172:288–296. doi: 10.1016/j. petrol.2018.09.069.

Zhang, W., Tang, X., Weisbrod, N., and Guan, Z. 2012. "A review of colloid transport in fractured rocks." *Journal of Mountain Science* 9:770–787. doi: 10.1007/s11629-012-2443-1.

12 Crystallizers for Brine Waste Treatment
Technologies and Design Heuristics

Sankaranarayanan Ayyakudi Ravichandran,
Jacob Hutfles, and John Pellegrino
University of Colorado Boulder

CONTENTS

12.1 INTRODUCTION

12.1.1 BRINES AND CRYSTALLIZATION

Supersaturated salt waters are encountered in RO/NF (reverse osmosis/nanofiltration) membrane-based water treatment processes, cooling water discharge, and produced water from oil and natural gas operations. This book chapter is intended to provide an

DOI: 10.1201/9781003091011-12

overview of concepts pertaining to the crystallization of sparingly soluble salts from supersaturated waters and recent developments geared toward the design of industrial crystallizers. Salt crystallization from supersaturated waters is a thermodynamically controlled process, such that the total crystal mass precipitated at equilibrium is fixed in relation to the initial level of supersaturation. While this total supersaturation depletion is fixed, the crystallization kinetics can be controlled via engineered interventions. According to classical theory, crystallization is thought to begin with the initial appearance of crystal nuclei (i.e. nucleation). Crystal nucleation can occur by a homogenous or heterogeneous process. Homogenous nucleation refers to the spontaneous appearance of nuclei in solution by virtue of supersaturation, and heterogeneous nucleation refers to the generation of nuclei on exposed surfaces or foreign particulates in the crystallizing liquor. Heterogeneous crystallization is more likely to occur than homogenous nucleation; this is due to the nucleating surface reducing the surface area of the crystal nuclei, thus lowering the free energy barrier of formation [1]. Figure 12.1 provides a simplistic illustration of heterogenous and homogenous nucleation, in which crystallization can simultaneously occur in the bulk of the crystallizing liquor, as well as on surfaces exposed to the crystallizing liquor. Some crystals adhere strongly to the surfaces in the form of mineral scale while loosely attached crystals may be removed from the surface via hydraulic flow. In industrial crystallizers, crystallization in the bulk liquor is the foremost mechanism desired, but heterogeneous nucleation and subsequent surface crystallization will occur on exposed surfaces. Since heterogeneous nucleation is thermodynamically favored, brine waste crystallizers typically contain "seed" crystals or foreign particulate media such as sand [2] to promote crystallization. This crystal growth on equipment surfaces, commonly known as scaling, is an important design obstacle as well. Surface scaling is responsible for plumbing blockage, decreased efficiency of heat transfer and separation operations (membrane scaling), degradation of pumps [3–7], etc. Due to the adverse effect of mineral scaling,

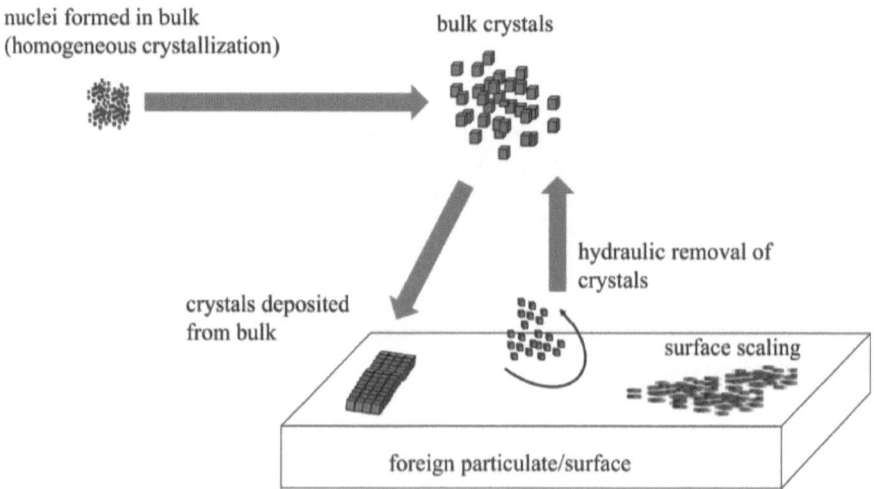

FIGURE 12.1 Heterogeneous and homogeneous crystallization, partitioning of crystals between the bulk and on surfaces. (Adapted from ref. [10] with permission from Elsevier B.V.)

reasonable techniques to quantify the extent of mineral scaling on different types of surfaces are available in the literature [8,9]. This chapter will not delve into the mechanisms of crystallization physics, as this has been adequately covered by previous literature [1], but the focus of this chapter is on the engineering ideas related to crystallization in supersaturated brines at industrial scales.

12.1.2 Current Brine Disposal and Challenges

Currently, brine is largely disposed of in the environment, there is a requirement to reduce the volume of waste (dewater the brine) and also increase the amount of reusable water. For example, in the United States, brine waste from municipal water treatment facilities is mostly disposed via surface discharge, sewer discharge, and injection wells [11]. Figure 12.2 gives a breakdown of the type and respective percentage of municipal treatment facilities that use the particular type of brine disposal.

Besides the previously discussed methods of disposal, evaporation crystallization is also used. Evaporation crystallization is the simplest type of industrial crystallization to minimize brine volume. This involves collecting brine waste in large ponds and allowing the water to evaporate utilizing solar heat. If the brine is already supersaturated, the evaporation of water will induce supersaturation resulting in crystallization; the dewatering process reduces the amount of waste that has to be disposed of [12]. Largely, the methods used currently for concentrate disposal may not always be environmentally sustainable and require local permitting. Certain jurisdictions have regulatory regimes that place limits on the concentration of brines that are disposed in sewers and through surface discharge [13]; consequently sewer discharge is limited by the extent of brine salinity [11]. Therefore, engineered solutions to dispose brine waste in an environmentally and economically sustainable manner are broadly sought.

In general, brine waste treatment is enabled by brine concentration and a zero liquid discharge (ZLD) type system for brine disposal is the most desired outcome. But brine concentration is currently expensive for several fundamental reasons. For example, as salinity increases, thermodynamics requires more energy to achieve the separation, which, in turn, increases temperature and pressure requirements that

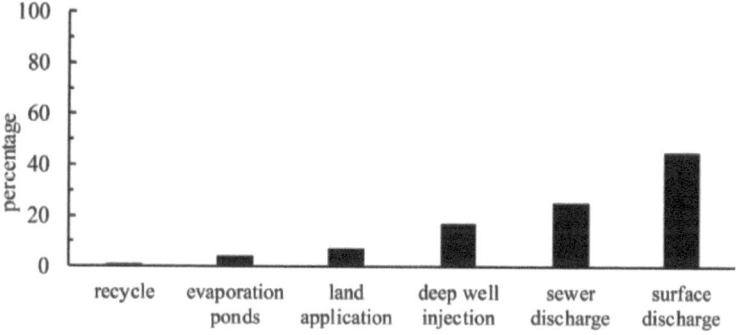

FIGURE 12.2 Methods of brine waste disposal by municipal MF/RO and BWRO facilities [11].

make equipment more expensive. In addition, most systems need custom engineering because of the variable chemistry that can be encountered, as well as equipment redundancy required for cleaning operations.

An example of a system involving multiple treatment steps is high-efficiency reverse osmosis (HERO). HERO is a patented technology [14] that has been applied at a commercial scale for brine treatment [15]. This method is used to remove hardness and overcome silica scaling on RO membranes [14]. The first step of the HERO process involves raising the pH (~10.5) of the brine feed by the addition of a base, followed by treatment in a weak acid cation resin exchange system. During cation exchange, H^+ ions replace cations, consequently the pH of the brine increases, and carbonate alkalinity is converted to carbonic acid ($CO_2 \cdot H_2O$). Next, acid is added to the brine to neutralize the remaining alkalinity. The acidified stream is then degasified to remove CO_2. The next treatment again involves increasing the pH to decrease the silica precipitation. Then, this high pH effluent is sent to the RO treatment system. The high pH of the effluent increases the removal of total organic carbon (TOC) and weakly ionized species such as boron and silica while simultaneously overcoming RO membrane performance loss by retarding silica scale formation. This system can allow for up to 90% water recovery from brackish waters [14]. Though HERO offers a path to a ZLD system, the various steps involved make it a very complex system to operate and the various process steps have to be specifically tailored to the brine chemistry.

A nominal technology and economics landscape for brine concentration/management is presented in Figure 12.3 Thus, we can see that incorporating brine crystallizers can add a simple, and more robust, physical solid–liquid separation component to all primary brine concentration schemes.

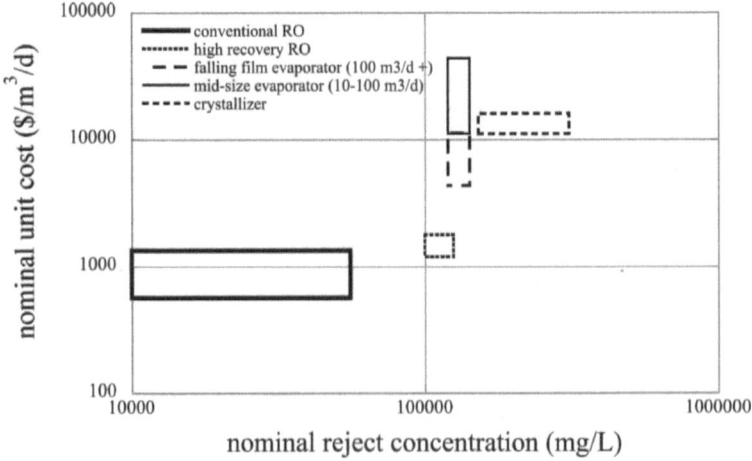

FIGURE 12.3　Range of installed cost for various brine concentration processes versus the total dissolved (and/or suspended) solids concentration produced. (Adapted from [16].)

12.2 USING CRYSTALLIZERS FOR BRINE TREATMENT

A brine crystallizer uses a solid–liquid separation technique to remove salts via precipitation from brines [9]. Figure 12.4 shows a schematic engineering process that incorporates crystallization. In this case, a crystallizer is used to treat brine waste from an RO membrane reject stream. The supernatant stream from the crystallizer is then cycled back to the membrane stage to increase water recovery. In such a process, increased crystallizer efficiency can improve overall process outcomes.

Crystallization of salts from brines can be accomplished by inducing supersaturation (if the brine stream is not already supersaturated) by using (1) a reactor with certain a residence time to allow crystallization to take place; this process can be aided by chemicals to induce/increase supersaturation; (2) by removing water by thermal evaporation to achieve supersaturation and then crystallization; and (3) dewatering the brine by using selective transport across membranes. To provide the reader with an understanding of how the three ideas described above are implemented in practicum, the operation of pellet softeners, evaporators, and membrane crystallizers are described in the successive sections.

12.2.1 Pellet Softeners

Pellet softeners are one of the most ubiquitous types of crystallizers used at the industrial scale for domestic water treatment, this process involves reducing water hardness (softening) by precipitating dissolved salts. Pellet softeners are a type of fluidized bed crystallizer whereby particulates are suspended in the crystallizing liquor (brine stream) to induce heterogeneous crystallization. If the brine stream is undersaturated or insufficiently supersaturated, chemicals are added to induce supersaturation or increase the degree of supersaturation. For example, if the brine is supersaturated with calcium, lime is added to the pellet softener to increase crystallization kinetics,

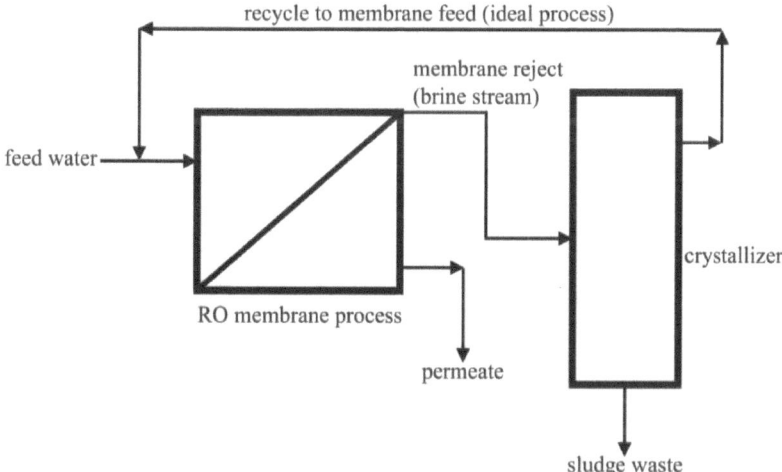

FIGURE 12.4 Schematic of how a crystallizer to treat brine streams can be incorporated with conventional RO.

thus inducing rapid precipitation. A diagram showing the typical operation of a pellet softener can be seen in Figure 12.5. In the fluidized bed pellet softener, small seed particles with high specific surface area (e.g. sand) are suspended in the upper region of the bed providing an amenable surface for heterogeneous crystallization. As the particles grow due to salt deposition, they develop a higher settling velocity and settle to the bottom of the bed. Typically, pellet softeners are operated as semi-batch reactors allowing for the removal of settled particles (pellets) and replenishment of seed particulates at staggered intervals. Pellet softeners can also be operated in continuous mode, allowing for the continuous removal of pellets and continuous replenishment of seed particulates, but this has been mechanically challenging to operate due to particle jamming. Usually, sand is used as the seed particulate [2,17,18], but other particulate types, such as heated iron oxide with a high surface-to-volume ratio, have been used to reduce the volume of seed particles required and reduce the footprint of pellet softeners[19].

12.2.2 EVAPORATIVE CRYSTALLIZERS

Evaporative crystallization uses an external thermal source to evaporate water from brine waste. As water is evaporated, the brine achieves supersaturation and the salts can crystallize out; then, the salt can be reused or disposed. Since evaporation is energy intensive, to increase the energy efficiency multiple-effect evaporators, i.e., evaporators-in-series, are used. A multiple-effect evaporator requires external heat only in the first vessel. Each successive stage is held at a lower pressure than the preceding stage, and at lower pressure, the boiling point of water decreases the steam from initial stages and can be used to heat the successive stages (see Figure 12.6). Besides being energy intensive, the capital costs are high as expensive materials, which can withstand being exposed to corrosive brines, are required for evaporator fabrication [20,21].

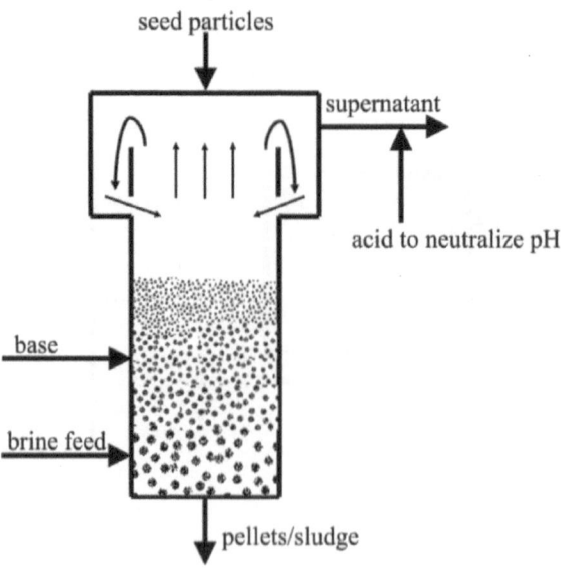

FIGURE 12.5 Schematic of pellet softener operation.

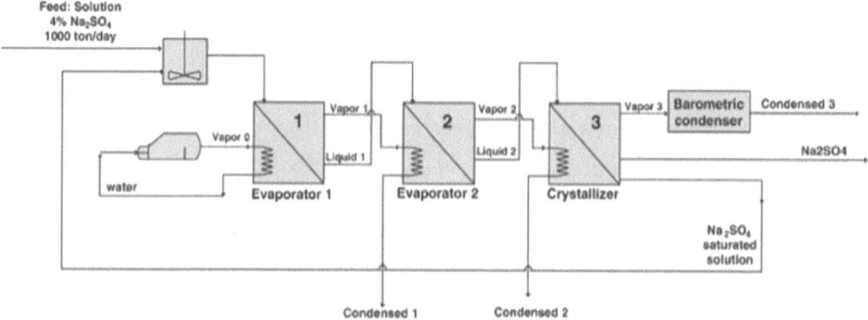

FIGURE 12.6 Schematic of a multiple-effect evaporator. (Reprinted from ref. [20] with permission from Elsevier B.V.)

12.2.3 MEMBRANE CRYSTALLIZATION

12.2.3.1 Membrane Distillation Crystallization

Membrane distillation (see Figure 12.7) crystallization involves using a hydrophobic membrane to separate concentrated brine feed from the pure water permeate. In this process, a thermal gradient is applied between the feed and permeate, and a pressure gradient exists across the membrane due to differential vapor pressure caused by the gradients in temperature and concentration. The solvent (water) vapor transports through the membrane from the brine side to the permeate side, resulting in the brine side liquid becoming rich in electrolytes. This concentrated and dewatered brine will

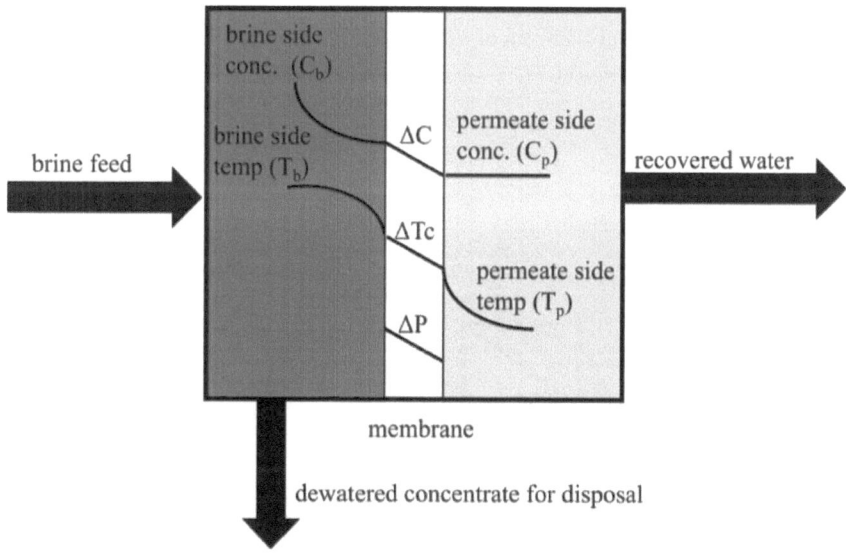

FIGURE 12.7 Schematic of membrane distillation crystallizer system adapted and modified. (Adapted from ref. [22] with permission from Taylor&Francis.)

consequently become supersaturated if not already supersaturated resulting in the crystallization of salts [22]. This process of dewatering reduces the amount of brine to be disposed and also allows for recovering water for reuse.

12.2.3.2 Pressure-Driven Membrane Crystallization

In the case of a pressure-driven membrane process, the brine feed would be processed by a series of polyamide thin-film composite (PA-TFC) RO membrane stages to dewater the brine (Figure 12.8). The brine from dewatering would achieve supersaturation and crystallization would take place. The approach to reaching a ZLD system would require high water recoveries from the brine. The realization of such a system is limited by the propensity of PA-TFC to mineral scaling [23] when subjected to sparingly soluble salt-rich brines; if, through engineering intervention, bulk partitioning of crystals is predominant over scale buildup on the membrane, favorable outcomes may be achieved. In theory, if we were to engineer the surface of PA-TFC membranes to resist mineral scaling and allow crystallization to take place in the bulk fluid, as opposed to on the membrane surface, the pressure-driven membrane crystallization could prove to be a viable strategy.

12.3 OVERCOMING CHALLENGES TO PRESSURE-DRIVEN MEMBRANE CRYSTALLIZERS

12.3.1 VIBRATORY SHEAR-ENHANCED FILTRATION PROCESS

Vibratory shear-enhanced filtration process (VSEP) is a patented technology [24] that was initially designed to overcome colloidal fouling in ultrafiltration (UF) membranes. VSEP involves applying vibration to the membrane to induce shear on the surface of the membrane to prevent buildup on the membranes [25]. VSEP technology has also been explored to combat the effects of mineral scale formation in the RO membrane

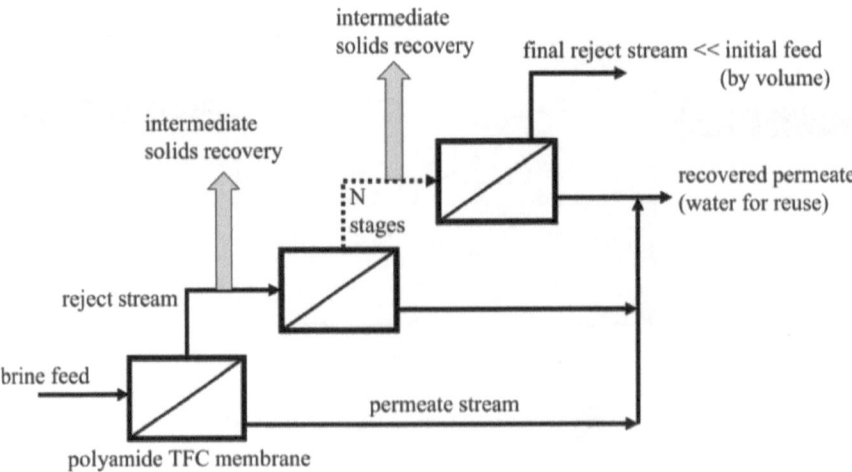

FIGURE 12.8 Theoretical application of scaling-resistant membranes in membrane crystallizers for brine treatment.

treatment process [23]. This technology may help overcome the effects of RO membrane performance loss due to mineral scale formation when membrane crystallizers are used to treat brines. A pilot-scale study [26] to treat brine waste from a brackish water facility showed that VSEP was effective in reducing silica deposition; nevertheless, there were problems with $BaSO_4$ precipitation, and frequent cleaning was required. Studies showed that vibration-induced shear decreased the extent of concentration polarization and changed the morphology of the mineral scale cake layer. Overall, membrane performance due to mineral scale formation can be mitigated using VSEP [26].

12.3.2 Scaling-Resistant Membranes for Brine Disposal

Mineral scaling on RO membranes is an obstacle to realize a membrane crystallizer system, as envisioned in Figure 12.8, that is devoid of complex process steps as in the case of VSEP and HERO. Many efforts have been geared toward developing scaling-resistant membranes [27–29]. On this front, we shall be discussing membrane patterning (engineered surface roughness) and surface chemistry modification to PA-TFC membranes.

12.3.2.1 Membrane Patterning

One approach is to modify the membrane surface by imprinting the membrane with nano/micro-scale features to influence the mass transfer characteristics at the boundary layer [30,27]. The physics behind VSEP and patterning is similar in the sense that both aim to increase the shear in the region above the membrane [31] to reduce concentration polarization and performance loss due to scaling. Whereas VSEP involves using an external actuator to induce increased shear, membrane patterning attempts to use a passive approach to modify flow characteristics at the surface of the membrane. Previous studies have shown that nanoscale patterns on membranes reduce scale deposition and flux decline in RO membranes [27]. The patterns present on the surface of the membrane may disrupt the boundary layer on top of the membrane surface, thereby disturbing the concentration profile in the feed side and consequently reduce the extent of concentration polarization. Carman et al. also showed that micron-scale patterned surfaces resisted settling of particulate zoospores versus unpatterned surfaces [30]. These ideas can be used to develop scaling-resistant PA-TFC membranes (Figure 12.9).

12.3.2.2 Membrane Surface Chemistry Modification

While patterning aims at developing a mechanical approach to scaling-resistant polyamide thin-film composite RO membranes, another method is to modify the surface chemistry of membranes to make mineral scales adhere less to the membrane. Tong et al. [32] consider that surface chemistry modifications offer a viable route to develop scaling-resistant RO membranes. In their extensive review, they support the view that understanding relationships between supersaturation/scalant chemistry and membrane surface chemistry is important for achieving scaling-resistant membranes. Various chemical modifications to PA-TFC membranes have been studied by researchers, such as, modifying the membrane to impart negative surface charge by grafting various functional groups to resist silica scaling [33,34]; graphene oxide

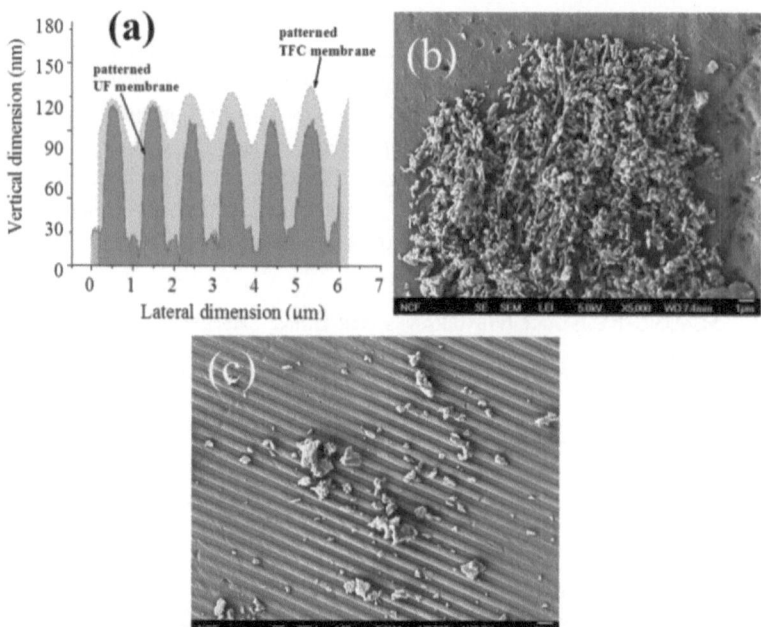

FIGURE 12.9 (a) AFM cross-sectional image of patterned polyamide RO membrane, (b) unpatterned RO membranes prone to mineral scaling, and (c) patterned membrane show scaling resistance. (Modified and adapted from ref. [27] with permission from Elsevier B.V.)

(GO) coating on the membrane to introduce -OH groups and reduce -COOH groups [35]; introducing poly(methacrylic acid) and poly(acrylamide) brush layers [36] to resist gypsum scaling; and surface patterning combined with carboxyl group surface functionalization to resist biofouling [37]. Patterning or modifying the surface chemistry of the PA-TFC layer or a combination of both would be key to realizing a RO/NF membrane-based crystallizer system.

12.4 BRINE CRYSTALLIZER DESIGN HEURISTICS

Crystallizer design is a well-studied field, numerous complex crystallizer designs exist and are used to produce commercially valuable crystals such as pharmaceutical products, and a prior paper by the authors of this chapter [38] contains a tabulation of various crystallizer designs that have been studied. However, in the case of crystallizers for brine disposal, the crystals are mostly not valuable products but waste that has to be disposed of [39,40]. Though there have been suggestions to recover valuable salts from specific brines [41,42], these apply only to specific brine chemistries, with valuable species, and fortuitous circumstances [43]. Therefore, complex crystallizer designs could be cost prohibitive for brine treatment. Thus, simple designs based on a few well-understood parameters affecting the kinetics, with simple and responsive methods to control them, are sought. Researchers have studied parameters that influence crystallization kinetics with a focus on improving brine crystallizer design [9].

There has also been work carried out to develop a bench-scale testing methodology to screen how different types of surfaces influence crystallization in terms of bulk crystallization and surface scale buildup. This methodology of screening a surface can allow for identifying surfaces that are less prone to mineral scaling [9]. These studies can help in evaluating the suitability of various crystallizer designs for the variable brine chemistries that result from diverse waters.

12.4.1 Effect of Mixing and Residence Time

Previous bench-scale studies using a series of CSTRs-in-series (continuously stirred tank reactor) studied the effect of residence time, the level of mixing energy applied to the crystallizing liquor, and the surface-to-volume ratio within regions where mixing was applied. The test setup consisted of a series of 6 CSTRs, and some mixing was applied using stirrers in each of the CSTRs, as well as, when the liquor was moved from one CSTR to another, due to peristaltic pumping. Further, the specific influence of peristaltic pumping (mixing) was investigated using a recirculation loop in the first CSTR (see Figure 12.10). As expected, increased crystallization was observed as residence time in each crystallizer was increased, but it was also interesting to note that faster crystallization was dependent on applying mixing in regions of high surface-to-volume ratio versus simply increasing mixing energy applied to the bulk crystallizing liquor [38].

The various operational parameters of the crystallizer that influence nuclei/crystal formation in the first tank of the CSTR-in-series system were formulated by the authors by using turbidity measurements, an indicator of nucleation. The expression relates the kinetics of nuclei formation to the Kolmogorov mixing length associated with the mixing input, the S/V (surface-to-volume ratio) of the region of mixing associated with a particular mixing length, the feed flow subjected to mixing, mass action factors, and the salt concentration in the brine. The variables in the expression are explained in Table 12.1. Readers may refer to prior literature [38] for more details.

$$\frac{dN_n}{dt} = k_0 \frac{D_x}{\eta_0^2} C_A^* C_C^* \left[1 + \frac{\eta_0^2}{D_x} \frac{1}{\eta_B^2} q_B S V_B \frac{k_B}{k_0} \phi_{S,B} + \frac{\eta_0^2}{D_x} \frac{1}{\eta_0^2} q_{tot} S V_T \frac{k_T}{k_0} \phi_{S,T} \right]$$

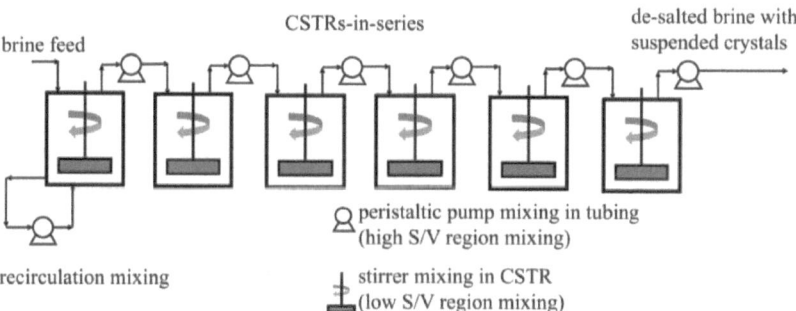

FIGURE 12.10 Schematic of CSTR-in-series crystallization system. (Adapted from ref.[38] with permission from Elsevier B.V.)

TABLE 12.1
Parameters Used in the Empirical Formulation Correlating Nuclei Generation to Crystallizer Operational Parameters

Variable	Units	Description
N_n	mol m^{-3}	No. of nuclei/vol
dN_n/dt	mol m^{-3}s^{-1}	No. of nuclei/vol/time
C_c^*, C_a^*	mol m^{-3}	Excess cation and excess anion concentration versus saturation, respectively
D_x	m^{2}s^{-1}	Ion pair diffusion coefficient $D_x \approx \dfrac{2}{\dfrac{1}{D_C} + \dfrac{1}{D_A}}$, 8.52×10^{-10}.
q_{tot}	m^{3}s^{-1}	Total feed rate into the first CSTR
q_B	m^{3}s^{-1}	Flowrate in the recirculation tubing
SV$_T$, SV$_B$	m^{-1}	Surface area to volume ratio of CSTR and recirculation tubing, respectively, e.g., SV$_T$ = S$_{a,T}$/V$_0$.
η_B	m	Kolmogorov mixing length scale associated with the volume of fluid within the tubing during its period of movement within the tubing for recirculation
η_0	m	Kolmogorov mixing length scale associated with the volume of fluid within the tank
k_0, k_B, k_T	m^{3}mol^{-1}	Mass action collision factors due to efficiency of "collisions" culminating in a nucleation event in the bulk, in the recirculation tubing, and CSTR
$\Phi_{S,T}, \Phi_{S,B}$	-	Fraction of the surface area with active sites favorable for nucleation events per unit area of the CSTR's surface and recirculation of tube's surface (material property of crystal and material of the surface of CSTR/ tube). $0 \leq \Phi_{S,T} \leq 1$, $0 \leq \Phi_{S,B} \leq 1$
τ	S	CSTR residence time (= V_0/q_{tot})
V_0, V_B	m^3	Wetted volume of CSTR and recirculation tube
$S_{a,T}, S_{a,B}$	m^2	Wetted surface area of CSTR and recirculation tubing

Reproduced from ref. [38] with permission from Elsevier B.V.

12.4.2 INTERACTIONS BETWEEN SURFACES AND CRYSTALLIZATION

Considering that heterogeneous crystallization is energetically favored compared to homogenous crystallization [1], the impact of surfaces exposed to the crystallizing liquor on crystallization and how these surfaces themselves are impacted by crystallization is important from the perspective of crystallizer design. Both surface chemistry and surface roughness can impact crystallization/mineral scale deposition on surfaces as seen from discussions in Sections 12.4.1. and 12.4.2. Studies on the ability of various materials to act as nucleating agents have been conducted extensively [44–51]; these studies performed under plethostatic conditions (constant temperature and pH) showed that materials with different functional group chemistry affect the kinetics of crystallization. If the material is very adept in providing nucleating sites, the kinetics of crystallization can be increased by introducing such materials into the crystallizer. On the other hand, there is also a value in identifying how materials that are exposed to the crystallizing liquor affect the partitioning of crystals into the

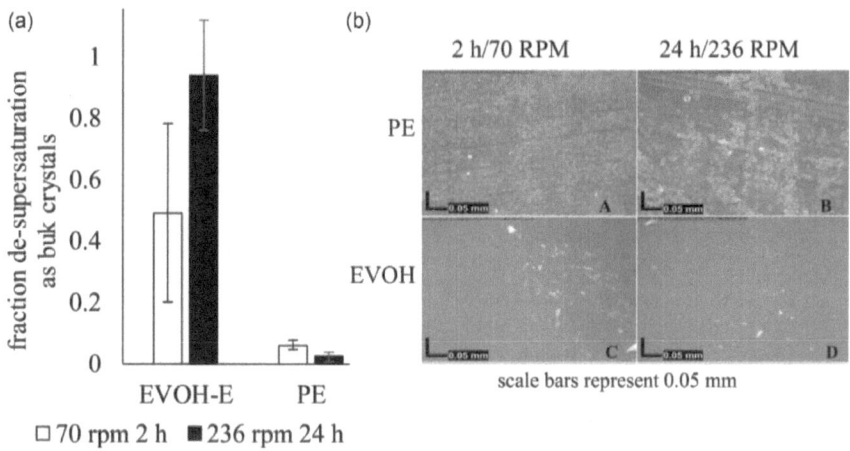

FIGURE 12.11 (a) Supersaturation depletion in the form of bulk crystals for solutions exposed to EVOH-E and PE surfaces at varied duration and mixing intensity/shaker RPM and (b) corresponding micrographs of exposed polymers. (Adapted from ref. [9] with permission from Elsevier B.V.)

bulk and on said surface. Researchers have recently [9] identified a method to macroscopically screen how mineral scales are formed on different polymers surfaces with different surface chemistries. The procedure involved placing circular coupons of different polymers on the lids of jars and sealing the lids onto the jars containing a supersaturated crystallizing liquor. Interactions between polymer surfaces with different functional groups and supersaturated brine were evaluated using these shaken jars, wherein, a polymer surface is exposed to synthetic brine solution under constant mixing. Polymer chemistry was found to have a significant effect on bulk and surface partitioning of crystals, likely due to functional group interactions with solvated ions. The results from these studies showed that a copolymer (EVOH-E) of vinyl alcohol (56 %) and ethylene (44 %) was much more resistant to mineral scale formation than polyethylene (PE), refer Figure 12.11. These types of surface crystallization studies can help in reasonable quantification of Φ terms shown in Table 12.1. Such studies also allow for understanding how surface chemistry modifications can help in developing scaling-resistant membranes and surfaces.

12.5 CONCLUSIONS

Effective brine waste management can be accomplished through engineered solid–liquid separations systems via crystallizers. The effective design of crystallizers for brine management requires an approach that is economically and environmentally sound. Understanding the physics of bulk and surface partitioning of crystals and developing engineering heuristics to tune the design parameters of crystallizers is key to develop efficient brine crystallizers. Whereas various non-thermal process techniques to treat brine waste such as VSEP and HERO have been developed, future research should be guided toward understanding complex relationships between crystallization and materials exposed to the crystallizing brine. A materials science approach can aid in developing scaling-resistant membranes for membrane crystallizers and plumbing that is resistant to scale buildup.

REFERENCES

1. Crystallisation, 4th Edition By J.W. Mullin. 2001. Butterworth Heinemann: Oxford, UK. 600 pp. £75.00. ISBN 075-064-833-3, *Organic Process Research & Development*, 6 (2002) 201–202.
2. G. van Houwelingen, R. Bond, T. Seacord, E. Fessler, Experiences with pellet reactor softening as pretreatment for inland desalination in the USA, *Desalination & Water Treatment*, 13 (2010) 259–266.
3. J. Glater, J.L. York, K.S. Campbell, Chapter 10 - Scale Formation and Prevention A2. In K.S. Spiegler, A.D.K. Laird, (Eds.) *Principles of Desalination* (Second Edition), Academic Press, Cambridge, MA, 1980, pp. 627–678.
4. C.J. Gabelich, M.D. Williams, A. Rahardianto, J.C. Franklin, Y. Cohen, High-recovery reverse osmosis desalination using intermediate chemical demineralization, *Journal of Membrane Science*, 301 (2007) 131–141.
5. M. Mickley, Brackish groundwater concentrate management. In: M. Associates (Ed.) State-of-Science White Paper, Boulder, CO, Prepared for NMSU and Consortium for High Technology Investigations on Water and Wastewater, presented at CHIWAWA Concentrate Management Workshop, El Paso, TX, May 2010.
6. P. Xu, T.Y. Cath, A.P. Robertson, M. Reinhard, J.O. Leckie, J.E. Drewes, Critical review of desalination concentrate management, treatment and beneficial use, *Environmental Engineering Science*, 30 (2013) 502–514.
7. B.K. Pramanik, L. Shu, V. Jegatheesan, A review of the management and treatment of brine solutions, *Environmental Science: Water Research & Technology*, 3 (2017) 625–658.
8. Y.N. Wang, J. Davidson, L. Francis, Scaling in polymer tubes and interpretation for use in solar water heating systems, *Journal of Solar Energy Engineering*, 127 (2005) 3–14.
9. J. Hutfles, S.A. Ravichandran, J. Pellegrino, Screening polymer surfaces in crystallization, *Colloids and Surfaces A: Physicochemical and Engineering Aspects*, 582 (2019) 123869.
10. S. Lee, J. Kim, C.-H. Lee, Analysis of CaSO4 scale formation mechanism in various nanofiltration modules, *Journal of Membrane Science*, 163 (1999) 63–74.
11. M. Mickley, Updated and Extended Survey of U.S. Municipal Desalination Plants, Mickley & Associates LLC (for) Bureau of Reclamation, USA, 2018.
12. H. Lu, J. Wang, T. Wang, N. Wang, Y. Bao, H. Hao, Crystallization techniques in wastewater treatment: An overview of applications, *Chemosphere*, 173 (2017) 474–484.
13. M. Mickley, Membrane Concentrate Disposal: Practices and Regulation (Second Edition) Mickley & Associates (for) Bureau of Reclamation, 2006.
14. D. Mukhopadhyay, Method and apparatus for high efficiency reverse osmosis operation, in: USPTO (Ed.) USA, 1999.
15. AquaTech, Duke Energy selects HERO™ Membrane Process for ZLD Application, 2014.
16. C. Gasson, The economics of desalination, www.nawihub.org/archive, Global Water Intelligence, 2020.
17. A.B. de Haan, H. Bosch, *Industrial Separations Processes: Fundamentals*, De Gruyter, Berlin, 2013.
18. J. Van Dijk, D. Wilms, Water treatment without waste material. Fundamentals and state of the art of pellet softening, *Aqua- Journal of Water Supply: Research and Technology*, 40 (1991) 263–280.

19. C.-W. Li, J.-C. Jian, J.-C. Liao, Integrating membrane filtration and a fluidized-bed pellet reactor for hardness removal, *Journal AWWA*, 96 (2004) 151–158.
20. M.J. Fernández-Torres, D.G. Randall, R. Melamu, H. von Blottnitz, A comparative life cycle assessment of eutectic freeze crystallisation and evaporative crystallisation for the treatment of saline wastewater, *Desalination*, 306 (2012) 17–23.
21. O. Lefebvre, R. Moletta, Treatment of organic pollution in industrial saline wastewater: A literature review, *Water Research*, 40 (2006) 3671–3682.
22. M. Gryta, Concentration of NaCl Solution by membrane distillation integrated with Crystallization, *Separation Science and Technology*, 37 (2002) 3535–3558.
23. L.S.G. Johnsona, M. Monroea, VSEP Treatment of RO Reject from Brackish Well Water A Comparison of Conventional Treatment Methods and VSEP, a Vibrating Membrane Filtration System. *El Paso Desalination Conference, El Paso, TX*, March 2006.
24. B. Culkin, *Device and method for filtering a colloidal suspension* (U.S. Patent 4,952,317). U. S. Patent and Trademark Office 1990.
25. W. Shi, M.M. Benjamin, Effect of shear rate on fouling in a Vibratory Shear Enhanced Processing (VSEP) RO system, *Journal of Membrane Science*, 366 (2011) 148–157.
26. A. Subramani, J. DeCarolis, W. Pearce, J.G. Jacangelo, Vibratory shear enhanced process (VSEP) for treating brackish water reverse osmosis concentrate with high silica content, *Desalination*, 291 (2012) 15–22.
27. S.H. Maruf, A.R. Greenberg, J. Pellegrino, Y. Ding, Fabrication and characterization of a surface-patterned thin film composite membrane, *Journal of Membrane Science*, 452 (2014) 11–19.
28. Y.-F. Guan, C. Boo, X. Lu, X. Zhou, H.-Q. Yu, M. Elimelech, Surface functionalization of reverse osmosis membranes with sulfonic groups for simultaneous mitigation of silica scaling and organic fouling, *Water Research*, 185 (2020) 116203.
29. J.R. Ray, W. Wong, Y.-S. Jun, Antiscaling efficacy of $CaCO_3$ and $CaSO_4$ on polyethylene glycol (PEG)-modified reverse osmosis membranes in the presence of humic acid: interplay of membrane surface properties and water chemistry, *Physical Chemistry Chemical Physics*, 19 (2017) 5647–5657.
30. M.L. Carman, T.G. Estes, A.W. Feinberg, J.F. Schumacher, W. Wilkerson, L.H. Wilson, M.E. Callow, J.A. Callow, A.B. Brennan, Engineered antifouling microtopographies – correlating wettability with cell attachment, *Biofouling*, 22 (2006) 11–21.
31. Y.K. Lee, Y.-J. Won, J.H. Yoo, K.H. Ahn, C.-H. Lee, Flow analysis and fouling on the patterned membrane surface, *Journal of Membrane Science*, 427 (2013) 320–325.
32. T. Tong, A.F. Wallace, S. Zhao, Z. Wang, Mineral scaling in membrane desalination: Mechanisms, mitigation strategies, and feasibility of scaling-resistant membranes, *Journal of Membrane Science*, 579 (2019) 52–69.
33. K.-G. Lu, H. Huang, Dependence of initial silica scaling on the surface physicochemical properties of reverse osmosis membranes during bench-scale brackish water desalination, *Water Research*, 150 (2019) 358–367.
34. T. Tong, S. Zhao, C. Boo, S.M. Hashmi, M. Elimelech, Relating silica scaling in reverse osmosis to membrane surface properties, *Environmental Science & Technology*, 51 (2017) 4396–4406.
35. B. Cao, A. Ansari, X. Yi, D.F. Rodrigues, Y. Hu, Gypsum scale formation on graphene oxide modified reverse osmosis membrane, *Journal of Membrane Science*, 552 (2018) 132–143.
36. N.H. Lin, M.-M. Kim, G.T. Lewis, Y. Cohen, Polymer surface nano-structuring of reverse osmosis membranes for fouling resistance and improved flux performance, *Journal of Materials Chemistry*, 20 (2010) 4642–4652.
37. S.T. Weinman, S.M. Husson, Influence of chemical coating combined with nanopatterning on alginate fouling during nanofiltration, *Journal of Membrane Science*, 513 (2016) 146–154.

38. S.A. Ravichandran, J. Krist, D. Edwards, S. Delagah, J. Pellegrino, Measuring sparingly-soluble, aqueous salt crystallization kinetics using CSTRs-in-series: Methodology development and CaCO3 studies, *Separation and Purification Technology*, 211 (2019) 408–420.
39. Innovative reclamation of membrane concentrates: conceptual evaluation of combining two innovative technologies Kennedy/Jenks Consultants, Irvine, CA, USA (www.etcc-ca.com/sites/default/files/OLD/images/stories/pdf/ETCC_Report_321.pdf), 2005, pp. 45.
40. N. Hantzsche, S. Itagaki, R. Ryder, September Ranch Subdivision Project Response to Comments to Draft REIR Appendix A: Water System Considerations -Background Technical Documents Appendix B: Condition Compliance and Mitigation Monitoring and Reporting Plan Kennedy/Jenks Consultants, Michael Brandman Associates, Questa Engineering Corporation., Monterey, CA (www.co.monterey.ca.us/home/showdocument?id=37238), 2006.
41. C.A. Quist-Jensen, F. Macedonio, E. Drioli, Membrane crystallization for salts recovery from brine—an experimental and theoretical analysis, *Desalination and Water Treatment*, 57 (2016) 7593–7603.
42. E. Curcio, G. Di Profio, Chapter 7 - Membrane crystallization, in: A. Basile, E. Curcio, Inamuddin (Eds.) *Current Trends and Future Developments on (Bio-) Membranes*, Elsevier, Amsterdam, 2019, pp. 175–198.
43. A. Shahmansouri, J. Min, L. Jin, C. Bellona, Feasibility of extracting valuable minerals from desalination concentrate: a comprehensive literature review, *Journal of Cleaner Production*, 100 (2015) 4–16.
44. J. Yang, Y.H. Liu, T. Wen, X.X. Wei, Z.Y. Li, Y.L. Cai, Y.L. Su, D.J. Wang, Controlled mineralization of calcium carbonate on the surface of nonpolar organic fibers, *Crystal Growth & Design*, 12 (2012) 29–32.
45. E. Dalas, J. Kallitsis, P.G. Koutsoukos, The crystallization of calcium-carbonate on polymeric substrates, *Journal of Crystal Growth*, 89 (1988) 287–294.
46. E. Dalas, J. Kallitsis, P.G. Koutsoukos, The growth of sparingly soluble salts on polymeric substrates, *Colloid Surface*, 53 (1991) 197–208.
47. E. Dalas, The overgrowth of calcium-carbonate hexahydrate and calcium-phosphate on new functionalized polymers, *Journal of Materials Science Letters*, 11 (1992) 1408–1410.
48. F. Manoli, E. Dalas, Calcium carbonate overgrowth on elastin substrate, *Journal of Crystal Growth*, 204 (1999) 369–375.
49. E. Dalas, Crystallization of sparingly soluble salts on functionalized polymers, *Journal of Materials Chemistry*, 1 (1991) 473–474.
50. F. Manoli, S. Koutsopoulos, E. Dalas, Crystallization of calcite on chitin, *Journal of Crystal Growth*, 182 (1997) 116–124.
51. N.H. Lin, W.Y. Shih, E. Lyster, Y. Cohen, Crystallization of calcium sulfate on polymeric surfaces, *Journal of Colloid and Interface Science*, 356 (2011) 790–797.

Index

Note: **Bold** page numbers refer to tables and *italic* page numbers refer to figures.